P9-BHX-905

ACS SYMPOSIUM SERIES **426**

Enhanced Biodegradation of Pesticides in the Environment

Kenneth D. Racke, EDITOR
DowElanco

Joel R. Coats, EDITOR
Iowa State University

Developed from a symposium sponsored
by the Division of Agrochemicals
at the 198th National Meeting
of the American Chemical Society,
Miami Beach, Florida
September 10-15, 1989

American Chemical Society, Washington, DC 1990

Library of Congress Cataloging-in-Publication Data

Enhanced Biodegradation of Pesticides in the Environment
Kenneth D. Racke, editor Joel R. Coats, editor

p. cm.—(ACS Symposium Series, 0065–6156; 426).

"Developed from a symposium sponsored by the Division of Agrochemicals at the 198th National Meeting of the American Chemical Society, Miami Beach, Florida, September 10–15, 1989."

Includes bibliograpical references.

ISBN 0–8412–1784–X
1. Pesticides—Biodegradation—Congresses. 2. Microbial metabolism—Congresses.

I. Racke, Kenneth D., 1959– . II. Coats, Joel R., 1948–
III. American Chemical Society. Division of Agrochemicals.
IV. American Chemical Society. Meeting (198th : 1989 : Miami Beach, Fla.) V. Series

SB951.145.B54E54 1990
632'.95042—dc20

90–34194
CIP

The paper used in this publication meets the minimum requirements of American National Standard for Information Sciences—Permanence of Paper for Printed Library Materials, ANSI Z39.48–1984.

∞

PRINTED IN THE UNITED STATES OF AMERICA

ACS Symposium Series

M. Joan Comstock, *Series Editor*

Foreword

The ACS SYMPOSIUM SERIES was founded in 1974 to provide a medium for publishing symposia quickly in book form. The format of the Series parallels that of the continuing ADVANCES IN CHEMISTRY SERIES except that, in order to save time, the papers are not typeset but are reproduced as they are submitted by the authors in camera-ready form. Papers are reviewed under the supervision of the Editors with the assistance of the Series Advisory Board and are selected to maintain the integrity of the symposia; however, verbatim reproductions of previously published papers are not accepted. Both reviews and reports of research are acceptable, because symposia may embrace both types of presentation.

Contents

Preface ix

1. **Pesticides in the Soil Microbial Ecosystem** 1
   Kenneth D. Racke

THE PHENOMENON CHARACTERIZED

2. **Effects of Long-Term Phenoxyalkanoic Acid Herbicide Field Applications on the Rate of Microbial Degradation** 14
   Allan E. Smith and Guy P. Lafond

3. **Enhanced Carbamothioate Herbicide Degradation: Research in Nebraska** 23
   Fred W. Roeth, Robert G. Wilson, Alex R. Martin, and Patrick J. Shea

4. **Enhanced Biodegradation of Carbamothioate Herbicides in South Carolina** 37
   Horace D. Skipper

5. **Enhanced Biodegradation of Dicarboximide Fungicides in Soil** 53
   Allan Walker and Sarah J. Welch

6. **Enhanced Biodegradation of Insecticides in Midwestern Corn Soils** 68
   Kenneth D. Racke and Joel R. Coats

7. **Enhanced Degradation of Insecticides in Soil: Factors Influencing the Development and Effects of Enhanced Microbial Activity** 82
   R. A. Chapman and C. R. Harris

PESTICIDES AND MICROBIAL ADAPTATION

8. **Enhanced Degradation of S-Ethyl *N,N*-Dipropylcarbamothioate in Soil and by an Isolated Soil Microorganism** 98
W. A. Dick, R. O. Ankumah, G. McClung, and N. Abou-Assaf

9. **The Role of Fungi and Bacteria in the Enhanced Degradation of the Fungicide Carbendazim and the Herbicide Diphenamid** 113
N. Aharonson, J. Katan, E. Avidov, and O. Yarden

10. **Influence of Pesticide Metabolites on the Development of Enhanced Biodegradation**...128
L. Somasundaram and Joel R. Coats

11. **Molecular Genetics of Pesticide Degradation by Soil Bacteria**...141
Jeffrey S. Karns

12. **Response of Microbial Populations to Carbofuran in Soils Enhanced for Its Degradation**..153
R. F. Turco and A. E. Konopka

13. **Adaptation of Microorganisms in Subsurface Environments: Significance to Pesticide Degradation**..167
Thomas B. Moorman

14. **Microbial Adaptation in Aquatic Ecosystems** 181
J. C. Spain

IMPLICATIONS AND SOLUTIONS

15. **Evaluation of Some Methods for Coping with Enhanced Biodegradation of Soil Insecticides** 192
A. S. Felsot and J. J. Tollefson

16. **Systems Allowing Continued Use of Carbamothioate Herbicides Despite Enhanced Biodegradation**..................................214
Robert G. Harvey

17. **Cultural Practices and Chemicals That Affect the Persistence of Carbamothioate Herbicides in Soil**..222
Dirk C. Drost, James E. Rodebush, and Joanna K. Hsu

18. **Spectrophotometric Methodologies for Predicting and Studying Enhanced Degradation** 240
Joseph P. Reed, Robert J. Kremer, and Armon J. Keaster

19. **Enhancing Biodegradation for Detoxification of Herbicide Waste in Soil** 249
A. S. Felsot and E. K. Dzantor

20. **Implications of Enhanced Biodegradation for the Use and Study of Pesticides in the Soil Environment** 269
Kenneth D. Racke

Appendix: Key Chemical Structures 283

Author Index 291

Affiliation Index 291

Subject Index 291

Preface

MICROBIAL DEGRADATION HAS LONG BEEN RECOGNIZED as a primary means of dissipating many pesticides in soil and water ecosystems, and recognition of this has prompted the development of biodegradable herbicides, insecticides, and fungicides. Because these biodegradable pesticides have predictable patterns of environmental persistence, they have become key components of agricultural and industrial pest management systems. Early research on the degradation of phenoxy herbicides in soil provided evidence that microbial adaptation could result in abnormally accelerated rates of pesticide breakdown. However, because of the lack of observable economic impact on pest control practices, the phenomenon of enhanced microbial pesticide degradation languished as an academic curiosity for many years. The contemporary occurrence of enhanced degradation and associated pest control failures has spurred reexamination of adapted microbial pesticide degradation as a critical environmental process. The recent flurry of research into microbial adaptation for pesticide degradation has spanned disciplines ranging from applied agronomy to molecular biology.

This book is the first collection of research results to focus exclusively on the adaptation of microorganisms for rapid pesticide degradation. After an introductory chapter, the book is organized into three sections, and it concludes with an appendix of key chemical structures. The first section contains general field and laboratory observations of enhanced herbicide, insecticide, and fungicide degradation. The second section focuses more specifically on the microbiological and biochemical aspects of adapted microbial pesticide metabolism in terrestrial and aquatic systems. Finally, the third section addresses management strategies and implications of enhanced pesticide degradation.

The editors thank the contributors to this volume, including authors, coauthors, and reviewers, for their time and insight. We also express our thanks to the Division of Agrochemicals, the sponsor of the symposium on which this book is based, and to the ACS Books Department staff for their hard work in presenting this volume.

Dedication

This book is dedicated to Dr. E. P. Lichtenstein of the University of Wisconsin and Dr. R. L. Metcalf of the University of Illinois, who conducted pioneering work on the environmental chemistry of pesticides and served as early mentors for the editors.

KENNETH D. RACKE
Environmental Chemistry Laboratory
DowElanco
Midland, MI 48841

JOEL R. COATS
Department of Entomology
Iowa State University
Ames, IA 50011–3140

January 19, 1990

Chapter 1

Pesticides in the Soil Microbial Ecosystem

Kenneth D. Racke

Environmental Chemistry Laboratory, DowElanco, 9001 Building, Midland, MI 48641–1706

The soil environment comprises a diverse ecosystem in which the recycling of organic matter provides a key link with global carbon and nutrient cycles. Soil contains populations of microorganisms that have exceptional degradative capabilities and the ability to adapt to utilize the variety of allochthonous organic materials that enter the soil carbon cycle. As man has come to rely upon the use of soil-applied pesticides in many agroecosystems, he has learned that the residual control provided by these compounds is modulated by both abiotic and biotic degradative processes. The ability of microorganisms to adapt for rapid catabolism of some soil pesticides has in some cases resulted in economically significant pest control failures. The phenomenon of enhanced pesticide degradation has become a significant concern requiring the development of management strategies. Yet, this unwelcome occurrence has led to valuable opportunities to investigate the remarkable adaptability and complexity of the soil microbial ecosystem.

In order to properly understand pesticides in the context of the soil microbial ecosystem, it is important to consider the relevant properties of the soil environment, the metabolic capabilities of the soil microbial community, and the uses and dissipation routes of soil-applied pesticides.

The Soil Environment

Soil has been defined as the "unconsolidated mineral material on the immediate surface of the earth that serves as a natural medium for the growth of land plants" (1). Although this sounds relatively

0097–6156/90/0426–0001$06.00/0

simple, in reality soil is a complex ecosystem, not merely the "dirt" of vernacular language. In a physical sense, soil is composed of mineral material (clay, silt, sand, stones), air or water-filled pore space, and organic matter. The soil organic matter is a very important and active component. Roughly three fractions can be delineated; a macroscopic component composed of particulate plant and animal debris in early stages of breakdown , a chemically well-defined assemblage of various simple organic compounds (carbohydrates, amino acids, proteins), and a complex, dark-colored component, largely aromatic and polymeric (i.e., "humus"), that is relatively resistant to degradation (2). Turnover times for organic debris, soluble compounds, and humus are on the order of 2-5 years, 5-25 years, and 250-2500 years, respectively (3). The physical components of soil are integrated into a heterogeneous matrix that forms the basis for the soil aggregate and particle. For example, clays and humic materials are often intimately associated to form organo-mineral complexes (4).

The soil should be understood as an ecosystem, with populations of microscopic (bacteria, fungi, protozoans, algae) and macroscopic (annelids, arthropods) organisms forming complex food web communities. The soil microflora play a key role in globally significant nutrient cycles. Important examples of such microbially-mediated processes are nitrification, denitrification, and carbon mineralization. The soil ecosystem constitutes an allochthonous one in an energetic sense, with plant and animal debris entering the soil to begin the degradative process. In a practical sense, the soil may be viewed as a vast recycling depot, continually receiving reduced compounds of carbon and continually oxidizing them to CO_2 and water (5). Of the soil organisms active in this recycling effort, the soil fungi are particularly effective in initiating the degradation of rather complex polymeric substrates (e.g., cellulose), whereas the soil bacteria have in general specialized in the degradation of simpler, soluble organics. Among the indigenous populations are many bacterial species that are largely quiescent, but flourish dramatically when readily available organic nutrients are added (6). Thus, although the soil has often been viewed as a rich, fertile ecosystem, in reality soil is considered to be an oligotrophic environment (7). The fierce nature of competition for the easily degradable organic compounds added to soil is exemplified by antibiotic production by soil microorganisms, which probably represents a form of chemical warfare over the defense of niches (6).

Pesticides Applied to the Soil Environment

Soil is the foundation of agriculture. It is the substrate in which the agronomic crops of the earth find required nutrients, water, microbial symbionts, and a physical anchor. Due to the presence in soil of crop competitors and competitors for the use of this plant biomass, pesticides have become indispensible tools in the management of weeds, insects, and fungi of economic significance. At present there are approximately 600 different pesticide active ingredients registered for use in the U.S., and millions of pounds are globally applied each year, with the world market estimated at greater than 20 billion dollars (8). Much of this use represents direct application to soil, and a considerable portion of that which is foliarly applied may reach the soil as spray drift, runoff, or wash-off. One consideration of the use of soil-applied pesticides is that they should persist long enough in soil to control the target pest(s), but not so long as to create environmental risks.

Herbicides which are applied directly to soil with the intent of providing residual weed control may be applied prior to crop planting (preplant) or after planting and prior to weed emergence (preemergence). Classes of herbicides that are applied preplant or preemergence include the triazines, substituted ureas, carbamates, carbamothioates, sulfonyl ureas, chloracetamides, and dinitroanilines (9). Many important crops, such as corn and soybeans, rely upon preplant or preemergence herbicides. A typical preemergent weed control scenario is provided by the occurrence of wild proso millet (*Panicum miliaceum)* in Midwestern U.S. corn. Wild proso millet is a pernicious grass species, and often exhibits both early and late season flushes (10). One of the herbicides that is effective against this weed is EPTC, which must be formulated with a metabolic safener (dichlormid) that allows corn to escape its herbicidal effects. EPTC is applied preplant-incorporated at 2.2-6.7 kg/ha and must persist in significant quantities for roughly 15-30 days for effective control of early season flushes of wild proso millet (11). It should be noted that in many agroecosystems, such as corn, herbicides or "tank mixes" of multiple herbicides are applied to control complexes of weed species.

Although the use of insecticides is not as heavily skewed toward soil application as that for herbicides, there are several significant crops such as corn, peanuts, and sugarbeets which rely heavily on successful control of soil insect pests. A typical example of soil insect pest control is provided by the corn rootworm (*Diabrotica spp.*). This key pest of corn overwinters in the egg stage in corn stubble, and after a variable hatching period the larvae actively damage corn roots from early June to late July (12). The most common method of corn rootworm control involves an at-planting (April-May) application of a granular insecticide, such as carbofuran, at 1 kg/ha. To provide effective rootworm control the carbofuran applied must persist in toxic concentrations for approximately 4-8 weeks after application (13). Insecticides of several classes have been employed for soil insect control, including organophosphorus, carbamate, and pyrethroid compounds.

Fungicides are applied to soil in a few specific instances for control of soil-borne plant pathogens. Some crops for which soil-applied fungicide applications are often necessary include onions, peanuts, and avocado. An example of this would be the use of iprodione for the control of white rot disease (*Sclerotium cepivorum*) of onions in the United Kingdom (14). Soil-applied fungicides are often systemic in nature, and may need to be taken up by plant roots and translocated to stem and foliage for efficacious pathogen control. Several commonly used soil fungicides are iprodione, carbendazim, and thiabendazole.

Mechanisms of Pesticide Dissipation in Soil

Pesticides which are applied to soil become enmeshed in the transport and degradation processes that affect all allochthonous organics added to this dynamic ecosystem (15).

Transport Processes. There are several processes that are active in transporting pesticide residues within or out of the soil ecosystem. The first class of processes involves aerial transport of pesticides from the soil into the atmosphere. Volatilization is an important process by which pesticides in the vapor phase can move from the soil surface into the atmosphere. Pesticides with vapor pressures greater than approximately 10^{-6} mm Hg at 25°C are most greatly affected, and losses from surface applications can approach greater than 50% in less than 48 hours (16). In addition to the vapor pressure of a pesticide, which increases with increasing temperature, several soil properties such as moisture and organic matter content can greatly modify the kinetics of volatilization. This is due to the fact that pesticides in soil partition between at least three phases: soil-sorbed, soil solution, and soil air (17). Although not intensively investigated, movement of soil particles and attached pesticides by wind may also play a role under some environmental conditions.

The second class of pesticide transport processes is that of movement of dissolved or particulate-sorbed pesticides in water. Leaching of pesticides has been recognized as a critical process, if not in the sense of the absolute percentage of applied pesticide leached into the soil profile, then with respect to the contamination of groundwater by trace quantities of pesticide. Several pathways of dissolved pesticide movement with leaching water are recognized (18). Convection and diffusion are often simultaneous processes that involve pesticide transport in water moving gravitationally or with a concentration gradient. Macropore flow involves pesticide transport in water flowing through the irregular network of channels created by earthworms, roots, insects, and burrowing animals. The extent of leaching is governed by partitioning processes between sorbed and solution phase pesticide, and often laboratory-determined sorption coefficients (e.g. Freundlich K_d) are used to predict field leaching potential (19). An understanding of the relationship between observed partitioning behavior and actual field leaching measurement is complicated by such factors as nonequilibrium partitioning resulting from slow-desorption kinetics (20). Transport of pesticides in water moving over the soil surface is also an important process that can impact surface-water quality. Pesticides with water solubilities greater than approximately 10 μg/g have been hypothesized to move largely in the solution phase, while less soluble pesticides are thought to move mainly sorbed to eroding soil particles (21).

Abiotic Transformation Processes. Pesticides that reside within or on the soil are subject to the abiotic and microbiological transformation processes that operate in this great recycling ecosystem. After numerous problems with the prolonged persistence of some early soil pesticides (e.g. aldrin, chlordane), degradability is viewed as a desirable attribute of modern soil-applied pesticides. Degradation processes in soil may result in accumulation of well-defined metabolites, incorporation of pesticide carbon or nitrogen into the soil organic matter fraction, or complete mineralization. Many of the abiotic degradation processes active in soil result in partial degradation of the pesticide to products that are further degraded microbially, accumulate in soil temporarily, or bind to soil organic components. Exposure of pesticide residues to ultraviolet radiation can result in considerable photodegradation of pesticides on the soil surface, although only residues on the extreme surface layer (<1mm) appear to be susceptible (22). More significant abiotic degradative reactions are those that occur between pesticides and reactive inorganic and organic compounds in solution and on surfaces within the soil profile. Hydrolytic reactions are very important transformation mechanisms of many pesticides in soil including organophosphorus and carbamate insecticides and herbicide esters (23,24). These hydrolytic reactions can be base or acid catalyzed in nature and often occur through interactions with reactive chemical groups exposed on clay mineral surfaces (25), reactive organics (e.g., enzymes) immobilized on or in soil aggregates (26), or inorganic species (e.g., Cu^{+2}) in soil solution (27). Hydrolysis of selected functional groups is often a necessary prelude for microbial degradation to commence (28). Various other reactions including oxidations (terbufos ⇒ terbufos sulfoxide) and reductions (parathion ⇒ amino-parathion) can also be important transformation processes for specific pesticides (29,30).

Microbiological Transformation Processes. The importance of the soil microbial community in mediating pesticide degradation has long been recognized. In fact, it was for one of the first synthetic organic pesticides, 2,4-D, that microbiological degradation was first shown to be an important factor (31). The conclusion of this early work was "...the detoxication of 2:4-dichlorophenoxyacetic acid in garden loam is due almost entirely to the activity of microorganisms". Since this early discovery there has been considerable appreciation for the role of the soil microbial community in pesticide transformations. In some cases the role of microorganisms has been stressed by comparison of degradation rates in natural and sterilized systems (32), while in others researchers have chosen to use isolation of pesticide-degrading microorganisms from soil as evidence of microbial involvement (33). It has also been recognized that environmental conditions that influence microbial activity (temperature, moisture,...) affect the microbial degradation of pesticides in soil (34). The importance of microbial involvement is demonstrated by the many reviews of pesticide/microbe interactions that have appeared (35-41).

The microbial metabolism of pesticides has often been subdivided into 2 distinct classes. The first of these is termed simply "catabolism". This process often results in the mineralization of some portion of an organic compound via enzymatic pathways to simple products of universal currency (CO_2, NH_3). In some cases, one portion of the molecule may be mineralized and another portion may accumulate in soil. This is true for the soil microbial degradation of carbofuran (42,43). Therefore, catabolism should not be equated with mineralization or complete destruction of a pesticide. It should be pointed out, however, that mineralization of a pesticide in soil *is* nearly always a consequence of microbial activity (44). The key to understanding catabolism is that it is primarily a process driven by the microbial quest for energy. Therefore, catabolism has come to be equated with utilization of a pesticide as an energy source and thus a growth substrate (40,41). Catabolism is most commonly linked to the conversion of pesticides into carbon skeleton feedstocks for existing metabolic pathways. However, in a few instances pure cultures of pesticide-degrading microorganisms isolated from soil have been capable of exploiting pesticides as sole sources of phosphorus (45,46), nitrogen (42), or sulfur (47).

A second recognized class of microbial pesticide metabolism, "cometabolism", is related to the observed uncoordination of microbial enzyme activities. As Dagley (48) has observed: "Most microbial degradative (catabolic) enzymes are not linked *directly* to processes that harness released energy for growth. Thus, enzymes that add water, or break chemical bonds by hydrolysis or aldol fission, simply serve the function of supplying substrates for the metabolic sequences that terminate in energy production". When a disjunction occurs, incomplete pesticide degradation results, and there is an accumulation of intermediate metabolic products. The phenomenon of cometabolism has been defined as the degradation "of substances without utilization of the energy derived from their oxidation to support growth..." (49). Microbiological studies to distinguish catabolism from cometabolism under natural conditions are not trivial, and most researchers have relied upon work with isolated pure cultures. The conclusions, appropriate or not, drawn from pure culture work in the laboratory depend on whether a microorganism capable of growing on the pesticide is isolated (i.e., catabolism) or not (i.e., cometabolism). A new emphasis is now being placed by researchers on understanding cometabolism in its ecological context. This reevaluation has been spurred by instances in which no single microbe can catabolize a pesticide, but only by the synergistic interactions of several members of a microbial consortium is mineralization and growth realized (50-53).

Microbial Adaptation for Degradation

The pattern of microbial population growth presents some interesting ecological implications. Because the most common mode of bacterial reproduction involves fission, with two daughter cells formed from each split parent, microbial populations exhibit the phenomenon of exponential growth. The classic growth curve for bacterial populations involves 4 recognized phases (54). The *lag phase* is a period during which no growth occurs upon presentation of a new carbonaceous energy source to the population. This lag is due to the time necessary to gear up the metabolic machinery through enzyme and coenzyme synthesis, a process which has been termed "induction". During the *exponential phase* the population grows exponentially as it exploits the growth substrate. The *stationary* phase is reached when population growth slows and halts due to limitations of nutrient depletion or toxin accumulation. Finally, the *death phase* is characterized by a decline in the viable population. Microbiologists have used this scenario as a paradigm of microbial population growth, yet it should be noted that this is almost entirely based on pure culture work in the laboratory. Although in some cases this type of growth pattern may be observed in soil (55), there is some doubt whether in the matrix of complex microbial community interactions in the harsh soil environment this pattern is as clearly displayed. Soil bacterial growth rates in nature seem to be far slower than those typical of soil bacteria grown in rich nutrient media in the laboratory (7). It has been concluded that microbial populations in soil are largely dormant, with only 15-30% of the bacterial population active even under the most favorable conditions (56,57). The majority of evidence portrays soil bacterial populations as operating on what is ecologically known as *R* population growth strategy: populations rapidly increase to exploit favorable growth conditions (i.e., "boom"), and maintain negligible population growth and activity during periods of unfavorable conditions (i.e., "bust") (7,58).

The soil ecosystem is primarily an allochthonous one in terms of energetic input, receiving its energy input primarily as organic carbon constituents of plant and animal debris. Plants alone introduce an incredible diversity of compounds into the soil: cellulose, hemicelluloses, lignin, starch, pectins, proteins, lipids, pigments, and defense chemicals (e.g., nicotine, rotenone) (59). As a consequence, analyses of soil reveal the presence of a myriad of biochemically identifiable components in addition to the more amorphous humic materials (3). As might be expected, the soil microbial community contains populations of fungi, actinomycetes, and bacteria that degrade this diverse nutritional offering. The fungal and actinomycete populations often act as primary degraders in soil, and are very active in the degradation of rather complex substrates such as cellulose, lignin, and chitin (6). The soil bacterial population contains some organisms which have become specialized to utilize some fairly simple aliphatic and aromatic substrates as growth substrates, some of which appear to present peculiar metabolic conundrums. For example, a survey of the *Arthrobacter spp.* of agronomic soils found a spectrum of nutritional subgroupings represented, that in addition to typical substrates such as glucose, butyric acid, and glycine, could utilize such compounds as benzoic acid, salicylic acid, histamine, and methylamine as sole carbon sources (60). The capabilities of several genera of soil bacteria for degradation of complex, soluble substrates are well known. For example, *Pseudomonas spp.* are common soil bacteria that can utilize many aromatic compounds for growth. Some of the compounds degraded include camphor, naphthalene, toluene, phthalic acid, and octane (61,62). The bacterial community is capable of degrading virtually any organic compound that is added to the soil environment, due to some microorganism or microbial consortium that can degrade and mineralize the compound and extract energy from it for growth (5).

How can some portions of the soil microbial community be so flexible in their nutritional requirements so as to exploit the variety of carbon sources that enter the soil environment? The answer to this question lies in the unique microevolutionary nature of microbial substrate adaptation. Characteristic of microbial metabolism is the adaptability of microorganisms through mutation and induction to develop the ability to degrade compounds which initially could not be degraded or were toxic (63). This can involve the evolution of suitable cell wall transport systems, of more efficient enzymes, or of enzymes with new activities (64). Any of these scenarios would confer a selective advantage on the microorganism involved, and would have significant ecological implications because each bacterial cell has the ability to form an entire population (65). One of the key factors in the metabolic adaptability of microorganisms is related to the possession and utilization of plasmids. Plasmids are extrachromosomal circlets of genetic material

(DNA) that in bacteria exist independently of the chromosome, and often encode for "nonessential" but useful characteristics such as conjugation factors, capsule formation, host specificity, and drug resistance (66). Metabolic genes residing on plasmids include the ability to degrade camphor, toluene, naphthalene, octane, and PCB's (61,67). Degradative plasmids can be transferred between individual bacteria of the same or different species (68). Possession of degradative plasmids allows microorganisms to "experiment" with auxiliary metabolic pathways via gene mutation or promiscuous genetic exchange without jeopardizing the irreplaceable genes on the central chromosome. The ultimate selective factor in determining whether a novel organic compound can provide the selective advantage necessary to induce catabolic adaptation appears to be its ability to support the synthesis of at least one molecule of ATP during its degradation (5,48). As Delwiche has observed "Few energy-yielding reactions have not been exploited by microorganisms, and some formidable obstacles to exploitation have been overcome by ingenious adaptations, some of which remain to be explained by the microbiologist" (5).

Enhanced Pesticide Biodegradation

In light of the metabolic adaptability of microorganisms residing in the energy-poor environment of the soil, it should not then be surprising that soil-applied pesticides can be degraded biologically and that microbial adaptation for pesticide catabolism would be possible. This fact became apparent not long after the introduction of the first synthetic organic pesticides (e.g., phenoxyacetic acid herbicides). Very early work demonstrated that 2,4-D dissipated much more slowly from autoclaved (i.e., sterilized) soil than from natural soil (69). Intrigued by the possibility of microbial degradation, L.J. Audus of the Botany Department of Bedford College University, London, began a series of investigations that first clearly elucidated the role of soil microorganisms in the degradation of pesticides applied to soil. In his experiments Audus recycled an aqueous solution of 2,4-D (10-1,000 μg/ml) through a column packed with garden soil. Using bioassay techniques he was able to demonstrate that after a defined lag period, during which little 2,4-D degradation was noted, there followed a period of very rapid pesticide degradation to negligible levels (31). A second key finding of this study was that a second dose of 2,4-D recycled through the soil column was immediately degraded with no lag period evident. These soil perfusion studies were quickly followed by other investigations which demonstrated that soil pretreated with 2,4-D displayed the ability to more rapidly degrade a subsequent dose of 2,4-D than previously untreated soil (70). The microbially-mediated nature of this process was later confirmed through the isolation of 2,4-D catabolizing microorganisms and the study of 2,4-D degrading populations of soil microorganisms (71-74).

During the 30 years following the discovery of microbial adaptation for pesticide degradation, there were intermittent reports of the same phenomenon affecting the persistence of other pesticides. However, due to the fact that virtually none of these reports concerned situations where reduced pesticide persistence affected pest control efficacy, microbial adaptation for pesticide degradation was largely relegated to the status of academic curiosity. In fact, the focus of environmental chemists during much of this period was on the recalcitrance and *lack* of degradation of many early pesticides. By the mid 1970's, the replacement of these materials with degradable, soil-applied pesticides in nearly all markets set the scene for the development of current interest in adapted microbial pesticide degradation.

Although agronomists periodically observe failures of soil-applied pesticides to control target pests, these have historically been attributed to improper application technique, unusual environmental conditions, or development of pest resistance. However, a pattern of pest control failures after application of normally efficacious soil pesticides emerged in the late 1970's. In New Zealand, Wisconsin, Nebraska, and Iowa the carbamothioate herbicide EPTC no longer provided effective control of certain weeds in some fields with historic use of EPTC (75-78). In Illinois, Iowa, Kansas, and Eastern Canada the carbamate insecticide carbofuran did not provide adequate protection against certain soil insect pests in some soils with historic carbofuran use (79-83). The failures of these and a few other pesticides were eventually recognized to occur due to rapid microbial degradation that manifested itself after a particular pesticide had been applied for two or more consecutive years to the same field. This phenomenon came to be termed "enhanced degradation" or "accelerated degradation", because a more rapid rate of pesticide degradation was observed in previously treated than in previously untreated fields. The term "enhanced degradation" has come to be synonymous with adapted microbial catabolism.

Since enhanced degradation has come to the forefront of environmental chemistry, much research effort has been directed at understanding the field implications, microbiology, and biochemistry of this phenomenon. The chapters which follow in this book have been organized so as to convey the depth of understanding we have achieved regarding microbial adaptation for pesticide degradation. These chapters reveal that we should not only view this phenomenon as an agricultural problem requiring management and solutions, but as a window into the remarkable adaptability and complexity of the soil microbial ecosystem.

Literature Cited

1. Foth, H.D. Fundamentals of Soil Science; John Wiley and Sons: New York, NY, 1978; p 423.
2. Stotzky, G.; Burns, R.G. In Experimental Microbial Ecology; Burns, R.G.; Slater, J.H., Eds.; Blackwell Scientific Publications: Oxford, 1982; Chapter 7.
3. Stevenson, F.J. Humus Chemistry: Genesis, Composition, Reactions; John Wiley and Sons: New York, NY, 1982; p 12.
4. Theng, B.K.G. Formation and Properties of Clay-Polymer Complexes; Elsevier: Amsterdam, 1979.
5. Delwiche, C.C. In Soil Biochemistry; McLaren, A.D.; Peterson, G.H., Eds.; Marcel Dekker: New York, NY, 1967; Vol. I, Chapter 7.
6. Alexander, M. Introduction to Soil Microbiology; John Wiley and Sons: New York, NY, 1977; p 17.
7. Williams, S.T. In Bacteria in Their Natural Environments; Academic Press: London, 1985; Chapter 3.

8. McDougall, J.; Woodburn, J. Agrochemical Review; County Natwest Woodmac, London, May 1989; p 7.
9. Ross, M.A.; Lembi, C.A. Applied Weed Science; Macmillan: New York, NY, 1985.
10. Cannell, A.M. Crops and Soils 1983, 36, 16-18.
11. Harvey, R.G.; McNevin, G.R.; Albright, J.W.; Kozak, M.E. Weed Sci. 1986, 34, 773-780.
12. Krysan, J.L.; Miller, T.A. Methods for the Study of Pest Diabrotica; Springer-Verlag: New York, NY, 1986.
13. Felsot, A.S.; Steffey, K.L.; Levine, E.; Wilson, J.G. J. Econ. Entomol. 1985, 78, 45-52.
14. Walker, A.; Brown, P.A.; Entwistle, A.R. Pestic. Sci. 1986, 17, 183-193.
15. Sawhney, B.L.; Brown, K., Eds., Reactions and Movements of Organic Chemicals in Soils; Soil Science Society of America: Madison, WI, 1989.
16. Spencer, W.F. In Fate of Pesticides in the Environment; Biggar, J.W.; Seiber, J.N., Eds.; University of California Publication 3320: Oakland, CA, 1987; Chapter 6.
17. Glotfelty, D.E.; Schomburg, C.J. In Reactions and Movements of Organic Chemicals in Soils; Sawhney, B.L.; Brown, K., Eds., Soil Science Society of America: Madison, WI, 1989; Chapter 7.
18. Green, R.E.; Khan, M.A. In Fate of Pesticides in the Environment; Biggar, J.W.; Seiber, J.N., Eds.; University of California Publication 3320: Oakland, CA, 1987; Chapter 9.
19. Cohen, S.Z.; Creeger, S.M.; Carsel, R.F.; Enfield, C.G. In Treatment and Disposal of Pesticide Wastes; American Chemical Society: Washington, DC, 1984; Chapter 18.
20. Pignatello, J.J. In Reactions and Movements of Organic Chemicals in Soils; Sawhney, B.L.; Brown, K., Eds., Soil Science Society of America: Madison, WI, 1989; Chapter 3.
21. Wauchope, R.D. J. Environ. Qual. 1978, 7, 459-472.
22. Miller, G.C.; Herbert, V.R. In Fate of Pesticides in the Environment; Biggar, J.W.; Seiber, J.N., Eds.; University of California Publication 3320: Oakland, CA, 1987; Chapter 8.
23. Tinsley, I.J. Chemical Concepts in Pollutant Behavior; John Wiley and Sons: New York, NY, 1979.
24. Drossman, H.; Johnson, H.; Mill, T. Chemosphere 1988, 17, 1509-1530.
25. Yaron, B. Soil Sci. 1978, 125, 210-216.
26. Skujins, J. CRC Critic. Rev. Microbiol. 1976, May, 383-421.
27. Mortland, M.M.; Raman, K.V. J. Agric. Food Chem. 1967, 15, 163-167.
28. Wolfe, N.L.; Metwally, M.E.S.; Moftah, A.E. In Reactions and Movements of Organic Chemicals in Soils; Sawhney, B.L.; Brown, K., Eds., Soil Science Society of America: Madison, WI, 1989; Chapter 9.
29. Wahid, P.A.; Ramakrishna, C.; Sethunathan, N. J. Environ. Qual. 1980, 9, 127-130.
30. Chapman, R.A.; Tu, C.M.; Harris, C.R.; Dubois, D. J. Econ. Entomol. 1982, 75, 955-960.
31. Audus, L.J. Plant Soil 1949, 2, 31-36.
32. Lichtenstein, E.P.; Schultz, K.R. J. Econ. Entomol. 1964, 57, 618-627.
33. Steenson, T.I.; Walker, N. J. Gen. Microbiol. 1957, 16, 146-155.

34. Chapman, R.A.; Harris, C.R.; Harris, C. J. Environ. Sci. Health 1986, B21, 125-141.
35. Bollen, W.B. Ann. Rev. Microbiol. 1961, 15, 69-92.
36. Cullimore, D.R. Residue Rev. 1971, 35, 65-80.
37. Bollag, J.M. Adv. Appl. Microbiol. 1974, 18, 75-130.
38. Tu, C.M.; Miles, J.R.W. Residue Rev. 1976, 64, 17-65.
39. Laveglia, J.; Dahm, P.A. Ann. Rev. Entomol. 1977, 22, 483-513.
40. Hill, I.R.; Wright, S.J.L., Eds. Pesticide Microbiology; Academic Press: London, 1978.
41. Lal, R., Ed. Insecticide Microbiology; Springer-Verlag: Berlin, 1984.
42. Karns, J.S.; Mulbry, W.W.; Nelson, J.O.; Kearney, P.C. Pestic. Biochem. Physiol. 1986, 25, 211-217.
43. Gorder, G.W. Ph.D. Dissertation, Iowa State University, IA, 1980.
44. Alexander, M. Science 1981, 211, 132-138.
45. Bhaskaran, B.; Kandasamy, D.; Oblisami, G.; Subramaniam, T.R. Current Sci. 1973, 42, 835-836.
46. Cook, A.M.; Daughton, C.G.; Alexander, M. Appl. Environ. Microbiol. 1978, 36, 668-672.
47. Cook, A.M.; Hutter, R. Appl. Environ. Microbiol. 1982, 43, 781-786.
48. Dagley, S. Residue Rev. 1983, 85, 127-137.
49. Horvath, R.S. Bacteriol. Rev. 1972, 36, 146-155.
50. Gunner, H.B.; Zuckerman, B.M. Nature 1968, 217, 1183-1184.
51. Senior, E.; Bull, A.T.; Slater, J.H. Nature 1976, 263, 476-479.
52. Ou, L.T.; Sikka, H.C. J. Agric. Food Chem. 1977, 25, 1336-1339.
53. Ou, L.T.; Sharma, A. J. Agric. Food Chem. 1989, 37, 1514-1518.
54. Brock, T.D.; Smith, D.W.; Madigan, M.T. Biology of Microorganisms; Prentice-Hall: Englewood Cliffs, NJ, 1984; Chapter 7.
55. Smith, O.L. Soil Microbiology: A Model of Decomposition and Recycling; CRC Press: Boca Raton, FL, 1982; Chapter 12.
56. Paul, E.A.; Voroney, R.P. In Contemporary Microbial Ecology; Ellwood, D.C.; Hedger, J.N.; Latham, J.; Lynch, J.M.; Slater, J.H., Eds., Academic Press: London, 1980; pp 215-237.
57. Clarholm, M.; Rosswall, T. Soil Biol. Biochem. 1980, 12, 49-57.
58. McLaren, A.D. Environ. Lett. 1973, 5, 143-154.
59. Raven, P.H.; Evert, R.F.; Curtis, H. Biology of Plants; Worth: New York, NY, 1976; Chapter 2.
60. Hagedorn, C.; Holt, J.G. Can. J. Microbiol. 1975, 21, 353-361.
61. Wheelis, M.L. Ann. Rev. Microbiol. 1975, 29, 505-524.
62. Palleroni, N.J. In Bergey's Manual of Systematic Bacteriology; Krieg, N.R., Ed., Williams and Wilkins: Baltimore, MD, 1984; Volume 1, pp 141-199.
63. Matsumura, F.; Benezet, H.J. In Pesticide Microbiology; Hill, I.R.; Wright, S.J.L., Eds., Academic Press: London; 1978, Chapter 10.
64. Clarke, P.H. In Evolution in the Microbial World; Cambridge University Press: Cambridge, 1974; pp 183-217.
65. Slater, J.H.; Godwin, D. In Contemporary Microbial Ecology; Ellwood, D.C.; Hedger, J.N.; Latham, J.; Lynch, J.M.; Slater, J.H., Eds., Academic Press: London, 1980; pp 137-160.

66. Williams, P.A. In Companion to Microbiology; Bull, A.T.; Meadow, A.T., Eds., Cambridge University Press: Cambridge, 1978; Chapter 4.
67. Furukawa, K.; Chakrabarty, A.M. Appl. Environ. Microbiol. 1982, 44, 619-626.
68. Slater, J.H. In Current Perspectives in Microbial Ecology; Klug, M.J.; Reddy, C.A., Eds., American Society for Microbiology: Washington, DC; 1984, pp 87-93.
69. Brown, J.W.; Mitchell, J.W. Bot. Gaz. 1948, 109, 314-323.
70. Newman, A.S.; Thomas, J.R. Soil Sci. Soc. Proc. 1949, 160-164.
71. Audus, L.J. J. Sci. Food Agric. 1952, 3, 268-274.
72. Walker, R.L.; Newman, A.S. Appl. Microbiol. 1956, 4, 201-206.
73. Loos, M.A.; Schlosser, I.F.; Mapham, W.R. Soil Biol. Biochem. 1979, 11, 377-385.
74. Ou, L.T. Soil Sci. 1984, 137, 100-107.
75. Rahman, A.; Atkinson, G.C.; Douglas, J.A. New Zealand J. Agric. 1979, 139, 47-49.
76. Martin, A.R.; Roeth, F.W. Proc. North Centr. Weed Contr. Conf. 1979, 34, 51-52.
77. Doersch, R.E.; Harvey, A.G. Proc. North Centr. Weed Contr. Conf. 1979, 34, 58-59.
78. Gunsolus, J.L.; Fawcett, R.S. Proc. North Centr. Weed Contr. Conf. 1980, 35, 18.
79. Felsot, A.; Maddox, J.V.; Bruce, W. Bull. Environ. Contam. Toxicol. 1981, 26, 781-788.
80. Felsot, A.S.; Wilson, J.G.; Kuhlman, D.E.; Steffey, K.L. J. Econ. Entomol. 1982, 75, 1098-1103.
81. Read, D.C. Agric. Ecosyst. Environ. 1983, 10, 37-46.
82. Wilde, G.; Mize, T. Environ. Entomol. 1984, 13, 1079-1082.
83. Tollefson, J.J. Solutions 1986, January, 48-55.

RECEIVED January 22, 1990

THE PHENOMENON CHARACTERIZED

Chapter 2

Effects of Long-Term Phenoxyalkanoic Acid Herbicide Field Applications on the Rate of Microbial Degradation

Allan E. Smith[1] and Guy P. Lafond[2]

[1]Agriculture Canada, Research Station, Regina, Saskatchewan S4P 3A2, Canada
[2]Agriculture Canada, Experimental Farm, Indian Head, Saskatchewan S0G 2K0, Canada

There is evidence from a relatively few field studies to indicate that repeated treatments of 2,4-D and MCPA result in enhanced herbicide degradation rates as a result of adaptation by soil microorganisms. There is also some evidence that cross-enhancement can occur under field conditions, whereby previous applications of either 2,4-D or MCPA, result in enhanced breakdown of both herbicides.

The phenoxyalkanoic acid herbicides 2,4-D (2,4-dichlorophenoxyacetic acid) and MCPA (4-chloro-2-methylphenoxyacetic acid), developed during the Second World War (1, 2), were the first truly selective herbicides for the control of weeds in crops and heralded the start of modern chemical weed control. Success was immediate and their use increased dramatically with the annual production of 2,4-D in the United States rising from 6 million kg in 1950 to over 20 million kg by 1964 (2). Other herbicides based on phenoxyacetic, phenoxypropionic, and phenoxybutyric acids were soon developed for agricultural use in Europe and North America. Thus, 2,4,5-T (2,4,5-trichlorophenoxyacetic acid) and silvex (2-(2,4,5-trichlorophenoxy)propionic acid) were introduced in 1945 and 1953, respectively, for shrub and brushwood control. Mecoprop (2-(4-chloro-2-methylphenoxy)propionic acid) became available in 1956 and dichlorprop (2-(2,4-dichlorophenoxy)propionic acid) in 1961, both proving active against many weed species not controlled by 2,4-D or MCPA. The phenoxybutyric acids 2,4-DB (4-(2,4-dichlorophenoxy)-butyric acid) and MCPB (4-(4-chloro-2-methylphenoxy)butyric acid), introduced during the late 1950s, were able to control weeds in legume crops normally damaged by low concentrations of other phenoxy herbicides. Because of their effectiveness against annual and perennial weeds and other unwanted vegetation, and because of their relative low cost, these eight chemicals, known as

0097–6156/90/0426–0014$06.00/0

substituted phenoxyalkanoic acid herbicides are still extensively used throughout the world. Of these chemicals, 2,4-D and MCPA are used to the greatest extent.

Shortly after their introduction, it was reported that the phenoxyacetic acids were biologically degraded in soil with the carbon atoms of the herbicides being used as energy sources by the degrading organisms. Soil microorganisms capable of degrading phenoxyalkanoic acid herbicides were subsequently isolated and identified. Over the years, a considerable scientific literature has appeared on the degradation of phenoxyalkanoic acid herbicides, especially 2,4-D and MCPA, by soil microorganisms and on the various factors that affect their persistence.

Soils previously treated with certain herbicides often exhibit an increased ability to degrade these chemicals, seemingly as a result of a proliferation of microorganisms adapted to the herbicides in question. Such soils are said to have become "enriched". The enhanced degradation of herbicides in soil as a result of repeated applications has recently been reviewed (3). The phenomenon of cross-enhancement, or the ability of soils pretreated with specific herbicides to degrade other structurally related herbicides more rapidly than in untreated control soils, has been known for nearly 40 years. Both enhanced degradation and cross-adaptation have been observed under field conditions for phenoxyalkanoic acid herbicides (3-5).

The objectives of this review are to briefly summarize the various factors influencing the breakdown of the phenoxyalkanoic acid herbicides in soils of North America and Europe, and then to discuss the effects of their repeated field usage on the rate of their microbial degradation.

Mechanisms Resulting in Soil Degradation

Abiotic. With absorption maxima in the range 280 to 290 nM (6) phenoxyalkanoic acids are able to absorb radiation found in sunlight and undergo photochemical decomposition. Their breakdown in irradiated aqueous solutions is well studied (5). However, photochemical breakdown of phenoxyalkanoic acids by sunlight at the soil surface is not considered to be a major source of herbicidal loss (5), since the chemicals are applied as post-emergence applications to growing crops, and residues reaching the soil will thus be protected from the sun's radiation by the crop canopy.

Phenoxyalkanoic acids are formulated as esters and amine salts and chemical hydrolytic mechanisms are considered responsible for the rapid conversion of the phenoxyalkanoic esters to the corresponding phenoxy-alkanoic anion in moist soils (5). The dimethylamine salt of 2,4-D similarly undergoes chemical conversion in the soil, by dissociation, to the acid anion (5). However, abiotic processes do not appear to result in any significant degradation of phenoxyalkanoic acids in the soil (5).

Microbiological. The importance of soil microorganisms for the degradation of 2,4-D, MCPA, and 2,4,5-T was demonstrated, shortly after their introduction, by soil perfusion experiments (7, 8). Soil columns were continuously percolated with aqueous solutions containing high concentrations of the herbicide and the eluate was monitored on a regular basis for herbicide remaining. An initial slight reduction in herbicide concentration due to adsorption by soil colloids was followed by a lag-phase, during which there was little apparent herbicide loss. The lag-phase, ranging from 2 weeks for 2,4-D to 40 weeks for 2,4,5-T, was followed by rapid breakdown (7, 8). Further applications of the phenoxyacetic acid to the soil columns were rapidly degraded with no lag-phase. It was considered (7, 8) that during the lag-phase there was a build-up of microorganisms capable of metabolizing the herbicides and using them as energy sources. Once a sufficiently large population had been reached rapid breakdown of the phenoxyacetic acids occurred, so that additional applications of the chemicals to the enriched soils were also rapidly metabolized. Subsequent studies confirmed that microbiological degradation was the mechanism by which phenoxyalkanoic acids are broken down in the soil, and many soil bacteria and actinomycetes capable of effecting their breakdown have been isolated. These early studies have been extensively reviewed (7-11).

Over the years much has been made of the so-called lag-phase first reported by Audus (7, 8) from soil perfusion experiments. Many of the herbicide-degrading organisms could have been initially killed with the high herbicide concentrations and water-saturated conditions used in the perfusion studies. Also analytical procedures used for measuring the herbicide concentrations in the perfusates, based on ultra-violet absorption and plant bioassay, were nonspecific and insensitive. Much of the early microbiological work on the breakdown of the phenoxyalkanoic acid herbicides was carried out with isolated microorganisms in culture solutions using high substrate concentrations. Such conditions are vastly different from those occurring under field situations and in laboratory soils where much lower herbicide concentrations are used. Thus, the lag-phase was most likely the result of experimental and analytical conditions of the early investigators.

The effects of concentration on the degradation of 2,4-D in soil have been studied using (14C)labeled herbicide with the release of (14C)carbon dioxide being used as a measure of the rate of breakdown of the 2,4-D. In some of these studies, the loss of 2,4-D was considered to be biphasic with a slow initial evolution of (14C)carbon dioxide being followed by a more rapid release (12-14). In contrast, other studies have shown the evolution of (14C)carbon dioxide from (14C)2,4-D treated systems to be uniform with time (13, 15, 16). A similar phenomenon has been reported for the breakdown of (14C)MCPA in soil (17). In all the above experiments no attempts were made to specifically analyze for (14C)2,4-D or (14C)MCPA actually remaining, and it has been cautioned (4, 5) that evolution of (14C)carbon dioxide is not a true measure of the decomposition rate of (14C)phenoxyalkanoic acids in soils.

Radioactive studies in which solvent-extractable (14C) was characterized by chromatographic assay have revealed no lag-phase when low concentrations (<10 ppm) of (14C)2,4-D and (14C)2,4,5-T were incubated with moist soils (18-21). Degradation of several phenoxyalkanoic acids at soil concentrations of <5 ppm using specific gas chromatographic analytical techniques have indicated that breakdown of these herbicides followed apparent first-order kinetics with no evidence of a lag-phase or biphasic degradation (22-25).

It has been concluded that degradation rate, in soil, is probably dependent on herbicide concentration (5) with the breakdown approximating first-order kinetics at soil concentrations below about 5 ppm, while at increased herbicide concentrations biphasic breakdown occurs.

It must be remembered, however, that laboratory studies with uniform herbicide applications are not equivalent to field applications where the herbicide initially comes into contact with the soil surface. A herbicide applied to the soil surface at a rate of 1.0 kg/ha is roughly equivalent to 2 ppm assuming distribution in the top 5 cm (23). This would be equivalent to 4 ppm if distributed in the top 2.5 cm of field soil and 32 ppm if contained in the surface 0.32 cm. Thus, under field conditions, concentrations of phenoxyalkanoic herbicides at the soil surface at the time of application could be considerably higher than those used in some laboratory studies, and biphasic breakdown may be possible.

Since the phenoxyalkanoic acid herbicides are degraded in the soil by biological processes, factors that affect microbial activity will directly affect their breakdown. Soil pH, soil type, soil organic matter, herbicide formulation, and herbicide concentration can all influence the rate of microbial decomposition (4, 5). Greater effects are experienced with moisture and temperature, since these factors have a profound influence on microbial activity and thus on herbicide breakdown (4, 5). It has been concluded (5), that soil temperature above 10°C and moistures above the wilting point are necessary for biological degradation of phenoxyalkanoic acids.

Soils from around the world have been examined for 2,4-D degrading organisms, and microbial populations have been shown to decompose phenoxyalkanoic acids by direct metabolism with the microorganisms receiving energy for growth from the breakdown products, as well as by co-metabolism, where the degrading organisms derive no energy from the breakdown. These and related subjects on the microbial degradation of 2,4-D have recently been reviewed (4).

Repeat Treatments and Enhanced Degradation

Laboratory persistence studies with single applications of individual phenoxyalkanoic acids under controlled laboratory conditions and using herbicide specific analytical techniques have shown that these herbicides are rapidly degraded in warm moist soils (5). The microbial degradation is also influenced by the rate and frequency of herbicide application, resulting in enhanced degradation. As noted, the early soil perfusion studies (7, 8) indicated that soils treated with large amounts of MCPA, 2,4-D, and 2,4,5-T retained their ability to rapidly degrade subsequent applications of the same herbicide. It was also observed (7, 8) that the 2,4-D perfused soils would rapidly degrade treatments of

MCPA as well as 2,4-D though the rate of degradation of 2,4,5-T was not enhanced. Similarly, MCPA perfused soils resulted in a more rapid degradation of both 2,4-D and MCPA treatments than in untreated control soils (7, 8). Other laboratory experiments have shown an increase in the rate of breakdown of 2,4-D as a result of soil retreatment (15, 26-28). It was also demonstrated (28) that pretreatment of laboratory soil with 2,4-D resulted in an increased breakdown of (14C)MCPA, as measured by evolution of (14C)carbon dioxide, when compared to control experiments.

Field studies with single applications of phenoxyalkanoic acid herbicides have indicated that breakdown is rapid under temperature and moisture conditions that favour microbiological activity (5). Enhanced degradation of these herbicides, under field conditions, was first noted in the late 1940s. The use of plant bioassay procedures, led to the discovery that the persistence of 2,4-D, but not 2,4,5-T, was decreased by pretreatment of the soil with 2,4-D (26, 27). This enhanced breakdown was later confirmed using (14C)2,4-D and radiochemical analytical techniques (29). The breakdown of the (14C)2,4-D being more rapid in soil from the treated plots, tested 8 months after the last field application, than in soil from plots treated for the first time.

It was demonstrated using plant bioassay procedures that soil taken from plots receiving applications of MCPA in the spring and fall of each year showed an increased rate of disappearance of MCPA in the third and fourth years of the experiment (30). After further applications of MCPA, persistence was further decreased with the time for MCPA treatments to reach the limit of detection being reduced from three weeks after three previous applications to four days after 10 previous treatments (31). Soil that had received two annual applications of MCPA for 7 years, retained enhanced ability to rapidly degrade fresh additions for at least 178 weeks (32). Soil from the same plots that had received 9 previous field treatments of MCPA and exhibited enhanced degradation was also able to degrade MCPB with equal facility, though the breakdown of dichlorprop, mecoprop, and silvex was unaffected by the previous MCPA treatments (33). MCPB is known to undergo a very rapid metabolism in soil by a beta-oxidation mechanism to MCPA (25). Thus, a more likely explanation for the seeming enhanced breakdown of MCPB would be for the chemical to undergo a very rapid conversion to MCPA that would then be rapidly degraded in the soils pretreated with MCPA. It can also be argued that 2,4-D, formed in the soil from 2,4-DB as a result of a very rapid beta-oxidation in soil (23, 34), would undergo a similar enhanced breakdown in soils pretreated with 2,4-D.

The breakdown of repeated 2,4-D and MCPA applications made to a field soil has been monitored using a white mustard bioassay procedure (35). Repeated applications of the two herbicides resulted in a reduction in degradation time from 10 weeks for 2,4-D and 20 weeks for MCPA, after one treatment, to 4 and 7 weeks, respectively, after 19 annual applications. From these pretreated soils microbial isolates were extracted and, in mineral salt medium, were found to degrade 2,4-D and MCPA more rapidly than isolates from untreated control soils (35). The microbial isolates from the 2,4-D treatments would also rapidly degrade MCPA added to the aqueous medium and vice versa. However, 9 months after the 19th annual application there was no significant change in the numbers of organisms capable of utilizing 2,4-D and MCPA as carbon sources in soil collected from plots versus soil from control plots (35). Plots in Canada have been receiving annual spring applications of high (~1.12 kg/ha) and low (~0.42 kg/ha) rates of amine and ester formulations of 2,4-D since 1947, and MCPA amine since 1953. In the fall of 1987, after 40 successive applications of 2,4-D and 34 annual treatments of MCPA, soil samples were taken from the 0- to 15-cm and 15- to 30-cm depths and analyzed gas chromatographically. Residual amounts of 2,4-D and MCPA were less than 0.02 kg/ha indicating complete degradation of the herbicides (36). Under laboratory conditions, the breakdown of added (14C)2,4-D to soils treated for 40 years with amine and ester formulations of 2,4-D was rapid with no significant difference between the high and low field application rates (Table I). In control soils breakdown was slower.

Table I. Breakdown of Added (14C)2,4-D During Incubation with Moist Soil from Field Plots Treated with 40 Annual Applications of 2,4-D (Data from Ref. 36)

	Applied (14C)2,4-D remaining (%)*	
Plot treatment	4 Days	8 Days
High amine	12a	6a
High ester	15a	6a
Low amine	19a	7a
Low ester	19a	7a
Control	55b	9b

*Means (from three replicates) followed by a common letter are not significantly different at the 0.05 level according to Duncan's multiple range test.

Loss of added (14C)MCPA in soil from the plots after 35 annual applications of MCPA was also rapid with no significant difference in soils treated at the high or low rates (Table II). Degradation was slower in the soil from the control plots.

Table II. Breakdown of Added (14C)MCPA During Incubation with Moist Soil from Field Plots Treated with 35 Annual Applications of MCPA (Data from Ref. 36)

	Applied (14C)MCPA remaining (%)*	
Plot treatment	4 Days	8 Days
High amine	15a	8a
Low amine	35a	6a
Control	60b	42b

*Means (from three replicates) followed by a common letter are not significantly different at the 0.05 level according to Duncan's multiple range test.

Soil samples taken from these long-term Saskatchewan plots after 32 annual applications of the high rate of 2,4-D were found to contain (37) greater numbers of 2,4-D degrading organisms

(5700/g soil) than similar plots receiving the lower rate of the herbicide (2700/g soil), or untreated control plots (2500 organisms/g soil). Similar studies (38) have indicated that the numbers of 2,4-D metabolizing microorganisms in soil from plots after 35 consecutive treatments with the low 2,4-D rates were 20-fold greater (7200 organisms/g soil) than in soil from untreated control plots (300 organisms/g soil). It was also noted (38) that there were greater numbers of 2,4-D co-metabolizing microorganisms in the treated soils (60,000/g soil) compared to the control plots (30,000/g soil). It can therefore be considered that the increased degradation of 2,4-D, and probably MCPA, in the Saskatchewan long-term studies (36) are an indication of soil microbial adaptation in response to repeated treatments of these herbicides.

Conclusions

There is evidence from a relatively few studies to show that repeated treatments of 2,4-D and MCPA result in enhanced degradation rates under field conditions as a result of adaptation of soil microorganisms. Even less is known about the phenomenon of cross-enhancement under field conditions, where previous applications of either 2,4-D and MCPA may result in enhanced breakdown of both phenoxyalkanoic herbicides.

Several mechanisms have been proposed to explain how a soil microbial population becomes adapted to a particular pesticide to result in enhanced degradation and also to explain how this ability can be retained for months and even years without further pesticide applications. Although a discussion of this subject is considered beyond the scope of this paper, the topic has been extensively reviewed (3, 4, 39, 40).

Despite phenoxyalkanoic acid herbicides being commonly used throughout the world, and their soil breakdown being investigated for over three decades, surprisingly little research has been conducted on the enhanced breakdown of these herbicides in response to previous applications though this has been known for about 40 years. The phenoxyalkanoic herbicides are applied as post-emergence treatments, and their soil residual activity is of little practical significance for weed control. Thus, for the phenoxyalkanoic acid herbicides, enhanced breakdown in soil seems to have been considered more as an academic curiosity than an agronomic problem. Enhanced degradation does, however, have environmental advantages in reducing the residence time of the herbicides in the soil and thus lessening the possibility of chemical movement by leaching and run-off processes.

Literature Cited

1. Kirby, C. Hormone Weed Killers; The British Crop Protection Council; Lavenham Press: Lavenham, U.K., 1980; pp 1-55.
2. Peterson, G. E. Agric. Hist. 1967, 41, 243-253.
3. Roeth, F. W. Rev. Weed Sci. 1986, 2, 45-65.
4. Sandmann, E. R. I. C.; Loos, M. A.; van Dyk, L. P. Rev. of Environ. Contam. Toxicol. 1988, 101, 1-53.

5. Smith, A. E. Rev. Weed Sci. 1989, 4, 1-24.
6. Sirons, G. J.; Chau, A. S. Y.; Smith, A. E. In Analysis of Pesticides in Water; Chau, A. S. Y.; Afgan, B. K., Eds.; CRC Press: Boca Raton, 1982, Vol. 2; Chapter 3, pp 155-227.
7. Audus, L. J. Plant Soil 1951, 3, 170-192.
8. Audus, L. J. In The Physiology and Biochemistry of Herbicides; Audus, L. J., Ed.; Academic Press: London, 1964; Chapter 5, pp 163-206.
9. Freed, V. H.; Montgomery, M. L. Res. Rev. 1963, 3, 1-18.
10. Kaufman, D. D.; Kearney, P. C. In Herbicides: Physiology, Biochemistry, Ecology; Audus, L. J., Ed.; Academic Press: London, 1976; Vol. 2, pp 29-64.
11. Loos, M. A. In Herbicides: Chemistry, Degradation, and Mode of Action; Kearney, P. C.; Kaufman, D. D., Eds.; Marcel Dekker: New York., 1975, 2nd edn.; Vol. 1, pp 1-128.
12. Ou, L. T.; Rothwell, D. F.; Wheeler, W. B.; Davidson, J. M. J. Environ. Qual. 1978, 7, 241-246.
13. Parker, L. W.; Doxtader, K. G. J. Environ. Qual. 1982, 11, 679-684.
14. Parker, L. W.; Doxtader, K. G. J. Environ. Qual. 1983, 12, 553-558.
15. Fournier, J. C.; Codaccioni, P.; Soulas, G.; Repiquet, C. Chemosphere 1981, 10, 977-984.
16. Wilson, R. G.; Cheng, H. H. J. Environ. Qual. 1978, 7, 281-286.
17. Helweg, A. Weed Res. 1987, 27, 287-296.
18. McCall, P. J.; Vrona, S. A.; Kelley, S. S. J. Agric. Food Chem. 1981, 29, 100-107.
19. Ou, L. T. Soil Sci. 1984, 137, 100-107.
20. Smith, A. E.; Hayden, B. J. Proc. of the Europ. Weed Res. Soc. Symposium on the Theory and Practice of the Use of Soil Applied Herbicides, 1981, pp 156-162.
21. Walker, A.; Smith, A. E. Pestic. Sci. 1979, 10, 151-157.
22. Altom, J. D.; Stritzke, J. F. Weed Sci. 1973, 21, 556-560.
23. Smith, A. E. Weed Res. 1978, 18, 275-279.
24. Smith, A. E.; Hayden, B. J. Bull. Environ. Contam. Toxicol. 1980, 25, 369-373.
25. Smith, A. E.; Hayden, B. J. Weed Res. 1981, 21, 179-183.
26. Newman, A. S.; Thomas, J. R. Soil Sci. Soc. Am. Proc. 1949, 14, 160-164.
27. Newman, A. S.; Thomas, J. R.; Walker, R. L. Soil Sci. Soc. Am. Proc. 1952, 16, 21-24.
28. Soulas, G. Proc. of the Europ. Weed Res. Soc. Symposium on the Theory and Practice of the Use of Soil Applied Herbicides, 1981, pp 129-136.
29. Hurle, K.; Rademacher, B. Weed Res. 1970, 10, 159-164.
30. Kirkland, K. Weed Res. 1967, 7, 364-367.
31. Fryer, J. D.; Kirkland, K. Weed Res. 1970, 10, 133-158.
32. Fryer, J. D.; Smith, P. D.; Hance, R. J. Weed Res. 1980, 20, 103-110.
33. Kirkland, K.; Fryer, J. D. Weed Res. 1972, 12, 90-95.
34. Gutenmann, W. H.; Loos, M. A.; Alexander, M.; Lisk, D. J. Soil Sci. Soc. Am. Proc. 1964, 28, 205-207.

35. Torstensson, N. T. L.; Stark, J.; Goransson, B. Weed Res. 1975, 15, 159-164.
36. Smith, A. E.; Aubin, A. J.; Biederbeck, V. O. J. Environ. Qual. 1989, 18, 299-302.
37. Cullimore, D. R. Weed Sci. 1981, 29, 440-443.
38. Fournier, J.-C. Docteur es Sciences Naturelles Thesis, Universite de Perpignan, 1989.
39. Torstensson, N. T. L. In Interactions Between Herbicides and the Soil; Hance, R. J., Ed.; Academic Press: London, 1980; Chapter 6, pp 159-178.
40. Torstensson, N. T. L. In Herbicides. Progress in Pesticide Biochemistry and Toxicology; Hutson, D. H.; Roberts, T. R., Eds.; John Wiley: Chichester, 1987; Chapter 8, pp 249-270.

RECEIVED January 24, 1990

Chapter 3

Enhanced Carbamothioate Herbicide Degradation

Research in Nebraska

Fred W. Roeth[1], Robert G. Wilson[2], Alex R. Martin[3], and Patrick J. Shea[3]

[1]Department of Agronomy, University of Nebraska, Clay Center, NE 68933–0066
[2]Department of Agronomy, University of Nebraska, Scottsbluff, NE, 69631–0224
[3]Department of Agronomy, University of Nebraska, Lincoln, NE 68583–0915

The half-life of EPTC was reduced about 50% by enhanced degradation which occurred on EPTC reapplication. Enhanced carbamothioate herbicide degradation was associated with unsatisfactory shattercane control in about 60% of the fields receiving a repeat treatment of EPTC or butylate. Though dietholate extender effectively prevented enhanced EPTC degradation, dietholate was ineffective on reapplication. SC-0058 extender was effective in slowing EPTC degradation in soils previously treated with EPTC, EPTC + dietholate, and EPTC + SC-0058. Dietholate and SC-0058 did not increase EPTC persistence in soil without prior EPTC treatment. Reversion of EPTC degradation from an enhanced rate to a normal rate took about 18 and 30 months in western and south central Nebraska, respectively, after two annual EPTC applications. Among the common carbamothioate herbicides, EPTC, vernolate, and butylate caused enhanced degradation and exhibited cross-adaptation; however, butylate was only partially susceptible to EPTC-enhanced degradation. Cycloate did not cause enhanced degradation nor show any susceptibility to EPTC-enhanced degradation.

Use of EPTC (S-ethyl dipropylcarbamothioate) herbicide for weed control in corn (*Zea mays* L.) increased markedly in the mid-1970's following the introduction of dichlormid (2,2-dichloro-N,N-di-2-propenylacetamide) safener. EPTC was particularly effective for controlling shattercane (*Sorghum bicolor* (L.) Moench), a very troublesome weed in continuous corn. A sugarbeet (*Beta vulgaris* L.) - fieldbean (*Phaseolus vulgaris* L.) - corn rotation is common in portions of western Nebraska, and EPTC is used for weed control in all three crops. As a result, EPTC was applied annually in many fields.

0097–6156/90/0426–0023$06.00/0

Grower complaints of poor shattercane control with EPTC became numerous by 1977. Loss of efficacy was initially attributed to misapplication, inadequate incorporation, and adverse environmental conditions. However, the continuing widespread reports of unsatisfactory control with EPTC suggested other unknown factors must be involved. Field and greenhouse studies were initiated in 1978 to identify the reasons for poor field performance (1-3). Interestingly the problem appeared to be associated with repeated annual application. Though the extent of EPTC failure was unknown at the time, a 1983-84 survey revealed that 60% and 45% of corn growers in south central Nebraska who used butylate [S-ethyl bis(2-methylpropyl)carbamothioate] or EPTC were dissatisfied with their weed control (4). This was probably representative of the situation in 1978.

One hypothesis for decreased control was that shattercane had developed resistance to EPTC. Shattercane seeds were collected from soils where EPTC had failed in 1977 as well as fields where control had been satisfactory. In greenhouse studies, these shattercane sources did not differ in EPTC response from each other nor from a standard forage sorghum (Sorghum bicolor (L.) Moench) (1,3). Thus, herbicide resistance did not explain diminished EPTC performance. Additional greenhouse studies revealed that as shattercane seed population density increased in the soil, the number of plants escaping control increased proportionately (1). While a large shattercane soil seedbank could contribute to reduced control, this factor did not seem responsible for total failure.

The effect of corn stover on EPTC performance was evaluated in the greenhouse (3). At 1 ppm EPTC, bioactivity was reduced at stover levels of 3 and 6 g per 500 g of soil. The influence of the stover could be overcome by increasing the application rate to 2 ppm. Field studies involving residue removal prior to EPTC application showed minimal differences at normal EPTC rates (2).

Shattercane control (2) was evaluated in 1979 in fields with previous EPTC use (Saronville) and without previous EPTC use (Douglas). At Saronville the treatments of EPTC + dichlormid reduced shattercane head production only 35% compared to the untreated control (Table I). At Douglas, EPTC provided 90 to 95% shattercane control. The inclusion of an experimental extender, dietholate (O,O-diethyl O-phenyl phosphorothioate), with EPTC improved control at Saronville but not at Douglas. An extender is a chemical additive which can prolong the persistence of a pesticide. Follow-up greenhouse work showed that the extender did not increase EPTC activity but did retard EPTC degradation in soil treated previously with EPTC (5-6).

Laboratory and field experiments conducted in 1980 (5, 7-9) demonstrated enhanced degradation of EPTC in soils with previous EPTC use while dietholate effectively counteracted this effect. The cause of rapid degradation was determined to be microbiological. Enhanced EPTC degradation had not been previously observed in our research because herbicide evaluations were not conducted on the same site in consecutive years. Several authors (10, 11) have reviewed the history of enhanced herbicide degradation in detail so this paper concentrates on research conducted in Nebraska. Concurrent research in other states and countries contributed essentially to progress in understanding this phenomenon and appreciating its impact.

Table I. Shattercane control in corn at Saronville, NE, in 1979. All herbicides were applied preplant and incorporated immediately†

Herbicide[a]	lb/A	Stems/ m^2	Heads/ m^2	Corn Bu/A
Untreated	...	124	55	6
Handweed + EPTC	4.0	32	7	73
EPTC	4.0	84	36	18
EPTC	6.0	154	36	20
EPTC + cyanazine	4.0 + 2.0	94	35	22
EPTC[b]	4.0	26	5	43
EPTC[b]	6.0	18	3	42
EPTC[b] + cyanazine	4.0 + 2.0	3	1	66
LSD (.05)		83	4	23

[a] All EPTC was formulated with dichlormid.
[b] Formulated with dietholate extender (6:1).

Confirmation of Enhanced Degradation. Obrigawitch (12) compared EPTC persistence in soils without previous EPTC treatment (non-history soil) to soils treated with EPTC (EPTC-history) within the previous year. The soils were a Tripp very fine sandy loam (Aridic Haplustoll, pH 8.0, 1.0% organic matter) from Scottsbluff and a Kennebec sil (Cumulic Hapludolls, pH 6.6, 2.4% organic matter) from Lincoln. Following incorporation of ^{14}C-carbonyl-EPTC, $^{14}CO_2$ evolution was measured during a 56-day incubation at 75% moisture holding capacity and 25C. Greater $^{14}CO_2$ evolution indicated more rapid EPTC degradation in EPTC-history soil (Figure 1). The three-day lag phase observed in non-history Kennebec soil was significantly reduced in the EPTC-history soil. The initial lag phase was not significantly reduced in the Tripp soil although the rate of $^{14}CO_2$ evolution was greater after 9 days. The reduction in lag phase and greater $^{14}CO_2$ evolution showed that previous EPTC exposure enabled soil microorganisms to more rapidly degrade EPTC in subsequent application.

In a field study (5, 12) EPTC was applied at 6.7 kg/ha and incorporated into EPTC-history and non-history Kennebec soil. Soil samples were analyzed over a 50-day period for residual EPTC. EPTC was degraded more rapidly in the EPTC-history soil after the first rainfall on day 16 (Figure 2). EPTC had an apparent half-life of approximately 9 days as compared to 18 days in the non-history soil. These data confirmed the laboratory experiment and indicated that reduced EPTC activity could be attributed to more rapid degradation when applied in consecutive years.

An additional study (13) was conducted to determine if enhanced degradation could be induced in the laboratory. EPTC and butylate were applied to Kennebec soil with no prior carbamothioate history and residual concentrations determined over a 48-day period. The application and analytical cycle was repeated three times. One application of EPTC was sufficient to cause the maximum degradation rate of a subsequent dose (Figure 3). Butylate degradation increased after each butylate reapplication but the degradation rate for butylate remained lower than for EPTC. The soil enhancement pattern differed between the two herbicides and suggested some difference in their susceptibility to cause or undergo enhanced degradation.

Cross-Adaptation. To determine if other carbamothioate herbicides would degrade more rapidly in soils exhibiting enhanced EPTC degradation, the degradation of EPTC, and two similar carbamothioates, butylate and vernolate (S-propyl dipropylcarbamothioate), were monitored for 44 days following applications to non-history and EPTC-history Kennebec soil (13). Both vernolate and EPTC had half-lives of 3 days in EPTC-history soil compared to 17 and 13 days, respectively, in non-history soil (Figure 4). A single carbon increase in the *S*-alkyl group (vernolate) did not appear to significantly influence microbial utilization. Butylate degradation was more rapid in EPTC-history soil than in non-history soil, but the difference was less than that for EPTC or vernolate. Substitution of the *N*-diisobutyl group for the EPTC *N*-dipropyl may have partially decreased accelerated utilization by EPTC-adapted microorganisms. Enhanced degradation specificity has been reported for carbamate insecticides also (14).

Carbamothioate Extender. Field observations (6, 15-17) indicated that dietholate extender increased the efficacy of EPTC and butylate in EPTC-history soil. The effect of dietholate on EPTC and butylate persistence was monitored for 50 days after application to two EPTC-history and non-history soils. The EPTC-history soils had three and eight years of previous EPTC application. EPTC and butylate were degraded more rapidly in the EPTC-history soil, but EPTC degraded faster than butylate (Figures 5 and 6). Dietholate increased EPTC persistence in EPTC-history soil but not in non-history soil. An increase in butylate persistence in the presence of dietholate in both EPTC-history and non-history soils suggested that dietholate had an effect that was not limited to the EPTC-adapted microorganisms. The mechanism of dietholate action was unclear since it did not extend EPTC persistence in the non-history soil.

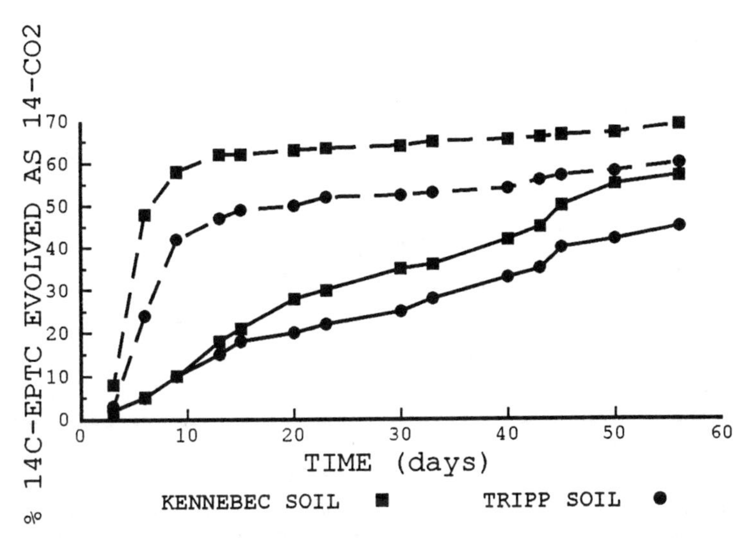

Figure 1. Degradation of 14C-EPTC in two soils with (dashed line) and without (solid line) previous EPTC application. (Reprinted with permission from ref. 12. Copyright 1982 Weed Science.)

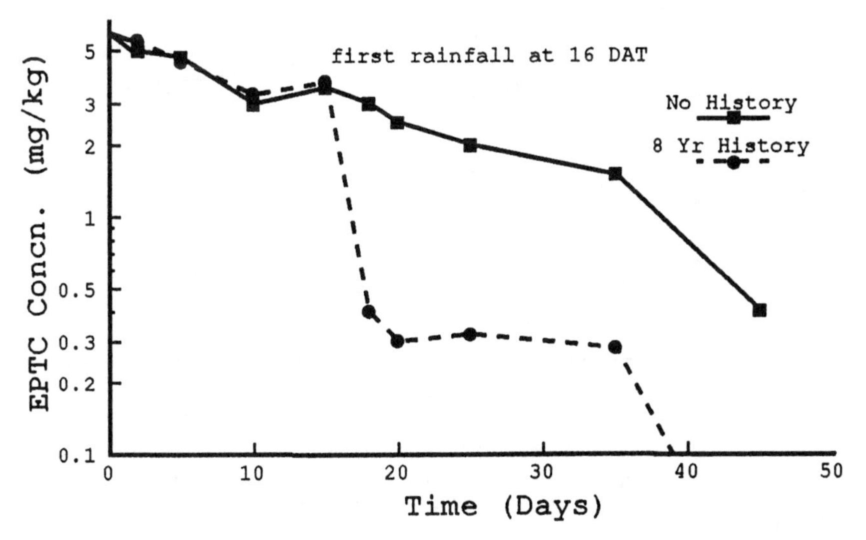

Figure 2. Field loss of EPTC in a Kennebec silt loam soil and without prior EPTC treatment.

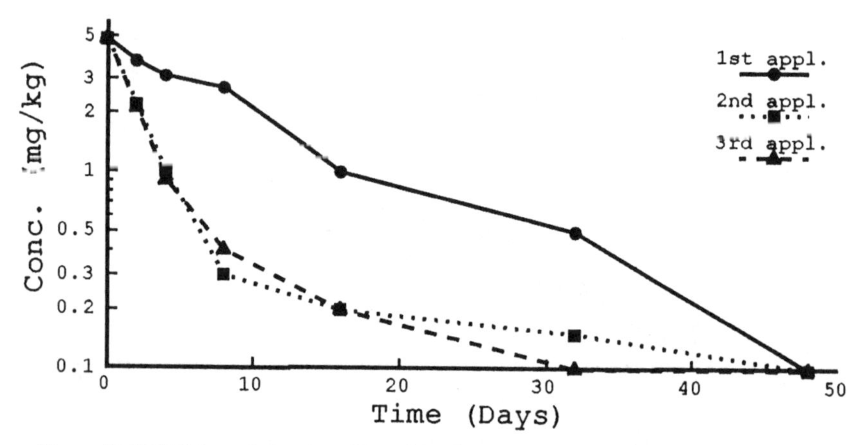

Figure 3. EPTC degradation in a Kennebec silt loam soil during the first, second, and third exposures to EPTC. The reapplication interval was 50 days. (Reprinted with permission from ref. 13. Copyright 1983 Weed Science.)

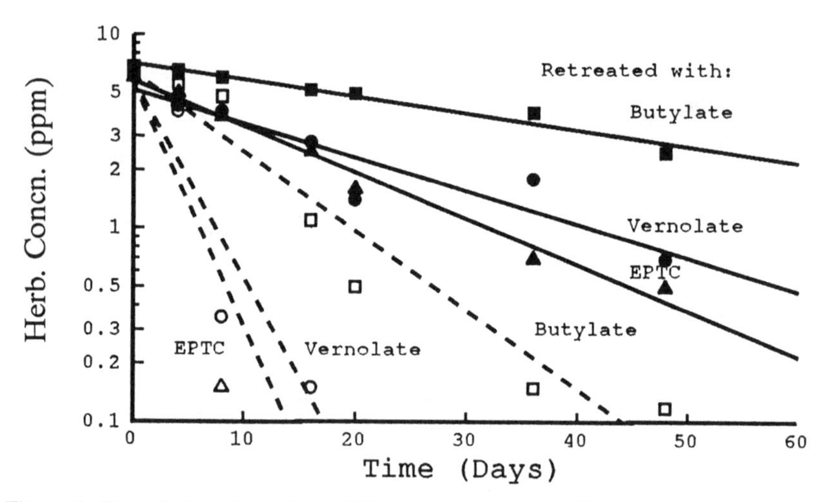

Figure 4. Degradation of butylate, EPTC, and vernolate in a Kennebec silt loam soil with eight annual applications of EPTC (dashed lines) or without previous EPTC applications (solid lines). (Reprinted with permission from ref. 13. Copyright 1983 Weed Science.)

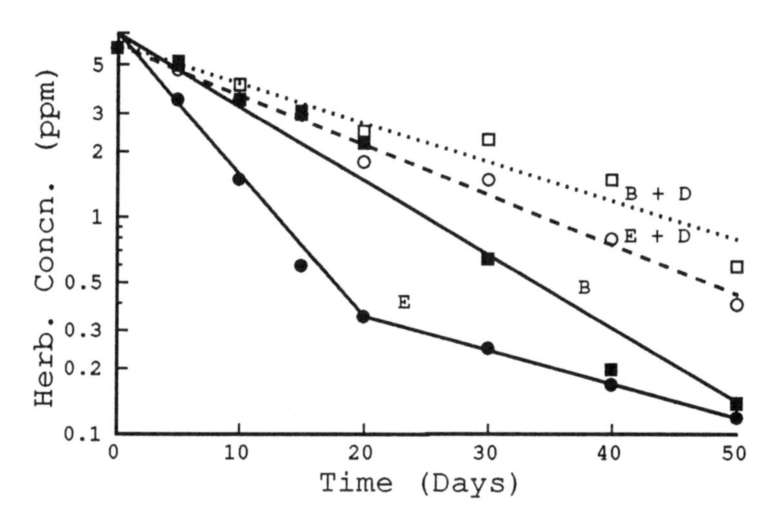

Figure 5. Dietholate(D) effect on butylate(B) and EPTC(E) degradation in a Hastings silt loam soil with three previous EPTC applications. (Reprinted with permission from ref. 13. Copyright 1983 Weed Science.)

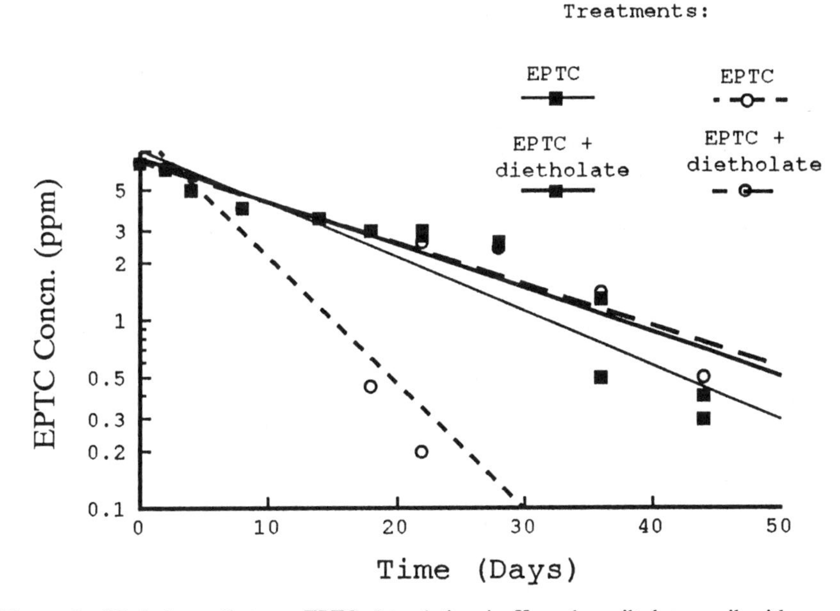

Figure 6. Dietholate effect on EPTC degradation in Kennebec silt loam soil with (dashed line) and without (solid linc) previous EPTC applications.

Controlled Studies. Three field sites were established in 1981 to more accurately examine enhanced carbamothioate degradation in soils with controlled herbicide histories (18-20). Soils and sites included a Sharpsburg silty clay loam at Mead (Typic Argiustoll, pH 6.4, 2.9% organic matter), Crete silty clay loam (Udic Argiustoll, pH 6.6, 2.7% organic matter) at Clay Center, and a Tripp very fine sandy loam at Scottsbluff. Soils at each location were annually cropped to corn and treated with the same annual pesticide treatments from 1981 through 1985. No carbamothioate pesticides had been previously applied.

One application of EPTC or butylate was sufficient to induce enhanced degradation of both herbicides applied the following year (19, 21). At Clay Center, EPTC degradation was most rapid in vernolate-history soil followed by EPTC-history soil (Figure 7). EPTC was degraded at an intermediate rate in butylate-history or carbofuran (2,3-dihydro-2,2-dimethyl-7 benzofuranyl methylcarbamate)-history soils. Previous carbofuran use did not reduce EPTC performance at Clay Center in 1984 (19). Prior soil treatment with atrazine (6-chloro-N-ethyl-N'-(1-methylethyl)-1,3,5-triazine-2,4-diamine), cyanazine (2-[[4-chloro-6-(ethylamino)-1,3,5-triazin-2-yl]] amino]-2-methylpropanenitrile), metolachlor (2-chloro-N-(2-ethyl-6-methylphenyl)-N-(2-methoxy-1-methylethyl)acetamide), alachlor (2-chloro-N-(2,6-diethylphenyl)-N-(methoxymethyl)acetamide) or cycloate (S-ethyl cyclohexylethylcarbamothioate) did not affect EPTC degradation. At Scottsbluff, EPTC degradation was equally enhanced in EPTC-history, vernolate-history, or butylate-history soils, but was not affected by the other pesticides.

The degree to which carbamothioate herbicides cause microbial adaptation depends on specific herbicide structure (19, 21, 22). Vernolate and EPTC have similar structures and soil exposure to either herbicide fully affected the degradation of the other. Butylate has a structure somewhat different than vernolate and EPTC and its degradation pattern was only moderately affected by prior use of EPTC and vernolate. Cycloate has a cyclohexane moiety that makes it unique from butylate, EPTC, and vernolate. The degradation rate of cycloate was not influenced by previous soil exposure to either EPTC, vernolate, or butylate (22), nor has prior cycloate treatment increased the degradation rate of the latter carbamothioates (19, 21).

EPTC and butylate sulfoxides are initial metabolites of EPTC and butylate degradation. EPTC and butylate sulfoxides induced enhanced degradation of their respective parent compounds (Figure 8) (19). Pretreatment of the soil with either EPTC sulfoxide or butylate sulfoxide caused cross-enhanced degradation of butylate or EPTC.

The degree and duration of enhanced EPTC degradation in soils varied with location and the number of years treated with EPTC (23). One prior EPTC application at Clay Center and Scottsbluff was sufficient to cause enhanced-EPTC degradation. At Scottsbluff, 18 months after the second annual EPTC application enhanced EPTC degradation did not occur. At Clay Center, soil with one previous EPTC treatment completely lost its ability to rapidly degrade EPTC after 24 months. Soil with two prior EPTC treatments did not fully

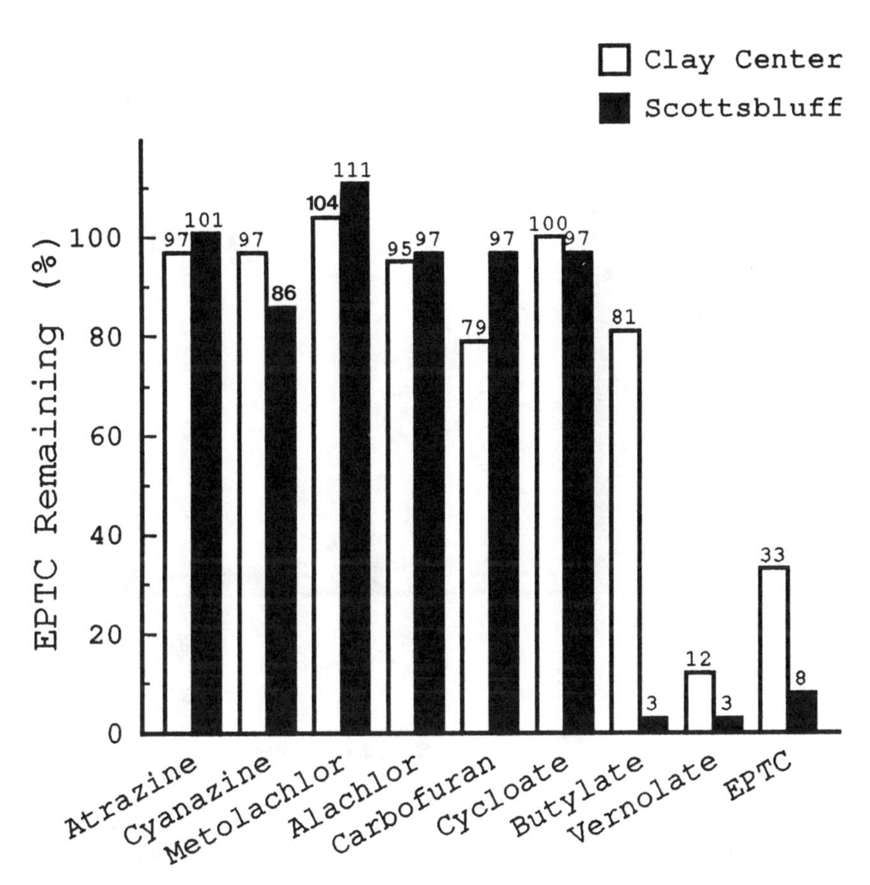

Figure 7. EPTC degradation in soil from two locations where various pesticides were applied for three consecutive years. Initial concentration was 6 ppm and samples were incubated for 3 days at CC and 4 days at SB. Non-history control=100%. Previous field treatment to retreatment interval was six months.

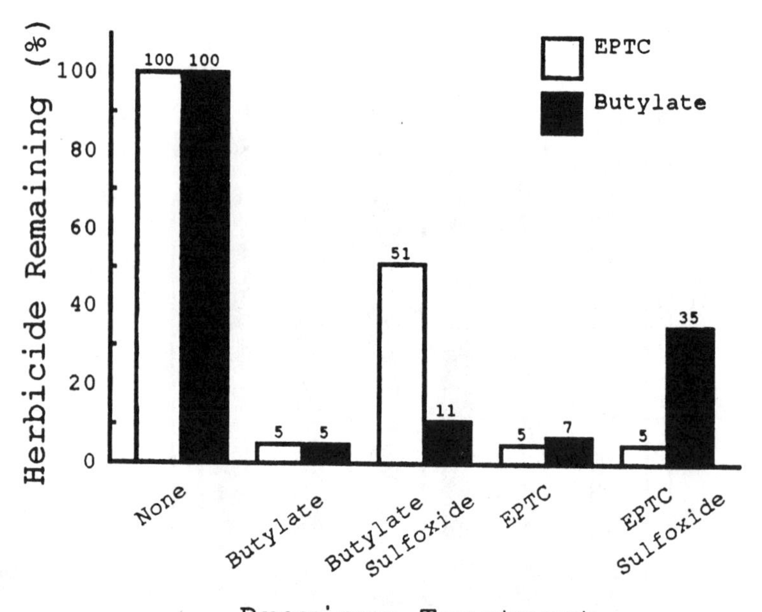

Figure 8. Butylate and EPTC degradation in Crete silty clay loam soil following previous exposure to butylate, EPTC, and their sulfoxides. Initial concentration was 6 ppm and samples were incubated for 3 days. Non-history control had 4 ppm remaining.

revert to a normal degradation rate until approximately 30 months after the last EPTC application.

The initial addition of dietholate to EPTC improved grass control and reduced the rate of EPTC degradation in soils exhibiting enhanced EPTC degradation (16). However, after multiple annual applications of dietholate plus EPTC, accelerated degradation of EPTC was again observed (24, 25). Several experiments have indicated that repeated application of dietholate did not induce enhanced dietholate degradation (11, 26), suggesting that degraders may have adopted an alternative degradation pathway.

Fonofos (O-ethyl-S-phenylether phosphonodithioate) insecticide and SC-0058 [S-ethyl di-(3-chlorallyl) carbamothioate] an experimental extender, have also reduced the degradation of EPTC in soils exhibiting enhanced EPTC degradation (27, 28). Fonofos was similar to dietholate in that enhanced degradation of EPTC was again observed after repeated annual applications of EPTC plus fonofos. SC-0058 was effective in slowing EPTC degradation in soils previously treated with EPTC, EPTC + dietholate, EPTC + fonofos, and EPTC + SC-0058. At Clay Center, EPTC degradation, though enhanced in all EPTC-history soils, was slower in soil with EPTC + SC-0058-history than in soils with EPTC or EPTC + dietholate histories (28).

Dietholate, fonofos, or SC-0058 did not extend the persistence of EPTC in non-history soils and were more effective in reducing enhanced EPTC degradation in soil from Scottsbluff than from Clay Center (28). There was also a tendency to have a reduced level of enhanced EPTC degradation in Scottsbluff soil when the treatment history included an extender with EPTC compared to EPTC without an extender (21).

A possible explanation for greater extender effectiveness at Scottsbluff than at Clay Center is that the EPTC-degrading microorganism population is probably lower at Scottsbluff (28). A higher population of EPTC-degraders may be able to degrade the extender faster. Tam et al. (29) reported that the inhibitory effect of dietholate was dependent on dietholate concentration and that an EPTC-degradation proficient bacteria could degrade dietholate.

Environmental factors such as water and temperature have an effect on EPTC persistence. The rate of EPTC degradation in Scottsbluff soil was moisture dependent below 25% moisture holding capacity but independent above that (9). EPTC degradation was more rapid at 15 and 25C than at 5C in a silt loam soil. If soil temperature and moisture were increased to 37C and field capacity for a week before EPTC was added to an EPTC-history soil, no enhanced degradation was observed. An EPTC-degrading bacterial isolate was shown to lose its degrading plasmid after incubation at 32° C for 16 hr. (30). The soil and its environment have considerable influence on EPTC degradation and are probably responsible for much of the differences in EPTC degradation rates observed at different locations in Nebraska and elsewhere (31, 32).

Coping with Enhanced Degradation. Combinations of atrazine with EPTC or butylate have been very effective for controlling grasses in corn even when enhanced degradation has rendered the carbamothioates ineffective when used alone (Table II). Since the carbamothioate herbicides are often combined with atrazine for maximum weed control, these herbicide combinations often mask enhanced degradation except where atrazine-tolerant weeds such as shattercane and wild proso millet (*Panicum milliaceum* L.) are present. The use of herbicide combinations is an effective strategy for coping with enhanced degradation as long as both herbicides are not mutually affected and provide complementary control.

Table II. Combinations of herbicides for grass control in corn at Clay Center, 1986-89. Evaluations were made about four weeks after planting

Herbicide	lb/A	% Weed Control[1]
Atrazine	3.0	94
EPTC	4.0	26
EPTC + atrazine	4.0 + 1.5	88
Butylate		4.020
Butylate + atrazine	4.0 + 1.5	91
lsd (.05)		15

[1] Weed was primarily green foxtail [*Setaria viridis* (L.) Beauv.]

Two-year herbicide rotations which include EPTC one year and another herbicide in the alternate year have kept EPTC more effective than when EPTC was applied every year (Table III). The particular alternate herbicide determines the degree of effectiveness. This may be related to the degree of weed control obtained in the alternate year (weed seed bank effect) rather than any direct effect on enhanced degradation, though the latter is possible. The carbamothioate cycloate, though the most effective alternate herbicide of the four tested, is not registered for use in corn. Rotation of herbicide combinations such as EPTC plus atrazine and cyanazine plus atrazine would be effective in controlling weeds in many situations.

Table III. Green foxtail control in corn with herbicides applied annually or in rotation at Clay Center, 1985-88. Evaluations were made about four weeks after planting and have been averaged across years

Two-Year Herbicide[1] Rotation[2]	Control With Each Herbicide[1]				
	E	C	M	B	A
			(control)		
Not rotated	23	74	96	17	40
E-C	72	83	-	-	-
E-M	87	-	98	-	-
E-B	50	-	-	-	-
E-A	64	-	-	-	58
First application in each year	86	-	-	-	48
lsd (.05) = 15					

1 E = EPTC, C = cyanazine, M = cycloate, B = butylate, A = alachlor
2 Treatments were initiated in 1981.

Acknowledgments

Special appreciation is extended to Brent Bean, Tim Obrigawitch, Gary Tuxhorn, and ICI Americas for their contributions to this research. We also acknowledge the help of many growers, extension agents, assistants, and colleagues who aided, counselled and shared their research results with us throughout these studies.

Literature Cited

1. Martin, A. R.; Roeth, F. W. 1978. No. Cent. Weed Cont. Conf. Proc. 33:108-109.
2. Martin, A.R.; Roeth, F.W. No. Cent. Weed Cont. Conf. Proc. 1979. 34, 51.
3. Roeth, F. W.; Martin, A. R. No. Cent. Weed Cont. Conf. Proc. 1979, 34, 51.
4. Roeth, F. W.; Wilson, R. G., Martin, A. R.; Shea, P. J. Weed Tech. 1989, 3, 24-29.
5. Obrigawitch; T. T., Martin, A. R.; Roeth, F. W. No. Cent. Weed Cont. Conf. Proc. 1980, 35, 20.
6. Obrigawitch; T. T.; Roeth, F. W.; Martin, A. R.; Wilson, R. G. Weed Sci. 1982, 30, 417-422.

7. Obrigawitch; T.T., Martin, A. R.; Roeth, F. W. No. Cent. Weed Cont. Conf. Res. Rep. 1980, 37, 196-198.
8. Roeth, F. W.; Obrigawitch, T. T.; Martin, A.R. No. Cent. Weed Cont. Conf. Res. Rep. 1980 37, 99-201.
9. Wilson, R. G.; Martin, A. R.; Roeth, F. W. Weed Cont Conf. Proc. 1980, 35, 77.
10. Harvey, R. G.; Dekken, J. H.; Fawcett, R. S.; Roeth, F. W.; Wilson, R. G. Weed Tech. 1987, 1, 341-49.
11. Roeth, F. W. Reviews of Weed Sci. 1986, 2, 45-65.
12. Obrigawitch; T. T., Wilson, R. G., Martin, A. R.; Roeth, F. W. Weed Sci. 1982, 30, 175-181.
13. Obrigawitch, T. T.; Martin, A. R.; Roeth, F. W. Weed Sci. 1983, 31, 187-192.
14. Racke, K. D.; Coats, J. R. J. Agric. Food Chem. 1988, 36, 1067-72.
15. Obrigawitch; T. T., Martin, A. R.; Roeth, F. W. No. Cent. Weed Cont. Conf. Proc. 1981, 36, 127.
16. Obrigawitch; T. T., Martin, A. R.; Roeth, F. W. No. Cent. Weed Cont. Conf. Res. Rep. 1981, 38, 166.
17. Roeth, F. W.; Obrigawitch, T. T.; Martin, A. R. No. Cent. Weed Cont. Conf. Res. Rep. 1981, 38, 164-165.
18. Bean, B. W.; Roeth, F. W.; Martin, A. R.; Wilson, R. G. No. Cent. Weed Cont. Conf. Proc. 1984, 39.
19. Bean, B. W.; Roeth, F. W.; Martin, A. R.; Wilson, R. G. Weed Sci. 1988, 36, 70-77.
20. Martin, A.R.; Roeth, F W. No. Cent. Weed Cont. Conf. Res. Rep. 1983. 40, 200.
21. Tuxhorn, G. L.; Roeth, F. W.; Martin, A. R.; Wilson, R. G. Weed Sci. 1986, 34, 961-965.
22. Wilson, R. G. Weed Sci. 1984, 32, 264-268.
23. Bean, B. W.; Roeth F. W.; Martin, A. R.; Wilson, R. G. Weed Sci. 1988 36, 524-530.
24. Harvey, R. G.; Mcnevis, G. R.; Albright, J. W.; Kozak, M. E. Weed Sci. 1986, 34, 773-80.
25. Rudyanski, W. J.; Fawcett, R. S.; McAllister, R. S. Weed Sci. 1987, 35, 68-74.
26. Wilson, R. G.; Rodebush, J. E. Weed Sci. 1987, 35, 289-94
27. Bean, B. W.; Roeth, F. W.; Martin, A. R.; Wilson, R. G. No. Cent. Weed Cont. Conf. Proc. 1985, 40, 11.
28. Bean, B. W.; Roeth, F. W.; Martin, A. R.; Wilson, R. G. Weed Science. 1990.
29. Tam, A.C.; Buhki, R. M.; Kahn, S. U. J. Agric. Food Chem. 1988, 36, 654-57.
30. Tam, A. C.; Buhki, R. M.; Khan, S. U. Appl. Environ. Microbiol. 1987, 53, 1088-93.
31. McCusker, V. W.; Skipper, H. D.; Zublema, J. P.; Gooden, D. T. Weed Sci. 1988, 36, 818,23.
32. Harvey, R. G. Weed Sci. 1987, 35, 683-89.

RECEIVED February 21, 1990

Chapter 4

Enhanced Biodegradation of Carbamothioate Herbicides in South Carolina

Horace D. Skipper

Department of Agronomy and Soils, Clemson University, Clemson, SC 29634–0359

Field and laboratory studies were conducted by an interdisciplinary team to: 1) confirm enhanced biodegradation, 2) ascertain microbial cross-adaptation, 3) determine microbial population shifts with repeated use of carbamothioates, 4) delineate the role of plasmids, and 5) investigate chemical/crop rotations and inhibitors to prevent or improve problem soils. Bioassays and evolution of $^{14}CO_2$ from ^{14}C-labeled herbicides confirmed accelerated biodegradation of butylate in soils with repeated use of butylate and cross-adaptation for EPTC, vernolate, and pebulate but not cycloate. Significant increases in bacteria or actinomycete populations were detected in soils treated with multiple applications of butylate, EPTC, or vernolate. Plasmids were associated with the microbial degradation of carbamothioate herbicides by a *Flavobacterium* sp. and a *Methylomonas* sp. isolated from herbicide-history soils. Chemical rotations were critical in the prevention of problem soils. A minimum of two years between applications of butylate or EPTC plus dietholate was essential to insure continued herbicidal efficacy. Enzyme inhibitors began to lose their effectiveness after 3 applications.

Enhanced biodegradation of pesticides has received considerable attention in recent years since it was first described by Audus (1) for the herbicide 2,4-D. Diphenamid, a soil applied herbicide, is also subject to enhanced biodegradation by soil microorganisms (2,3). Fungicides (4,5) and insecticides (6-9) are also subject to enhanced degradation by soil microorganisms. The carbamothioate herbicides are readily degraded by microbes (10,11) and especially after repeated applications (12-19).

0097–6156/90/0426–0037$06.00/0

One approach to restoration of herbicidal activity in problem soils has been the use of microbial/enzymatic inhibitors (20). Stauffer Chemical Company (now ICI Americas) has provided leadership for the agrochemical industry in this area and provided potential inhibitors to university scientists for validation. Kaufman et al. (21) had earlier suggested the use of inhibitors to improve the performance of herbicides with short residual lives in soils. Kaufman et al. (3) discussed the role of pesticides as inducers, substrates, and/or inhibitors of degradative enzymes in soil microorganisms and gave selected examples of specific pesticides as inducers and others as inhibitors of certain enzymes. They also considered the potential of "multiproblem" soils.

In this review, a number of questions were addressed since 1982 to resolve the role of enhanced biodegradation of carbamothioate herbicides in South Carolina soils: 1) Does enhanced biodegradation of carbamothioate herbicides result in performance failures? 2) Do environmental factors control, regulate, or modify the role of enhanced biodegradation in the performance of herbicides? Can a critical factor such as soil moisture negate or mask enhanced biodegradation of pesticides? 3) How does cross-adaptation fit into the puzzle? The carbamothioate herbicides are essential management tools for farmers in the southeastern United States. They are environmentally friendly, effective against a broad range of annual weeds and certain hard-to-control perennials, and are safe across multiple crops. If a soil becomes enhanced for the degradation of butylate in corn, can producers use EPTC or vernolate in their crop rotation in the same field? 4) Which soil microorganisms are responsible for enhanced biodegradation? If pesticide-degraders are identified and isolated in pure culture, how can we prove they are really functional in soils? 5) Are the genes responsible for degradation located on chromosomes or plasmids? Are the degradative genes stable within specific microorganisms or are they readily transferred within the microbial community? 6) Are there solutions to enhanced biodegradation? Can inhibitors be used to preserve environmentally friendly pesticides? Can organic/inorganic amendments be added to soils to retard enhanced biodegradation and insure an adequate concentration of the pesticide to control the specific pest(s) during the first 4 to 6 weeks that are critical to crop growth? Have standard crop/pesticide rotations been overlooked as potential solutions? 7) When we are given a lemon in life, can we make lemonade? Is there a positive use(s) of enhanced biodegradation--perhaps the microorganisms (enzymes/genes) could be used to clean up polluted sites?

To answer these questions, a number of techniques were employed by an interdisciplinary team which used applied and basic research, field and laboratory studies, bioassays and ^{14}C assays, whole plants and DNA aspects in a systems approach to enhanced biodegradation.

Field Research

In South Carolina, the first evidence of enhanced biodegradation of carbamothioate herbicides in Coastal Plains soils was generated in a Wagram loamy sand (loamy, siliceous, thermic Arenic Paleudults) and a Varina sandy loam (clayey, kaolinitic, thermic Plinthic Paleuldults) with a 7- and 2-year history, respectively, of annual applications of butylate at 6.7 kg/ha in corn. In these butylate-history soils, concurrent use of fonofos (O-ethyl-S-phenylether phosphonodithioate) and dietholate (O,O-diethyl-O-phenylphosphorothioate) significantly improved the control of large crabgrass (*Digitaria sanguinalis*) with butylate and EPTC, respectively. The increased efficacy was greater for butylate than for EPTC and generally greater for the Varina soil than the Wagram soil. These results agree with the specific herbicide use and length of continuous herbicide application. When soybeans were planted in these butylate-history soils, dietholate did not improve the performance of vernolate. Data indicated the butylate-adapted microorganisms were cross-adapted to EPTC but not to vernolate (19).

In Georgia, Dowler et al. (22) reported enhanced biodegradation of butylate in butylate-history soils, but dietholate effectively reduced the microbial degradation of butylate in their soils. In contrast, enhanced biodegradation of alachlor or metolachlor was not detected after six consecutive annual applications at 2.2 and 1.7 kg/ha, respectively.

In Piedmont soils of South Carolina with at least 3-years of butylate at an annual rate of 6.7 kg/ha, enhanced biodegradation of butylate was detected in the Congaree sandy loam (fine-loamy, mixed, nonacid, thermic Typic Udifluvents) but not in the Chewacla sandy loam (fine-loamy, mixed, thermic Aquic Fluventic Dystrochrepts). Deep tillage (turn-plow) of the Chewacla soil apparently diluted the population of butylate-degraders sufficiently to overcome a problem soil. The addition of dietholate to butylate restored the efficacy of butylate to a commercially acceptable level. EPTC plus dietholate (Table I) also gave excellent control of johnsongrass (*Sorghum halepense*). Tal et al. (23) and Tam et al. (24) found dietholate to be an

Table I. Performance of butylate, butylate plus dietholate, and EPTC plus dietholate 4 weeks after planting in two butylate-history soils near Clemson, SC

Treatment	Chemical rate	Control[a] of Johnsongrass	
		Chewacla	Congaree
	(kg/ha)	(%)	
1. Butylate	6.7	93 b	43 b
2. Butylate + dietholate	6.7 + 1.1	94 b	93 a
3. EPTC + dietholate	6.7 + 1.1	98 a	92 a

[a]Means within a column followed by different letters are significantly different at the 5% level by the LSD test.

effective inhibitor of microbial degradation of EPTC in soils or pure cultures, respectively.

For control of yellow nutsedge (*Cyperus esculentus*) in soybeans and peanuts, vernolate is a key component in growers' management programs. However, after 2-years of continuous use at 2.2 kg/ha, vernolate alone gave marginal control of yellow nutsedge. Dietholate significantly improved the weed control with vernolate in peanuts (Table II). Fonofos, an insecticide, was also effective in restoring the efficacy of vernolate (25).

When herbicides are used in crop rotations, the grower is concerned about the effects of potential herbicide residues on subsequent crops. However, a key question associated with enhanced biodegradation is the potential impact of one herbicide on the subsequent performance of a different herbicide, especially if they are structurally similar. If soil microorganisms adapt to one herbicide and produce a problem soil, would these same microorganisms be cross-adapted to a second herbicide and render it ineffective? Thus, a problem soil might be generated for not just one but two herbicides.

To address the issue of cross-adaptation, two sites in South Carolina and one in North Carolina with butylate-history soils were used to investigate cross-adaptation to EPTC, vernolate, pebulate, and cycloate. The herbicides were preplant incorporated with and without the addition of dietholate in a 6:1 ratio. Grain sorghum was then planted at 1, 8, or 12 days after treatment (DAT) as a bioassay to measure the performance of the

Table II. Performance of vernolate, vernolate plus dietholate, and other herbicides 4 and 6 weeks after planting in a two year vernolate-history soil near Sumter, SC

Treatment	Chemical rate	Control[a] of Yellow Nutsedge	
		4 WAP	6 WAP
	(kg/ha)	(%)	
1. Vernolate	2.2	71	30
2. Vernolate + dietholate	2.2 + 0.4	87	77 a
3. Vernolate	2.8	62	32
4. Vernolate + dietholate	2.8 + 0.5	80 a	81 a
5. Vernolate + benefin	2.8 + 1.3	74	42
6. Vernolate + dietholate + benefin	2.8 + 0.5 + 1.3	87	82 a
7. Benefin	1.3	0	3
8. Vernolate + napropamide	2.8 + 1.1	72	30
9. Vernolate + dietholate napropamide	2.8 + 0.5 + 1.1	86	87 a
10. Vernolate + fonofos	2.8 + 4.5	84	75

[a]Indicates the vernolate versus vernolate plus dietholate means were significantly different at the 5% level (Trt 1 vs Trt 2 or Trt 3 vs Trt 4). Modified from Ref. 25

carbamothioate herbicides. Control of the grain sorghum and ragweed (*Ambrosia artemisiifolia*) was evaluated four and seven weeks after treatment (WAT), respectively. As shown in Table III, butylate and EPTC along failed to control the grain sorghum even at 1 DAT and butylate was less effective than EPTC. The addition of dietholate significantly improved the performance of both butylate and EPTC. The improved efficacy associated with dietholate use was relatively greater for butylate than EPTC, and this may reflect the butylate-history of this soil. Thus, enhanced biodegradation of butylate and cross-adaptation for EPTC was readily apparent under these conditions. In contrast, dietholate did not alter the performance of vernolate, pebulate, or cycloate to indicate the butylate-degrading microorganisms were not cross-adapted to these related herbicides.

Field experiments were conducted to determine the effect of multiple applications of butylate alone or with enzyme inhibitors on the efficacy of butylate. A Dothan loamy sand (fine, loamy, siliceous, thermic Plinthic Paleudults) not previously treated with butylate was used for these studies. Dietholate and the experimental inhibitors, SC-0058 [S-ethyl di(3-chloroallyl)thiocarbamate), SC-0520 (chemistry not disclosed), and SC-7432 (diallylamine) were evaluated for their efficacy as inhibitors of microbial degradation (26). Treatments of butylate were applied at six-week intervals to the same plots for a total of four cycles in 1984. In 1985, the sequence was repeated using only dietholate and SC-0058 for a total of eight cycles.

Table III. Performance of carbamothioate herbicides in a butylate-history soil near Dobson, NC

Treatment	Chemical rate	Control[a] of Grain Sorghum-4 WAT 1 DAT	Control[a] of Grain Sorghum-4 WAT 12 DAT	Control[a] of Ragweed-7 WAT 0 DAT
	(kg/ha)	(%)	(%)	(%)
1. Butylate	4.5	21	5	0
2. Butylate + dietholate	4.5 + 0.7	90 a	30 a	25 a
3. EPTC	4.5	58	6	8
4. EPTC + dietholate	4.5 + 0.7	98 a	95 a	71 a
5. Vernolate	2.2	96	61	38
6. Vernolate + dietholate	2.2 + 0.4	100	62	48
7. Pebulate	4.5	100	80	52
8. Pebulate + dietholate	4.5 + 0.7	99	90	66
9. Cycloate	4.5	76	41	85
10. Cycloate + dietholate	4.5 + 0.7	78	32	74

[a]Indicates the herbicide versus herbicide plus dietholate means were significantly different at the 5% level (Trt 1 vs Trt 2 or Trt 3 vs Trt 4) Adapted from Lawrence, E. G. et al., Weed Science, in press.

Bioassays were planted at weekly intervals after each application cycle. Non-treated plots were included to evaluate butylate performance on a first-time application basis in each cycle.

Results indicated that enhanced biodegradation of butylate was initiated with a second application after just six weeks and that inhibitors were effective in restoring or exceeding initial control levels. Inhibitory effects of dietholate decreased in Cycle 5 while SC-0058 began to lose its efficacy in Cycle 7 (Figure 1). SC-0520 and SC-7432 were not effective as inhibitors of biodegradation (26). Dietholate was an effective inhibitor for various periods of time in other soils (27,28).

Another approach to the prevention or reduction of enhanced biodegradation of pesticides involves the use of chemical/crop rotations. In 1984, a five-year study with herbicides and crops was established in a Dothan soil to evaluate the effect of these rotations on enhanced biodegradation. Sixteen treatments were evaluated including continuous corn and herbicide, continuous corn with herbicide rotation, and crop rotation (corn/soybeans) with herbicide rotation. Herbicides investigated included butylate, EPTC plus dietholate, alachlor, vernolate, and trifluralin.

The efficacy of EPTC plus dietholate began to decline after two annual applications and was totally ineffective after five annual applications (Table IV). By inclusion of alachlor in corn or alachlor in soybeans in the rotation, a high degree of large crabgrass control was maintained. Inclusion of vernolate in soybeans in the rotation was detrimental to the subsequent performance of EPTC plus dietholate (29). Two or three years rotation away from butylate or EPTC plus dietholate were required to maintain adequate performance of these products and to prevent the development of enhanced biodegradation (Table V). It was critical to maintain the herbicide rotation because consecutive

Table IV. Control of large crabgrass by EPTC plus dietholate in corn as affected by crop and herbicide rotations

Rotations		Control[a] of large crabgrass		
Crop	Herbicide	1984	1986	1988
		(%)		
Corn	EPTC[b]	95 a	73 b	8 c
Corn	EPTC/alachlor	78 ab	95 a	70 a
Corn/soybean	EPTC/alachlor	94 a	100 a	79 a
Corn/soybean	EPTC/vernolate	69 ab	94 a	48 b

[a] Means within a column followed by different letters are significantly different at the 5% level by the LSD test.
[b] EPTC plus dietholate. Modified from Ref. 29.

applications of butylate or EPTC plus dietholate gave poor weed control. Rotations that included alachlor or trifluralin resulted in better performance of butylate and EPTC plus dietholate than rotations with vernolate, a closely related carbamothioate herbicide. Rotation of the crop was less effective than rotation of the herbicides in the prevention or retardation of enhanced biodegradation. Rotations of herbicides also was important under midwestern United States conditions (13,18,27,30) to prevent or alleviate enhanced biodegradation.

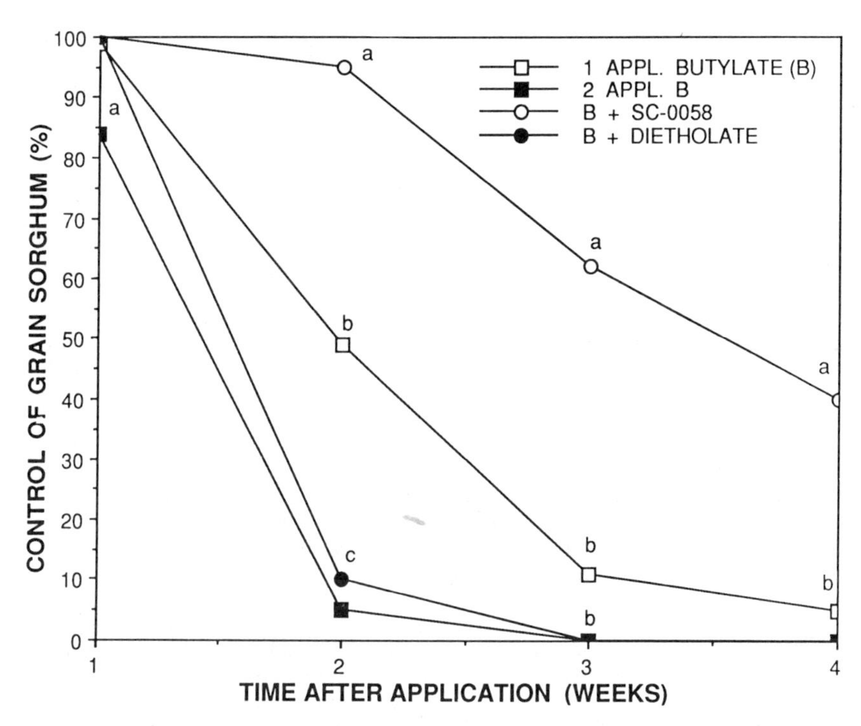

Figure 1. Control of grain sorghum by butylate (B) alone or combined with dietholate or SC-0058 in Cycle 6. Means followed by the same letter within each time are not significantly different at the 5% level by the LSD test. (Modified from Ref. 26).

Table V. Control of large crabgrass in corn with rotations of butylate (B) or EPTC plus dietholate (E) with alachlor (A) or trifluralin (T) in soybeans

Herbicide sequence					Control[a] of large crabgrass		
84	85	86	87	88	84	87	88
					(%)		
B	A	A	B		95 a	100 a	
B	T	T	B		95 a	100 a	
E	A	A	E		95 a	100 a	
B	A	A	A	B	95 a		92 a
B	T	T	T	B	95 a		97 a
E	A	A	A	E	95 a		95 a

[a]Means within a column followed by different letters are significanlty different at the 5% level by the LSD test. Modified from Ref. 29.

Laboratory Research

Degradation Studies. Some laboratory studies on enhanced biodegradation have involved use of gas chromatographic analyses while other studies have included measurement of $^{14}CO_2$ resulting from ^{14}C-labeled pesticide degradation to assess the severity of the problem, delineate the impact of environmental factors, or determine cross-adaptation (31). In our soil microbiology/herbicide research program at Clemson University, the use of $^{14}CO_2$ assays has been a major analytical tool (32,33). Cross-adaptation is defined as greater degradation of another herbicide in the butylate-history soil than degradation in a corresponding non-butylate-history soil (15). A similar definition for enhanced biodegradation could be applied to soils in which ^{14}C-butylate is applied to butylate-history soils. In general, the increased degradation was 5 to 20% greater for history vs non-history soils.

Butylate was degraded more rapidly in butylate-history soils (Dothan, Varina, and Wagram) than in each respective non-butylate-history soil. The soil microorganisms adapted so rapidly to butylate in the Dothan soil that after 22 days, there were no difference detectable in $^{14}CO_2$ evolved between the Dothan control soil and the Dothan butylate-history soil (Figure 2). From $^{14}CO_2$ studies in butylate-history soils, the butylate-adapted microorganisms in all three soils were cross-adapted to EPTC. Whereas, the butylate-adapted microbes in the Dothan soil were also cross-adapted to vernolate and pebulate. There was no cross-adaptation for cycloate in any of the butylate-history soils (15). An explanation for these differences (Table VI) is not readily apparent, but may be related to qualitative or quantitative differences in the soil microbial populations as suggested previously for biodegradation of atrazine (34).

In vernolate-history soils, the vernolate-adapted microorganisms were cross-adpated for EPTC and pebulate in both Dothan and Vaucluse soils with a history of vernolate use (25). The vernolate-adapted microorganisms were not cross-adapted for butylate or cycloate as measured by evolution of $^{14}CO_2$ (Table

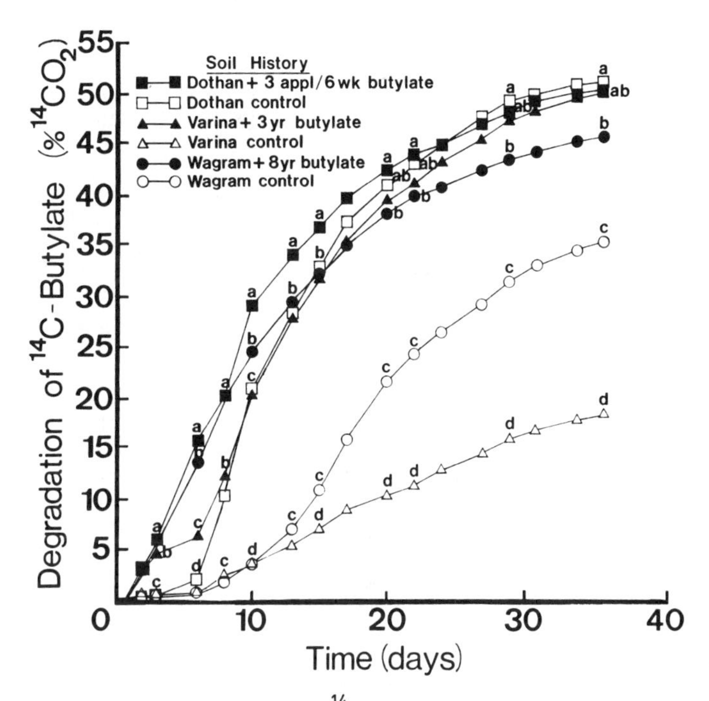

Figure 2. Accumulation of $^{14}CO_2$ evolved from soils with and without prior butylate use and treated with ^{14}C-butylate. Means followed by the same letter within a day are not significantly different at the 5% level by the LSD test. (Reproduced with permission from Ref. 15. Copyright 1988 Weed Sci. Soc. Am.).

VI). Overall, there has been a good agreement between the field and laboratory assessments for cross-adaptation. However, the cross-adaptation of butylate-adapted microoganisms in Dothan soil for vernolate and pebulate based on $^{14}CO_2$ assays has not been confirmed under field conditions.

At various times, poor incorporation or adverse environmental conditions have been blamed as causes for performance failures by

Table VI. Cross-adaptation[a] for carbamothioate herbicides in butylate- or vernolate-history soils

Soil herbicide history	Soil	Herbicide				
		Butylate	Cycloate	EPTC	Pebulate	Vernolate
Butylate	Dothan	NA[b]	--	X[a]	X	X
Butylate	Varina	NA	--	X	--	--
Butylate	Wagram	NA	--	X	--	--
Vernolate	Dothan	--	--	X	X	NA
Vernolate	Vaucluse	--	--	X	X	NA

[a]Cross-adaptation is defined as greater degradation of another herbicide in the butylate (or vernolate)-history soil than degradation in a corresponding non-herbicide-history soil.
[b]Enhanced biodegradation of butylate or vernolate.

pesticides when enhanced biodegradation was the culprit. Environmental factors may indeed temporarily inactivate the microorganisms responsible for enhanced biodegradation. Thus, although sufficient numbers of adapted microorganisms are present, due to their inactivity enhanced rates of herbicide degradation are not observed and the herbicide provides the expected weed control. Even under conditions more suitable for microbial activity and rapid pesticide degradation, a pesticide can provide apparent satisfactory pest (weed, insect, disease, nematode) control if the pest population is below the threshold level needed for a performance failure.

Soil moisture is a key environmental factor that regulates or modifies the efficacy of pesticides. Under dry field conditions, the performance of butylate in a Wagram butylate-history soil approached that of butylate in a non-butylate-history soil (22). To investigate the influence of soil moisture on enhanced biodegradation, a Varina butylate-history soil was adjusted to 10, 25, 50, or 100% of field capacity (FC; 0.05 bars or -5 kPa). At 10 and 25% FC, evolved $^{14}CO_2$ from ^{14}C-butylate was reduced by 80 to 90%. There were no differences in total $^{14}CO_2$ between 50 and 100% FC after 20 days (35). Similar moisture effects were observed for the degradation of vernolate in a vernolate-history soil (25). Degradation rate of EPTC was dependent on soil moisture below 3% and independent of soil moisture above 3% (36).

Pesticide-treated soils may be stored for various lengths of time before they are used in laboratory studies. With problem soils, one wonders if they lose their degradative potential upon storage. To address this question, butylate-history soils were stored at 4°C for up to two years in plastic bags. Even after two years of storage, neither the enhanced biodegradation of butylate nor the cross-adaptation to EPTC were significantly affected (35). Lee et al. (37) stored EPTC-history soils for one year at ambient temperatures without altering the rapid decomposition rate. Storage of Wisconsin soils for one year at 15 or 25°C was detrimental to enhanced biodegradation; whereas, soils stored at 5°C retained their enhanced degradative ability (27).

In general, increasing the number of consecutive applications of a pesticide susceptible to enhanced biodegradation increases the severity of the problem under field conditions. Thus, one would expect a greater amount of evolved $^{14}CO_2$ from ^{14}C-butylate-treated soil with more applications of butylate. As shown in Figure 3, there was good agreement between the amount of $^{14}CO_2$ evolved and the number of applications of butylate at six week intervals in a Dothan soil (38).

Ecological and Genetic Studies. If one observes a decrease in pesticide performance or a failure due to enhanced biodegradation, then an increase in pesticide-adapted microbial numbers or enzymatic activity would be expected. To enumerate and isolate herbicide-degraders from enhanced biodegradative soils, a minimal salts (MS) medium (39) containing 200 mg/L herbicide as the carbon source and 25 mg/L TTC (2,3,5-triphenyltetrazolium chloride; 40) as an indicator of organisms capable of utilizing the herbicide was employed. When active metabolic degradation of a carbon source (pesticide) occurs, excess electrons from metabolism reduce the TTC from colorless to red. Thus, an active pesticide-degrader appears as a red colony.

Butylate-history soil contained more TTC-red actinomycetes on the MS + TTC + butylate medium and vernolate-history soil contained more TTC-red bacteria than the corresponding non-herbicide-history soils (Table VII; 41). Some cross-adaptation between the carbamothioate herbicides was also reported where the butylate-history soil contained more TTC-red actinomycetes on the MS + TTC + vernolate medium than the control soil. Although there were good agreements between populations of herbicide-degraders, respired $^{14}CO_2$, and cross-adaptation (15,41), there are still some unanswered questions. EPTC-degraders could not be detected among populations of bacteria or actinomycetes. Thus, EPTC-degraders may be soil fungi or anaerobic bacteria (41) or more likely isolates that were unable to grow on the chosen medium. Moorman (42) recently developed a ^{14}C-MPN technique to enumerate EPTC-degraders. He concluded that increased rates of metabolism were apparently responsible for increased rates of degradation rather than increased populations of EPTC-degraders. Utilizing enriched soils, Lee (43) isolated 29 fungi and 9 bacteria that could degrade EPTC. Ghani and Allbrook (44) reported more rapid degradation of EPTC in soils of low pH and high C:N ratio which are generally suited to more fungal activity. In contrast, Tal et al. (23) indicated the degradation of EPTC was linked to the activity of soil bacteria that could be controlled by sterilization and chemical treatments. Additional

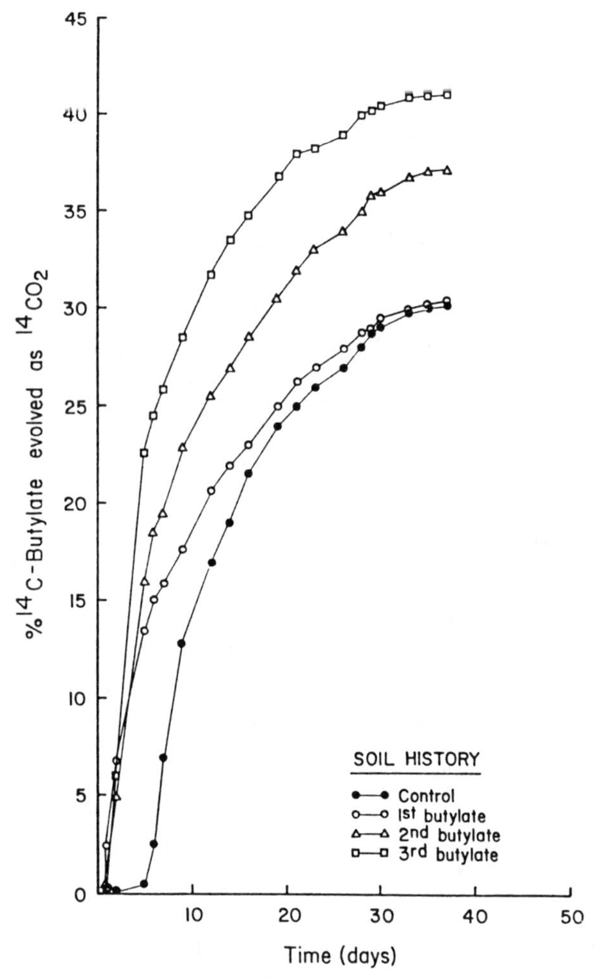

Figure 3. Accumulation of $^{14}CO_2$ evolved from Dothan soil with 1, 2, or 3 applications of butylate and without prior butylate use and treated with ^{14}C-butylate. (Modified from Ref. 38).

ecological research is needed on carbamothioate-degraders in soils.

From the population studies, several microbial isolates were selected to delineate the role of plasmids in enhanced biodegradation. A modification of the Birnboim and Doly (45) procedure was used to extract and demonstrate the presence of plasmid DNA in a *Flavobacterium* sp. (VI.15). A differential medium with 70 mg/L of yeast extract was developed to distinguish between butylate-degraders (But^+) and isolates unable to utilize

Table VII. Effects of previous soil exposure to butylate or vernolate on the populations[a] of butylate- or vernolate-degraders in a Dothan loamy sand

Previous herbicide exposure	Actinomycete MSB[b]	Actinomycete MSV[b]	Bacteria MSB[b]	Bacteria MSV[b]
	(cfu · 10^4/g)			
Butylate	13.3 a	8.8 a	2.0 b	1.5 b
Vernolate	3.7 b	2.8 b	6.6 a	8.5 a
None	1.6 c	1.0 b	0.6 b	0.4 b

[a]Means followed by different letters are significantly different at the 5% level by the LSD test.
[b]MSB = minimal salts (MS) + TTC + butylate; MSV = MS + TTC + vernolate. (Reproduced in part with permission from Ref. 41. Copyright 1989 Weed Sci. Soc. Am.)

butylate ($\underline{but}^-$). These latter isolates were obtained via heat or acridine orange curing (46). Plasmid profiles of But^+ derivatives revealed a plasmid band (pSMB 2) of approximately 100 kb that was not present in the $\underline{but}^-$ derivatives. Loss of pSMB 2 was associated with the loss of butylate-utilizing ability on the differential medium. Spontaneous loss of pSMB 2 upon cold storage suggests that close, careful monitoring, and handling of herbicide-degraders with instable plasmids is required (47). A 76.5-kb plasmid was identified in a soil bacterium by Tam et al. (48) for degradation of EPTC.

A *Methylomonas* sp (BI.10) capable of utilizing butylate also contained a plasmid DNA band (pSMB 4) that was estimated to be 50 kb. This plasmid was very stable and was not cured by hot or cold temperatures and had to be cured with acridine orange (46). Conjugal matings between $Herb^+$ donor (Sm^r) and $\underline{herb}^-$ recipient ($Tc^r Km^r$) cells produced $Herb^+$ transconjugants (Sm^s Tc^r Km^r) that contained pSMB 4 and turned red on the butylate-TTC medium (46) to demonstrate transfer of butylate-degrading ability.

Conclusions

Based on field and laboratory data, enhanced biodegradation of carbamothioate herbicides was confirmed under South Carolina conditions. A good correlation was observed between enhanced biodegradation and performance failure. Laboratory analyses of "problem" soils could be used to predict potential problems or weak efficacy under field situations. However, lack of soil moisture and thus a low microbial degradative potential could mask or alleviate enhanced biodegradation. Dietholate and SC-0058 were effective inhibitors to overcome enhanced biodegradation but lost their effectiveness after extensive, continuous use. Herbicide rotation also was an effective management tool to prevent or ameloriate soils conditioned for enhanced biodegradation. Soil microorganisms adapted to one herbicide also were cross-adapted to degrade certain other structurally related herbicides. Thus, extension agents, consultants, and farmers must be educated as to which ones can be used in effective management programs.

Population shifts in herbicide-degraders were noted in herbicide-history soils vs non-history soils. Actinomycetes, bacteria, and fungi have been implicated in the degradative process of carbamothioate herbicides; however, definitive linkages have yet to be established across soils. Although several microorganisms have been identified as carbamothioate-degraders, more efficient degraders (Dick, W. A. et al., 1990, this symposium) are needed for genetic engineering of super-degraders that can survive and function in the soil environment to clean up polluted sites. Plasmids associated with microbial degradation of carbamothioates were identified in several bacteria. Conjugal transfer of the plasmid, pSMB 4, was achieved with a *Methylomonas* sp. Additional ecological and genetic studies are needed to better understand population shifts associated with the use of pesticides and possible transfer of the degradative genes via plasmids within the soil biomass.

Acknowledgments

Special appreciation is expressed to DeWitt Gooden and Joe Zublena. Other faculty members who cooperated in this research were Mildred Amakiri, Ed Murdock, Dwight Camper, Ellis Kline, Ernie Lawrence, Larry Grimes, and John Evans. Graduate students involved were Vicki McCusker, Jim Mueller, Jim Struble, Joe Varn, Michelle Fleming, and Steve Wagner. Thanks to Patsy Ellis for typing and assistance with revisions. Financial support was received from the Stauffer Chem. Co. (ICI Americas), Southern Regional Pesticide Impact Assessment Program, and USDA Competitive Grants Program. Tech. Contribution No. 3036 of the South Carolina Agric. Exp. Stn., Clemson Univ.

Literature Cited

1. Audus, L. J. Plant Soil 1949, 2, 31-36.
2. Avidov, E.; Aharonson, N.; Katan, J. Weed Sci. 1988, 36, 519-523.
3. Kaufman, D. D.; Katan, J.; Edwards, D. F.; Jordan, E. G. In Agricultural Chemicals of the Future; Hilton, J. L., Ed.; Rowman and Allanheld, New Jersey, 1985; p 437-451.
4. Bailey, A. M.; Coffey, M. D. Phytopath. 1985, 75, 135-137.
5. Yarden, O.; Aharonson, N.; Katan, J. Soil Biol. Biochem. 1987, 19, 735-739.
6. Camper, N. D.; Fleming, M. M.; Skipper, H. D. Bull. Environ. Contam. Toxicol. 1987, 39, 571-578.

7. Felsot, A.; Maddox, J. V.; Bruce, W. Bull. Environ. Contam. Toxicol. 1981, 26, 781-788.
8. Gauger, W. K.; McDonald, J. M.; Adrian, N. R.; Matthees, D. P.; Walgenbach, D. D. Arch. Environ. Contam. Toxicol. 1986, 15, 137-141.
9. Sethunathan, N.; Pathak, M. D. Can. J. Microbiol. 1971, 17, 699-702.
10. Kaufman, D. D. J. Agri. Food. Chem. 1967, 15, 582-591.
11. Sheets, T. J. Weeds 1959, 7, 442-448.
12. Appleby, A. P.; Brewster, B. D.; Coleman, Q. C.; McAuliffe, D. Proc. West. Soc. Weed Sci., 1981, 34, 56-57.
13. Bean, B. W.; Roeth, F. W.; Martin, A. R.; Wilson, R. G. Weed Sci. 1988, 36, 524-530.
14. Harvey, R. G.; McNevin, G. R.; Albright, J. W.; Kozak, M. E. Weed Sci. 1986, 34, 773-780.
15. McCusker V. W.; Skipper, H. D.; Zublena, J. P.; Gooden, D. T. Weed Sci. 1988, 36, 818-823.
16. Rahman, A.; Atkinson, B. C.; Douglas, J. A.; Sinclair, D. P. New Zealand J. Agric. 1979, 139, 47-49.
17. Rahman, A.; James, T. K. Weed Sci. 1983, 31, 783-789.
18. Rudyanski, W. J.; Fawcett, R. S.; McAllister, R. S. Weed Sci. 1987, 35, 68-74.
19. Skipper, H. D.; Murdock, E. C.; Gooden, D. T.; Zublena, J. P.; Amakiri, M. A. Weed Sci. 1986. 34, 558-563.
20. Miaullis, B.; Nohynek, G. J.; Pereiro, F. Proc. Br. Crop Protect. Confer. - Weeds, 1982, p. 205-210.
21. Kaufman, D. D.; Kearney, P. C.; von Endt, D. W.; Miller, D. E. J. Agri. Food Chem. 1970, 18, 513-518.
22. Dowler, C. C.; Marti, L. R.; Kvien, C. S.; Skipper, H. D.; Gooden, D. T.; Zublena, J. P. Weed Technol. 1987, 1, 350-358.
23. Tal, A.; Rubin, B.; Katan, J.; Aharonson, N. Weed Sci. 1989, 37, 434-439.
24. Tam, A. C.; Behki, R. M.; Khan, S. U. J. Agric. Food Chem. 1987, 36, 654-657.
25. Varn, J. E., Jr.; Gooden, D. T.; Skipper, H. D.; Zublena, J. P. Proceed. South. Weed Sci. Soc., 1986, 39, 48.
26. Zublena, J. P.; Skipper, H. D.; Gooden, D. T.; Camors, F. B. Weed Sci.Soc. Am. Abst., 1986, 26, 95.
27. Harvey, R. G. Weed Sci. 1987, 35, 583-589.
28. Wilson, R. G.; Rodebush, J. E. Weed Sci. 1987, 35, 289-294.
29. Gooden, D. T.; Skipper, H. D.; Zublena, J. P. Am. Soc. Agron. Abstr., 1989, p. 216.
30. Gray, R. A.; Joo, G. K. Weed Sci. 1985, 33, 698-702.
31. Subba-Rao, R. V.; Cromartie, T. H.; Gray, R. A. Weed Tech. 1987, 1, 333-340.
32. Bartha, R.; Pramer, D. Soil Sci. 1965, 100, 68-70.
33. Skipper, H. D.; Mueller, J. G.; Ward, V. L.; Wagner, S. C. In Research Methods in Weed Science, 3rd Edition; N. D. Camper, Ed.; South. Weed Sci. Soc., Champaign, IL, 1986.; Chapter 23.
34. Skipper, H. D.; Volk, V. V. Weed Sci. 1972, 20, 344-347.
35. Ward, V. L. M.S. Thesis, Clemson University, Clemson, South Carolina, 1986.

36. Obrigawitch, T.; Wilson, R. G.; Martin, A. R.; Roeth, F. R. Weed Sci. 1982, 30, 175-181.
37. Lee, A.; Rahman, A.; Holland, P. T. New Zealand J. Agri. Res. 1984, 27, 201-206.
38. Skipper, H. D.; Gooden, D. T.; Zublena, J. P.; Amakiri, M. A.; Struble, J. E. Weed Sci. Soc. Am. Abst., 1985, 25, 91.
39. Kaufman, D. D.; Kearney, P. C. Appl. Microbiol. 1965, 65, 443-446.
40. Bochner, B. R.; Savageau, M. A. Appl. Environ. Microbiol. 1977, 33, 434-444.
41. Mueller, J. G.; Skipper, H. D.; Lawrence, E. G; Kline, E. L. Weed Sci. 1989, 37, 424-427.
42. Moorman, T. Weed Sci. 1988, 36, 96-101.
43. Lee, A. Soil Biol. Biochem. 1984, 16, 529-531.
44. Ghani, A.; Allbrook, R. F. New Zealand J. Agric. Res. 1986, 29, 469-474.
45. Birnboim, H. C.; Doly, J. Nucleic Acid Res. 1979, 7, 1513-1519.
46. Mueller, J. G. Ph.D. Dissertation, Clemson University, Clemson, South Carolina, 1988.
47. Mueller, J. G.; Skipper, H. D.; Kline, E. L. Pest. Biochem. & Physiol. 1988, 32, 189-196.
48. Tam, A. C.; Behki, R. M.; Khan, S. U. Appl. Environ. Microbiol. 1987, 53, 1088-1091.

RECEIVED January 22, 1990

Chapter 5

Enhanced Biodegradation of Dicarboximide Fungicides in Soil

Allan Walker and Sarah J. Welch

Institute of Horticultural Research, Wellesbourne, Warwick, United Kingdom

Experiments in the mid-1970's demonstrated excellent control of white rot disease of onions with iprodione. There is increasing evidence that in commercial practice the fungicide becomes less effective with repeated use and this has been linked with enhanced degradation in the soil. In a survey involving 33 soils, a clear relationship was established between the rate of degradation and the frequency of prior application. The related compound vinclozolin is also prone to enhanced degradation and there is evidence for cross-enhancement between iprodione and vinclozolin. Rapid degrading ability can be readily transferred to slow degrading soils by inoculation with small amounts of rapid-degrading soil, and rapid degradation can be inhibited by soil sterilization or treatment with anti-bacterial antibiotics. Studies of the pathway of degradation have led to development of a simple colorimetric test for identification of soils with rapid-degrading ability. Mixed cultures of bacteria capable of degrading the fungicides have been isolated and there is preliminary evidence of plasmid involvement in degradation.

The phenomenon of enhanced biodegradation of pesticides following repeated application to soil was first observed in the late 1940's with the phenoxyalkanoic acid herbicides 2,4-D, 2,4,5-T and MCPA (1,2). Although similar behaviour has been observed with other pesticides (3,4), it is only recently that consequences of direct practical importance have become apparent. Poor biological performance of several soil-applied insecticides, fungicides and herbicides following repeated use at the same site has now been correlated with enhanced degradation of their residues in the soil (see reviews by Roeth (5); Kaufman (6); Suett and Walker (7)). A minor but none the less important example of this problem in the United Kingdom is the apparent failure of dicarboximide fungicides

0097–6156/90/0426–0053$06.00/0

to control white rot disease (Sclerotium cepivorum [Berk.]) of onions. The only commercially available fungicides capable of adequately controlling this disease in the U.K. are iprodione (3-(3,5-dichlorophenyl)-N-isopropyl-2,4-dioxoimidazolidine-carboxamide) and vinclozolin ((RS)-3-(3,5-dichlorophenyl)-5-ethenyl-5-methyl-2,4-oxazolidinedione). Entwistle (8) reported failure of iprodione to control the disease at a site where the fungicide had been used repeatedly, and Entwistle (9) reported similar observations with vinclozolin. This paper summarises the results of experiments made at Wellesbourne over the past few years to investigate the apparent enhanced biodegradation of these dicarboximide fungicides in soil. Full details of experimental methods can be found in Walker, Entwistle and Dearnaley (10), Walker, Brown and Entwistle (11), and Walker (12,13).

Field Performance and Preliminary Observations with Iprodione

Before discussing fungicide degradation, it is worthwhile looking at the changes in performance against onion white rot that have been observed in the field. The data in Table I summarise results from experiments in a quarantine facility at the Institute of Horticultural Research (IHR), Wellesbourne over a number of years. Iprodione has been ineffective at this site since 1981 and vinclozolin since 1983, although it gave some control in the experimental plots during 1984. Myclozolin (3-(3,5-dichlorophenyl)-5-methoxymethyl-5-methyl-1,3-oxazolidine-2,4-dione) and procymidone (3-(3,5-dichlorophenyl)-1,5-dimethyl-3-azabicyclo[3.1.0]hexane-2,4-dione) are experimental fungicides only and not available commercially for white rot control. The former has given good control over a number of years; the latter gave good control in one experiment in 1984 and has not been tested subsequently.

Table I. Changes in Effectiveness of Dicarboximide Fungicides

Fungicide	% reduction in white rot infection in year:							
	1978	1979	1981	1982	1983	1984	1985	1986
Iprodione	93	99	31	0	0	13	0	0
Vinclozolin	-	-	86	85	0	76	5	7
Myclozolin	-	-	-	97	100	100	97	97
Procymidone	-	-	-	-	-	100	-	-

As part of a series of experiments to investigate the reasons for the changes in fungicide performance, laboratory degradation studies were made initially with iprodione. Carbon-14 labelled iprodione was incubated in soil from an area of the IHR quarantine field in which the fungicide was ineffective, and the rate of degradation was compared with that in soil from a grassed area of the same field that had not previously been treated with the fungicide (10). The results indicated more rapid degradation in the previously-treated than in the previously-untreated soil. The times to 50% loss were approximately 10 and 30 days, respectively. Similar differences in degradation rates were observed in soil samples from four commercial farms where iprodione had been used on several occasions to control the pathogen. The fungicide consistently degraded more quickly in soils from the previously-treated areas of the fields than from untreated areas. One problem with these early experiments was that the 'untreated' soils were generally taken from field headlands and boundaries and they were consistently of lower pH than their respective pre-treated samples. Published evidence indicated that iprodione degradation is pH dependent (14) hence the results, although consistent with enhanced degradation, were not conclusive.

Further Observations with Iprodione and Vinclozolin

In order to obtain more definitive evidence of enhanced degradation of iprodione, experiments were made using a soil with no known prior treatment with dicarboximide fungicides. Sequential applications of iprodione were made in both laboratory and field experiments and changes in rate of degradation were monitored. Similar experiments were made with the related compound vinclozolin. A full description of the experimental procedures can be found in Walker, Brown and Entwistle (11). The results gave a clear indication that both iprodione and vinclozolin are susceptible to enhanced degradation and those from the laboratory experiments are reproduced in Figure 1. In soil treated at time 0, the time for 50% loss of iprodione was about 23 days and less than 2% of the applied dose was recovered after 49 days. In soil treated for the second time (50 days), the time for 50% loss was reduced to 5 days. When treated for the third time (100 days), only 10% of the applied dose remained 2 days later, and none was recovered after 7 days. The results for vinclozolin were similar to those for iprodione (Figure 1).

To examine further the influence of soil pH on degradation of iprodione and vinclozolin, laboratory experiments were made with samples of soil from the top 10 cm of plots of a long-term liming experiment started in 1961 in Sawyers field at Rothamsted Experimental Station, Harpenden, England. The soils were of similar texture and organic matter content but pH measured in a 1:2.5 suspension of soil in distilled water was 4.3, 5.0, 5.7 and 6.5 in the different samples used for the fungicide degradation experiments. The experiments involved carbon-14 labelled fungicides and appropriate radioassay procedures which were described in detail by Walker (12). The results indicated little degradation of either fungicide at pH 4.3 or 5.0 during an 80-day incubation period. Degradation was more rapid at pH 5.7 and the

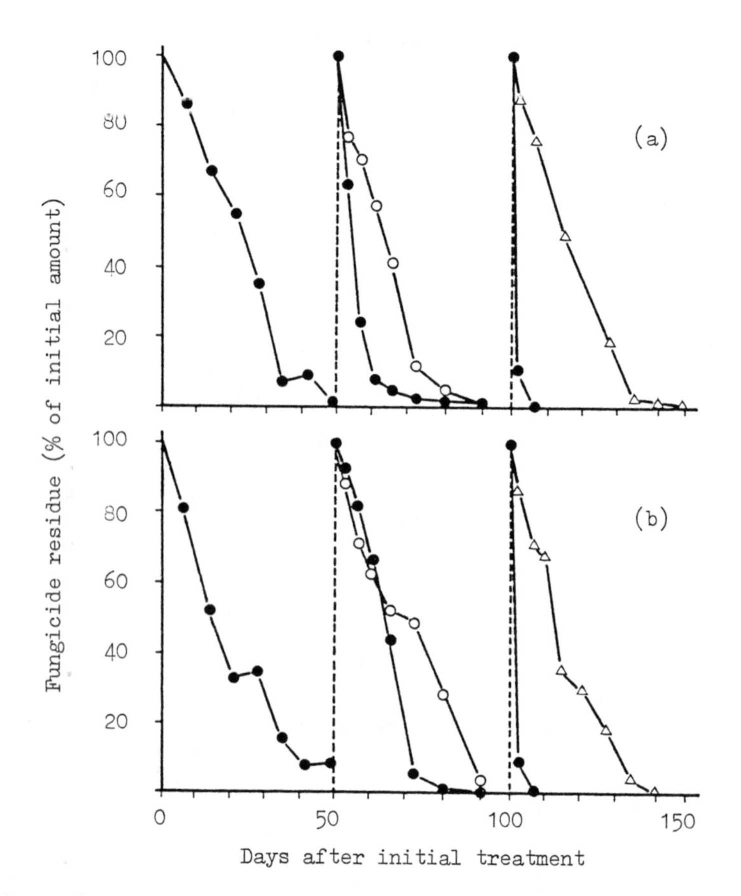

Figure 1. Degradation of (a) iprodione and (b) vinclozolin in soil. ● residues following sequential treatments at time 0, 50 and 100 days; ○ treated once only after pre-incubation for 50 days; △ treated once only after pre-incubation for 100 days. (Reproduced with permission from Ref. 11. Copyright 1986 Society of Chemical Industry).

times for 50% loss of iprodione and vinclozolin were about 60 and 75 days respectively. Degradation was even more rapid at pH 6.5 with times to 50% loss of 30 to 35 days for both compounds. When repeat applications were made to the soils 100 and 200 days after the initial incubation was started, there was still only limited degradation in the two more acid soils. However in the soils at pH 5.7 and 6.5, there was a progressive increase in degradation rate with successive doses of either fungicide. The results from the soil with pH 6.5 are shown in Figure 2a for iprodione and 2b for vinclozolin. The times for 50% loss of iprodione were 30, 12 and 4 days for the first, second and third doses respectively, and for vinclozolin, they were approximately 30, 22 and 7 days.

Evidence for Enhanced Degradation in Commercial Fields

Although enhanced degradation of iprodione and vinclozolin can be induced by repeated application to soil in the laboratory and field, problems with lack of disease control in commercial practice are not widespread. In order to determine the extent to which enhanced degradation has been induced in commercial field soils, and hence
the potential for a change in biological performance, samples of soil with known histories of iprodione use were collected from 33 different sites and the rates of iprodione degradation measured under standard laboratory conditions (20°C; soil moisture at a tension of -33kPa). The experiment was restricted to iprodione because, when the study was made, this fungicide had been available commercially for several years, whereas vinclozolin was a relatively new treatment and hence the opportunity for repeated application was much less. The degradation data were evaluated using curve-fitting procedures (12) and the time to 90% degradation (DT90) was estimated for each soil.

Table II. Summary of Iprodione Degradation Data for 33 Soils (Reproduced with permission from Ref. 12. Copyright 1987 Society of Chemical Industry)

Previous treatments	Number of soils	Time for 90% loss (days) minimum	maximum	mean	± S.D.
None	12	22	93	50	23.7
One	5	16	28	20	4.9
Two	4	5.2	23	17	8.2
Three	3	3.8	15	7.8	5.83
More than three	9	4.8	13	8.1	3.21

The results are summarised in Table II where the maximum, minimum and mean DT90 is presented for each group of similar pretreatment history soils. The mean time to 90% disappearance varied from 50 days in the soils that had not been treated previously with iprodione in the field, to about 8 days in soils which had been treated previously on three or more occasions. These results therefore demonstrate that enhanced degradation of iprodione in field soils following repeated use is a common phenomenon and in the soils showing the most rapid degradation (DT90 4-5 days), there is a clear potential for loss of biological activity against the target organism.

A Simple Color Test to Identify Rapid-Degrading Soils

Several of the experiments summarized above involved the use of carbon-14 labelled iprodione and vinclozolin. The analytical techniques used involved solvent extraction of the soils and thin-layer chromatographic separation of parent fungicide from its degradation products (10,12). The main degradation product from both iprodione and vinclozolin chromatographed in an identical manner to pure, unlabelled 3,5-dichloroaniline. In further experiments, Walker (13) confirmed by gas-liquid and high performance liquid chromatography that 3,5-dichloroaniline was an important degradation product of iprodione in soil, and demonstrated that it could be extracted in relatively high concentrations when iprodione degradation was rapid. A colorimetric test based on production of a magenta diazo color complex was shown to differentiate between amounts of 3,5-dichloroaniline in the range from 2 to 15 ug. The test was used to examine 3,5-dichloroaniline production from iprodione incubated at 50 mg/kg in the 33 soils from the field survey described above. A positive color reaction in acetone extracts made after 3 days was obtained with those soils in which the time to 90% degradation of iprodione was less than 6 days. A positive reaction after 7 days was obtained when the DT90 was less than 14 days, and a positive reaction after 10 days was associated with DT90 of less than 22 days. The type of result obtained is illustrated by the data in Table III. The test clearly has the potential to indicate soils in which enhanced degradation has been induced and hence to identify those in which efficacy problems might be encountered.

Studies with Other Dicarboximides and the Possibility of Cross Enhancement

As mentioned previously, the two dicarboximide fungicides approved in the U.K. for control of white rot disease of onions are iprodione and vinclozolin. Other related fungicides can give good control (e.g. Table I) and experiments were therefore made to determine whether they may be prone to enhanced degradation. The possibility of cross-enhancement of degradation between different dicarboximides has also been investigated. Field plots were established in Wharf Ground, a field at IHR Wellesbourne in which the soil is a sandy clay loam with 2.1% organic matter and pH of 6.9. Duplicate plots were sprayed with iprodione, vinclozolin, myclozolin or procymidone at 4.0 kg/ha on three occasions - 23

May, 3 July and 5 August 1986. There were also two unsprayed control plots. Samples of soil were taken from the top 5 cm of each plot on 2 April 1987 and separate subsamples were incubated with the different fungicides as described previously (12). The

Table III. Reaction of Different Soils in the Colorimetric Test

Time for 90% loss	Color response after:		
(days)	3 days	7 days	10 days
3.8	***	***	***
4.9	***	***	***
6.0	***	***	***
10.4	+	*	**
11.1	*	**	***
14.5	0	**	***
15.8	0	+	
17.6	0	+	*
19.1	0	+	*
22.0	0	+	**
24.6	0	0	0
33.1	0	0	0

0 indicates no color change; + indicates a slight pink coloration;
*, ** and *** indicate levels of magenta color equivalent to 5, 5-10 and >10 ug 3,5-dichloroaniline respectively.

decline in fungicide residues was measured by gas-liquid or high performance liquid chromatography as before (11,12). The results with myclozolin and procymidone (Figure 3) provided no evidence for enhanced degradation of either fungicide. Myclozolin was extremely stable in this particular soil and only 15-20% degradation occurred within the 60 day incubation period. Procymidone was also stable, and between 40 and 50% of the initial amount remained after 60 days irrespective of whether the soil had been treated previously in the field or not. These data are generally consistent with those in Table I. There is insufficient evidence to show whether biological performance of procymidone may be affected by repeated application. However, myclozolin gave good control in all experiments which is consistent with its high stability in the soil.

Examples of the data from the cross-enhancement experiment are shown in Table IV. These are the residues measured 14 and 28

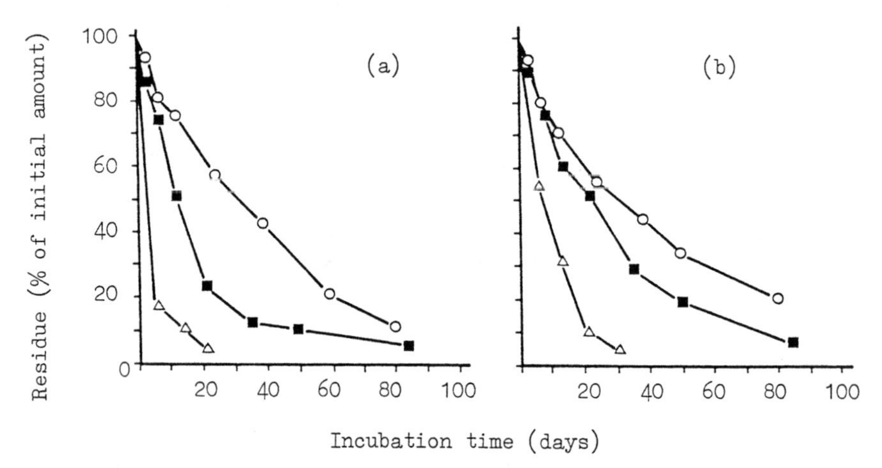

Figure 2. Degradation of sequential applications of (a) iprodione and (b) vinclozolin in soil with pH 6.5. O first treatment; ■ second treatment; Δ third treatment. (Reproduced with permission from Ref. 12. Copyright 1987 Society of Chemical Industry).

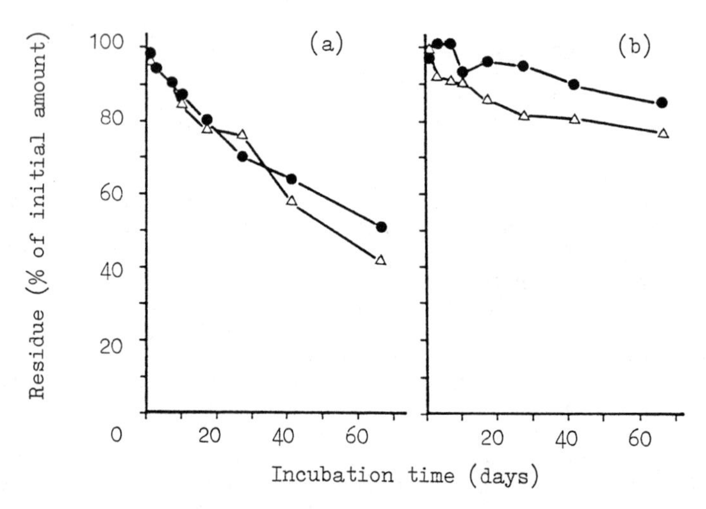

Figure 3. Degradation of (a) procymidone and (b) myclozolin in previously treated (Δ) and previously untreated (●) soil.

days after the start of the incubation experiment. They indicate that degradation of both iprodione and vinclozolin is enhanced by prior treatment of the soil with iprodione or vinclozolin respectively. As discussed above, pre-treatment of soil with myclozolin or procymidone did not have a major effect on the degradation rates of subsequent additions of the same chemicals, although procymidone degradation was affected to a small extent. The only evidence of significant cross-enhancement involved iprodione and vinclozolin; iprodione degradation was apparently enhanced by pre-treatment of the soil with vinclozolin. Vinclozolin degradation was affected only slightly by pretreatment of the soil with iprodione, which agrees with previous observations (12), and also affected to some extent by prior treatment with myclozolin. There are, therefore, no simple relationships where cross-enhancement is concerned, and it is probable that the only way to gain a better understanding of such phenomena will be to isolate the specific organisms and enzyme systems responsible for degradation.

Table IV. Cross Enhancement between Different Dicarboximide Fungicides

Pretreatment	Concentration (mg kg^{-1}) of fungicide:			
Fungicide	Iprodione	Vinclozolin	Myclozolin	Procymidone
	After 14 days incubation			
None	6.00	5.45	5.85	6.82
Iprodione	0.15	5.50	5.90	6.81
Vinclozolin	2.30	0.05	4.27	7.48
Myclozolin	7.93	5.39	5.44	6.72
Procymidone	4.97	5.45	5.55	4.86
	After 28 days incubation			
None	3.75	2.44	5.20	5.71
Iprodione	0.00	1.89	5.46	5.80
Vinclozolin	0.18	0.00	3.96	6.25
Myclozolin	3.96	2.37	5.15	6.10
Procymidone	2.88	2.31	5.08	3.12

Ease of Induction and Spread

The loss of control of white rot disease by iprodione in the IHR quarantine field summarized in Table I is widespread and the chemical is apparently inactive in all parts of the field.

However, the results in Table I refer to small scale field trials so that in any one year only a limited area of the field will have been treated with the fungicide. This raises the question of how soil throughout the field has developed the capacity for rapid degradation. It may be that degrading ability is readily transferred with soil that is blown with the wind or moved during routine cultivation operations. It is also possible that only trace amounts of chemical are required to bring about a change in subsequent degradation rates so that drift during application leads to effects in parts of the field not treated deliberately. Experiments were therefore made to examine the ease of transfer of degrading ability, and how the induction of rapid degradation is influenced by pre-treatment concentration of fungicide.

Samples of untreated and iprodione pretreated soil were collected from the Wharf Ground site described above. Iprodione was incorporated into subsamples of both soils at a concentration of 8 mg/kg. The soils were then mixed to give experimental treatments containing 0, 0.1, 0.5, 1.0, 5.0, 10.0 and 100% of the pre-treated rapid-degrading soil. All of the soils were incubated as before and iprodione residues were measured at intervals by high performance liquid chromatography (12). The results (Figure 4) indicate a progressive increase in iprodione degradation rate as the proportion of rapid-degrading soil in the mixture was increased. Incorporation of just 0.1% of the pre-treated soil was sufficient to alter the degrading-ability of the control soil indicating a very ready transfer of degrading ability. Incorporation of 5% or more of the pre-treated soil into the control soil resulted in a degradation rate identical with that in the pre-treated rapid-degrading soil.

Further samples of control soil from the Wharf Ground site were incubated under standard conditions as before (12). Duplicate amounts (500 g) contained iprodione at initial concentrations of 0, 0.05, 0.10, 0.50, 1.0, 5.0 or 10.0 mg/kg. After 70 days, further iprodione was added to all of the samples to give a concentration in soil of 10.0 mg/kg. The samples were re-incubated under the same standard conditions and residual concentrations of the fungicide were measured during the subsequent 21 days. The results (Figure 5) indicate little difference in degradation rate between soil samples pre-incubated with 0, 0.05 or 0.10 mg/kg iprodione, but pre-treatment concentrations greater than this had a significant effect on subsequent rates of loss. Following pre-incubation with 0.50 mg iprodione/kg soil or more, the residual concentrations 13 days after re-incubation were less than 5% of the initial amounts. In the other treatments, residual concentrations at this time were equivalent to more than 85% of the initial dose. The results therefore demonstrate that low soil concentrations of iprodione are able to enhance the biodegradation of subsequent additions of the fungicide and, taken in conjunction with those from the 'soil mixing' experiment described above, help to explain why enhanced

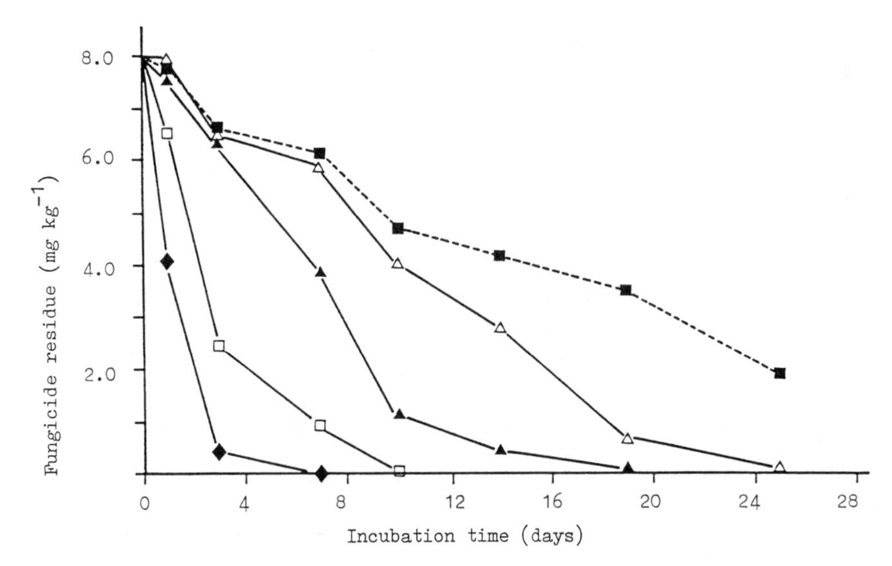

Figure 4. Transfer of iprodione degrading ability. ■ control soil; ◆ degrading soil; △ , ▲ and □ refer to 0.1, 0.5 and 1.0% of degrading soil mixed with control soil. 5 and 10% degrading soil in control soil gave results identical with degrading soil (◆).

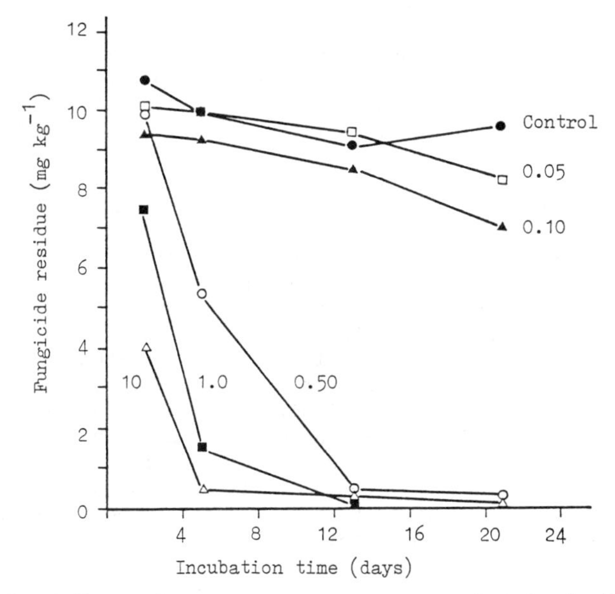

Figure 5. Effect of pre-treatment concentration (mg kg^{-1}) on degradation of subsequent addition of iprodione. Results for initial concentration of 5 mg kg^{-1} identical with those shown for 10 mg $kg-^{-1}$.

degradation of iprodione is easily induced and spread in treated field soils.

Effect of Microbial Inhibitors on Degradation

In order to confirm that soil microorganisms contribute to degradation of iprodione and vinclozolin and to make a preliminary assessment of the groups of organisms that may be involved, experiments were made to determine the influence of some microbial inhibitors on rates of loss. The experiments once more involved samples of soil from the Wharf Ground site pre-treated with iprodione or vinclozolin in the field as described above. Subsamples of soil (1 kg) were treated with sodium azide (800 mg/kg dry soil), cycloheximide (anti-fungal antibiotic, 500 mg/kg dry soil) or chloramphenicol (anti-bacterial antibiotic, 500 mg/kg dry soil) and incubated at 20°C for 7 days. The soils were then allowed to lightly air dry overnight and the water lost was replaced with an iprodione or vinclozolin suspension in sterile distilled water so that the concentration of fungicide was 8 mg/kg dry soil. Further samples of the same soils were sterilised by gamma-irradiation (Isotron plc, Swindon, U.K.; dose, 50 kGy) and then incubated with iprodione or vinclozolin as appropriate. Duplicate amounts of all the soils were incubated at 20°C with soil moisture equivalent to an applied pressure of 33kPa (0.33 bar) and subsamples (100 g) were removed for analysis at intervals during the subsequent 25 days. Analysis of fungicide residues was by HPLC as before (12) with the results summarised in Figure 6. All the sterilization procedures inhibited degradation of both fungicides. With any one fungicide, there was little difference in inhibitory effects between sodium azide, gamma-irradiation or chloramphenicol. Cycloheximide was less inhibitory, particularly with vinclozolin. Treatment of soil with cycloheximide markedly reduced the degradation rate of iprodione although there is no indication of the reasons for this. The antibiotic, although predominantly anti-fungal, may also inhibit growth of some soil bacteria, or removal of soil fungi may cause significant population changes in bacterial communities. In general, the results in Figure 6 indicate that soil bacteria may be primarily responsible for degradation of iprodione and vinclozolin in soil. Results similar to those in Figure 6 have been reported previously in studies of enhanced biodegradation (15). The rate of carbon-14 dioxide evolution from labelled isofenphos incubated in soil treated with cycloheximide was similar to that from labelled insecticide incubated in control soil. Carbon-14 dioxide evolution from the insecticide in soil treated with chloramphenicol was very low and similar to that from soil sterilized by autoclaving. On the basis of these observations, these authors also concluded that soil bacteria were responsible for degradation of the insecticide.

Preliminary Studies of Microbial Relationships

More detailed studies of the microbial relationships involved in enhanced degradation of the dicarboximide fungicides in soils are currently being undertaken in the Department of the Environmental

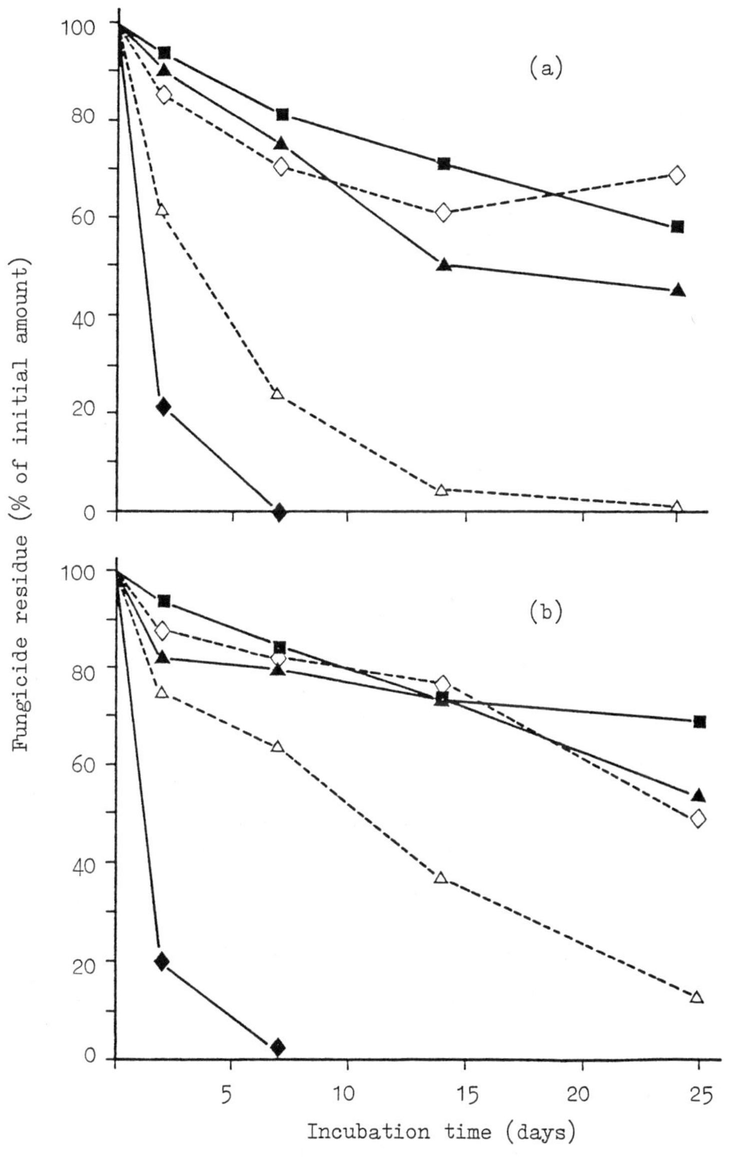

Figure 6. Degradation of (a) vinclozolin and (b) iprodione in a rapid degrading soil (◆) and in the same soil treated with chloramphenicol (■), cycloheximide (△), sodium azide (◇) or gamma irradiation (▲).

Microbiology, University of Newcastle upon Tyne, U.K. Some preliminary observations from these experiments were reported by Head, Cain, Suett and Walker (16). Bacterial cultures with the ability to degrade iprodione and vinclozolin in the absence of any exogenous carbon or nitrogen source have been isolated from samples of soil from the Wharf Ground experiments described above. These enrichment cultures were able to convert iprodione stoichiometrically to 3,5-dichloroaniline within 50 h and vinclozolin within 30 h. The cultures were not able to degrade the fungicides when the pH was less than 5.5, and the optimum pH for their activity was around 6.5. At pH 7.5 and above, degradation of iprodione and vinclozolin proceeded even in sterile culture thus confirming that chemical hydrolysis is important under alkaline conditions. Plasmid screening has revealed high molecular weight extrachromasomal DNA in fungicide-degrading cultures, and association of these plasmids with pesticide degrading ability is currently under investigation.

General Conclusions

The results summarized above demonstrate that enhanced degradation of iprodione in field soils is a common occurrence whenever the fungicide has been used repeatedly at the same site. They indicate that vinclozolin is also prone to enhanced degradation in soil. Some other dicarboximide fungicides are more stable in soil and the limited evidence available suggests that soil microorganisms do not adapt to degrade them so readily. The ability to rapidly degrade iprodione can be induced with a single pre-treatment at a very low initial concentration and the rapid-degrading ability is easily spread from one soil to another. There is clear evidence for the involvement of soil bacteria in degradation of both iprodione and vinclozolin. The very fast rates of loss observed in some soils which have received regular applications of iprodione indicate a clear potential for loss of biological activity against the target organism. In the absence of effective methods to counteract enhanced degradation in field soils, it is essential to avoid repeated use of either iprodione or vinclozolin at the same site in successive seasons, and it is important that good agricultural practices such as rotation of crops and pesticides should be encouraged. In addition to the well-known agronomic benefits from crop rotation, there may also be advantages in minimising the possible development of enhanced degradation.

Literature Cited

1. Audus, L.J. Plant Soil 1949, 2, 31-36.
2. Audus, L.J. Plant Soil 1951, 3, 170-192.
3. Engvild, K.C.; Jensen, H.L. Soil Biol. Biochem. 1969, 1, 295-300.
4. Groves, K.; Chough, K.S. J. Agric. Food Chem. 1979, 18, 1127-1128.

5. Roeth, F.W. Reviews of Weed Science 1986, 2, 45-65.
6. Kaufman, D.D. Proc. Br. Crop Prot. Conf.-Weeds 1987, pp. 5-5-522.
7. Suett, D.L.; Walker, A. Aspects of Applied Biology 1988, 17, 213-222.
8. Entwistle, A.R. Phytopathology 1983, 73, 800.
9. Entwistle, A.R. Aspect of Applied Biology, 1986, 12, 201-209.
10. Walker, A.; Entwistle, A.R.; Dearnaley, N.J. In Soils and Crop Protection Chemicals; Hance, R.J. Ed.; British Crop Protection Council Monograph No. 27; British Crop Protection Council, Croydon, 1984; pp 117-123.
11. Walker, A.; Brown, P.A.; Entwistle, A.R. Pestic. Sci. 1986, 17, 183-193.
12. Walker, A. Pestic. Sci. 1987a, 21, 219-231.
13. Walker, A. Pestic. Sci. 1987b, 21, 233-240.
14. Cayley, G.R.; Hide, G.A. Pestic. Sci, 1980, 11, 15-19.
15. Racke, K.D.; Coats, J.R. J. Agric. Food Chem. 1987, 35, 94-99.
16. Head, I.M.; Cain, R.B.; Suett, D.L.; Walker, A. Proc. Br. Crop. Prot. Conf. - Pests and Diseases 1988, pp. 699-704.

RECEIVED January 22, 1990

Chapter 6

Enhanced Biodegradation of Insecticides in Midwestern Corn Soils

Kenneth D. Racke[1] and Joel R. Coats[2]

[1]Environmental Chemistry Laboratory, DowElanco, 9001 Building, Midland, MI 48641–1706
[2]Department of Entomology, Iowa State University, Ames, IA 50011

An experimental strategy for the study of enhanced degradation is described based on its occurrence in Midwestern corn soils. The shift from recalcitrant chlorinated hydrocarbons to biodegradable organo-phosphorus and carbamate insecticides has resulted in the failure of some compounds, notably carbofuran and isofenphos, to provide adequate pest control following repeated use. Enhanced degradation of an insecticide involves its rapid degradation by a population of soil microorganisms that has adapted to beneficially catabolize it following exposure to it or a similar insecticide. For enhanced degradation to be thoroughly investigated studies must be carried out to demonstrate an increased rate of degradation in soils with prior insecticide exposure, to identify the rates and products of degradation in similar soils under controlled conditions, and to elucidate the microbiological mechanisms.

Agriculture in the Midwestern United States relies heavily on the cultivation of corn (*Zea mays*). Management of this crop in turn hinges on the successful control of a panoply of destructive pests, chief of which are soil-dwelling insect larvae of corn rootworms (*Diabrotica spp.*), cutworms (*Agrotis spp.*) wireworms (Elateridae), and grubs (*Phyllophaga spp.*). The corn rootworm is the key pest posing the greatest potential for yield reduction. It is worth noting that although corn is planted in April or May, rootworm larvae hatch and are active during an extended period from May to July. Because corn rootworms prefer to inhabit continuous corn, early control practices focused on crop rotations as a means of limiting populations (1).

Insecticidal Control of Corn Soil Pests

With the advent of synthetic organic insecticides following World War II, a new chemical control paradigm took root in corn agriculture. In the late 1940's it was discovered that soil application of

0097–6156/90/0426–0068$06.00/0

chlorinated hydrocarbon insecticides provided excellent control of these pests (2). The cyclodiene aldrin became the most widely used compound by the mid 1950's as use of these soil-applied insecticides blossomed. For example, soil insecticide use in Iowa went from nothing in 1950 to 237,000 acres in 1954 (6% of corn acres) (3). The control was so effective that a broadcast soil application in many cases provided control both that year and the next!

The perspective that "The persistent chlorinated hydrocarbons appear to be uniquely well adapted for soil use" (3) rather quickly collided with the reality of insect microevolution. The first evidence of rootworm control failures occurred in Nebraska in 1959, and by 1962 resistance of rootworms to the cyclodienes had been conclusively documented (4). The spread of this resistance during the 1960's, along with environmental scrutiny of the ecological impacts of these persistent compounds, created a demand for more environmentally sound insecticides.

Beginning in the mid to late 1960's the biodegradable organophosphorus and carbamate insecticides began to replace the chlorinated hydrocarbons as the rootworm control chemicals of choice. Diazinon, oxydisulfoton, and phorate were the first, and were quickly followed by bufencarb, trimethacarb, fonofos, and fensulfothion. In the mid to late 1970's the introduction of new chemicals continued as ethoprop, carbofuran, chlorpyrifos and terbufos came into use. Because rootworm insecticides are most commonly applied at planting, 4-8 weeks prior to rootworm egg hatch, it was early recognized that the inherently rapid degradation rate of some of these insecticides precluded their effective use (5). However, most provided effective control without the undesirable environmental impact of the chlorinated hydrocarbons. By 1973 roughly 20-30 million acres of midwest corn were treated annually with soil insecticides (6), and corn had become the leading crop for use of soil insecticides.

Enhanced Insecticide Degradation and Control Failure

Although it was realized that various environmental factors could influence rootworm control with soil-applied insecticides (7), the failures of several reliable insecticides wrought some degree of consternation within the agricultural community. As various compounds fell from common use during the 1960's and 1970's due to poor performance (diazinon, oxydisulfoton, bufencarb, trimethacarb, fensulfothion) resistance was examined as a possibility and found not to prove sufficent for explaining the observed failures (6,8). The continuing introduction of new insecticides for the corn rootworm market tended to balance out the overall effect of the losses of failure-prone products.

Research on control failures during the late 1970's and early 1980's focused on carbofuran, a compound that had earlier provided excellent control and had captured a considerable share of the rootworm insecticide market (e.g. 20% of treated acres in Iowa in 1977)(9). Although some studies found no correlation between prior use of carbofuran and an enhanced rate of carbofuran degradation (10,11) both monitoring of field residues (12-14) and laboratory degradation tests (15-17) conclusively demonstrated enhanced microbial degradation as the cause of decreased carbofuran persistence in fields with histories of carbofuran use. Thus, a phenomenon that had first been noted with the phenoxyalkanoic herbicides during the 1940's and had languished as an academic curiosity became a primary concern in the Midwestern corn belt. See Felsot (18) for a more complete account of the recognition of carbofuran enhanced degradation.

As control failures of other insecticides for the rootworm market (isofenphos, terbufos, fonofos, cloethocarb, bendiocarb) occurred, enhanced degradation began to be blithely invoked as the universal cause for many of these failures (19). With this assumption many extension agents began recommending rotations between rootworm insecticides as a possible means of avoiding this perceived problem (e.g.,20). The inference of enhanced degradation became a marketing weapon of some agrochemical companies to raise questions regarding the performance of competitor's products. It was obvious that a rational, experimental approach was needed to address the question of which insecticides had actually undergone enhanced degradation.

An Experimental Approach for the Study of Enhanced Degradation

In an effort to determine the criteria that should be used to invoke cases of enhanced degradation, an experimental approach for its study was developed that focused on laboratory investigations with field-collected soils. It was obvious that insecticide control failures were common occurrences and certainly not all due to enhanced degradation, as investigations of faulty application methods and unusual environmental conditions have shown (18). The ideal approach to the study of enhanced degradation would involve controlled field research in which pesticide persistence and control efficacy were both measured at many locations over a number of years. However, the tremendous cost in time and effort and confounding of results by environmental variables make a controlled laboratory approach desirable. The limitation of laboratory efforts focused exclusively on the soil-insecticide interaction is that they cannot fully address the additional insect-insecticide and insect-crop interactions present in the field. This means that caution must be excercised when proof of enhanced degradation is discovered in the laboratory, for this does not necessarily mean that insect control and crop yield will be adversely affected under field conditions. With this caution in mind, results of laboratory investigations can be kept in proper perspective for what they can best provide: a mechanistic understanding of soil-microbe-insecticide interactions.

Rapid Insecticide Degradation Assay

The first part of the laboratory methodology for study of enhanced degradation involved use of a rapid degradation assay by which an idea of the rate of insecticide degradation could be obtained for a large number of soils. The soils for this assay were collected during the fall from Iowa cornfields with known histories of insecticide use. In some cases the soils collected were specifically from fields in which an insecticide had been used for several years and no longer provided suitable control of soil insect pests. In these cases a separate soil was also collected from an adjacent field or from the field fencerow.

A 25-g portion of each soil was placed in a glass jar and treated with ^{14}C-insecticide at 5 µg/g. Insecticides used for this assay included those for which fields with some treatment histories could be identified: ^{14}C-carbonyl-carbofuran, ^{14}C-ring-isofenphos, ^{14}C-ring-fonofos, ^{14}C-ethoxy-terbufos, ^{14}C-ethoxy-phorate, and ^{14}C-ethyl-ethoprop. After each soil was moistened to 0.3 bar soil-moisture tension, glass vials containing 0.1 N NaOH were placed inside each jar to serve as CO_2 traps. The jars were incubated in the dark at 25°C for 1 week, during which the traps were analyzed daily for trapped $^{14}CO_2$.

Results of the rapid degradation assay (Table I) are expressed for each insecticide as cumulative mineralization in history soils (treated with that pesticide at least one previous year) and non-history soils (untreated or treated with other insecticides). All soils from fields previously treated with carbofuran exhibited a much higher mineralization rate than soils from non-history fields. Previous reports have confirmed the enhanced degradation of carbofuran, and its rapid mineralization in each carbofuran-history soil is indicative of its great susceptibility to enhanced degradation. Even the fencerow soils surrounding carbofuran-history fields exhibited increased carbofuran mineralization rates, and had apparently been contaminated by carbofuran or carbofuran-treated soil. Up to 5 years had elapsed since the last carbofuran soil application to history soils.

Table I. Effect of Soil Insecticide Use History on the Mineralization of ^{14}C-Insecticides to $^{14}CO_2$ in Soils During a 1-Week Laboratory Assay

	% Mineralization ± SE	No. of Soils
Carbofuran		
History	84.0 ± 8.0	8
Nonhistory	18.1 ± 7.6	11
History Fencerow	39.7 ± 8.2	2
Isofenphos		
History	35.9 ± 22.4	6
Nonhistory	2.4 ± 1.0	19
Fonofos		
History	20.8 ± 13.6	13
Nonhistory	9.1 ± 3.5	16
Terbufos		
History	12.5 ± 4.0	11
Nonhistory	9.7 ± 3.2	5
Phorate		
History	13.1 ± 4.1	4
Nonhistory	8.0 ± 1.3	12
Ethoprop		
History	32.7 ± 18.9	2
Nonhistory	19.9 ± 11.6	8

For isofenphos, there also was an increased mineralization rate in isofenphos-history soils. Very little mineralization occurred in nonhistory soils, and in all cases the rate of mineralization was greater in history soils, but varied in magnitude from 5.1 to 64.4%. Some difference in the mineralization of fonofos in the history and nonhistory soils was evident, but the behavior of fonofos was much more variable in the fonofos-history soils. There seemed to be two groups of history soils: those in which the fonofos mineralization rate was quite rapid (5 soils, 24.2-46.6%) and others in which it was comparable to the non-history soils (8 soils, 8.3-14.9%).

The mineralization of terbufos and phorate was not significantly greater in the corresponding history soils. Ethoprop exhibited quite variable mineralization behavior in both history and nonhistory soils, and from the small number of soils available with ethoprop history it is not possible to draw any meaningful conclusions. This corroborates earlier studies that have reported no evidence of enhanced terbufos or phorate degradation. Other reports have investigated the possible enhanced degradation of terbufos and phorate and also have found no evidence of an accelerated rate of degradation (21,22). Therefore, more intensive soil degradation studies focused on carbofuran, isofenphos, fonofos, and several related insecticides.

Cumulative plots of carbofuran mineralization in a history and a nonhistory soil are presented in Figure 1. The initially accelerating rate of $^{14}CO_2$ production, indicative of a microbial response (e.g., enzyme induction, population growth), is characteristic of enhanced degradation. Comparison of the mineralization of ^{14}C-carbonyl and ^{14}C-ring labelled carbofuran demonstrates an important consideration in this type of assay: the location of the ^{14}C label in the insecticide is critical for this type of assay to provide useful information. Although the carbonyl ^{14}C was almost completely evolved as $^{14}CO_2$, the ring ^{14}C was only slowly mineralized.

There are several complementary rapid degradation screening assays that have been effectively used in the laboratory. These include bioassay (16,23), gas chromatographic assay of persistence (17), colorimetric assay (24), and liquid medium assay (15,25). All can be useful in screening large numbers of soils for gross differences in insecticide-degrading behavior. However, it is extremely important that such assays are not used to provide the sole basis for explaining potential instances of enhanced degradation (26). In order for enhanced degradation to be truly confirmed for a given pesticide a more thorough laboratory and/or field investigation in which the microbiological aspects are clarified is required. Only then can enhanced degradation be distinguished from natural variations in soil insecticide degradation due to variation in soil properties (e.g., biomass, pH, organic content). The most useful aspect of these rapid degradation assays is in identifying suspect soils and pesticides for further investigation.

Pesticide Degradation in Companion Soils

The second component of the laboratory methodology for study of enhanced degradation involved a more intense investigation of the soil degradation of insecticides identified as suspect from the rapid degradation assay (carbofuran, isofenphos, fonofos) as well as 2 additional insecticides (cloethocarb, chlorpyrifos). The degradation

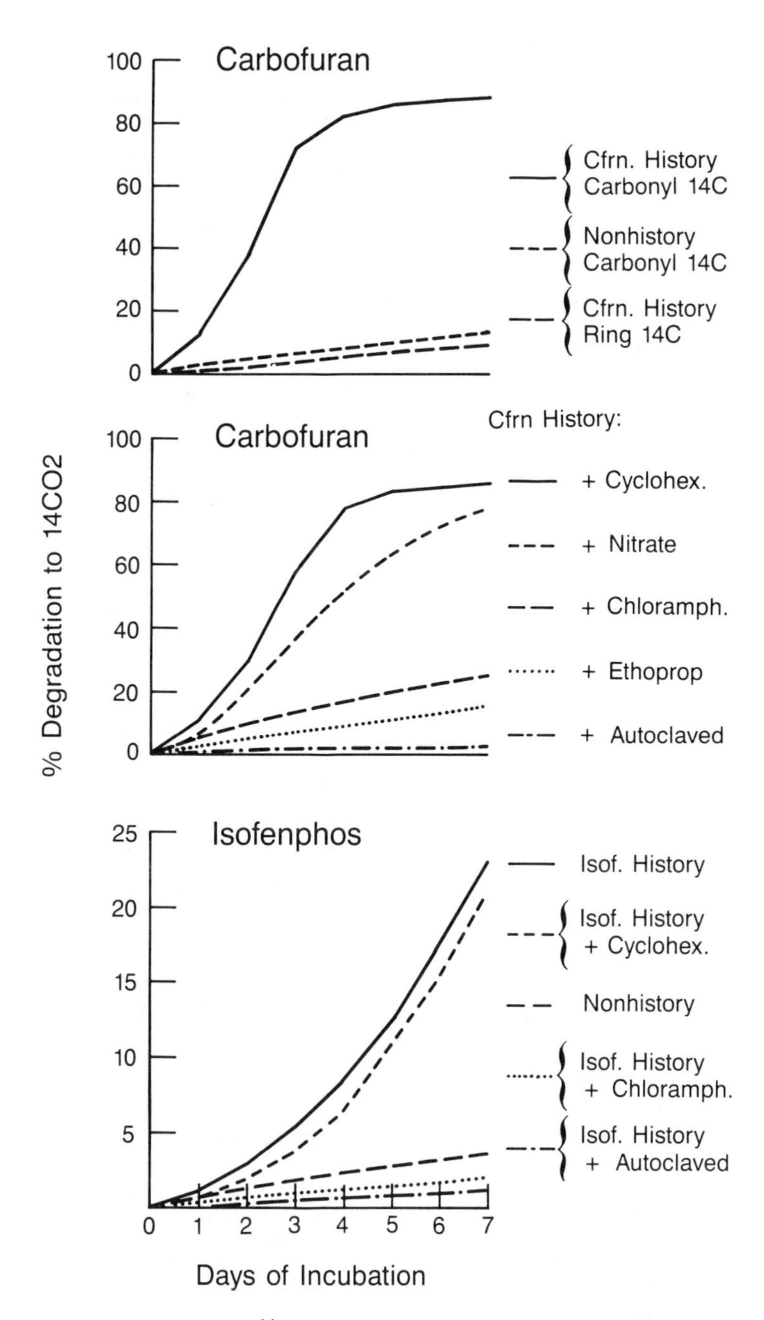

Figure 1. Mineralization of ^{14}C–labeled carbofuran and isofenphos to $^{14}CO_2$ during a one-week laboratory assay.

of these insecticides was examined in 'companion' soils. For a given insecticide this included a soil with a history of application of that insecticide and a similar soil from an untreated adjacent experimental plot or field. The history soils for carbofuran, isofenphos, and fonofos had displayed evidence of rapid insecticide degradation in the rapid degradation assay. The strategy was to examine the rates and products of degradation in treated and untreated companion soils to determine if a shift in degradative behavior had occurred.

For this study 100 g samples of soil were placed in glass jars and treated with either ^{14}C-ring-carbofuran, ^{14}C-ring-cloethocarb, ^{14}C-ring-isofenphos, ^{14}C-ring-fonofos, or ^{14}C-ring-chlorpyrifos at 5 µg/g. Soils were moistened to 0.3 bar soil moisture tension and incubated for 4 weeks at 25°C by using a flow-through incubation system in which air was periodically passed from the jars and through 0.1 N NaOH traps for monitoring of evolved $^{14}CO_2$ and maintenance of aerobic conditions (27). Soils were then extracted with organic solvents to remove extractable metabolites and the extract analyzed by thin-layer or high-pressure liquid chromatography for determining distribution of metabolites. Samples of extracted soil were combusted to determine soil-bound ^{14}C-residues.

Results of the soil pesticide metabolism study are presented in Table II, and the companion soils are listed first for each pesticide investigated. For carbofuran, there was a tremendous difference in degradation rate between the history and nonhistory soils. As oppposed to the mineralization assay, in which ^{14}C-carbonyl-carbofuran was used, this study employed ^{14}C-ring-carbofuran. Therefore, although little mineralization of the carbofuran ring occurred in any soil, there was a tremendous accumulation of soil-bound residues in the soil in which carbofuran was rapidly degraded. The enhanced degradation of carbofuran has been extensively documented (18). The behavior of cloethocarb was similar to that of carbofuran, and it too was extensively degraded in the history versus the nonhistory soil with substantial accompanying production of soil-bound residues. Cloethocarb is a carbamate insecticide that was under development for the rootworm insecticide market but was withdrawn about the same time that decreased persistence in soil after repeated use was noted (28).

Both isofenphos and fonofos were much less persistent in the history versus the nonhistory soils. A major difference between these insecticides and the carbamates was that considerable mineralization of the aromatic ring portion of the organophosphorus compounds occurred. For isofenphos, considerable quantities (15.2%) of isofenphos oxon accumulated only in the nonhistory, whereas no such accumulation was noted in the isofenphos-history soil. Several reports of the reduced persistence of isofenphos in field plots following repeated use have appeared (29,30), and isofenphos was withdrawn from the rootworm insecticide market after widespread experiences of control failure following second year applications. This is especially ironic because upon first application to soil isofenphos behaves as one of the most persistent organophosphorus insecticides (31).

In contrast to the other insecticides intensely investigated, chlorpyrifos showed no decrease in persistence in soil with a history of chlorpyrifos use. This corroborates earlier laboratory and field

Table II. Recovery of Parent Insecticides and Degradation Products From Companion Insecticide-History and Untreated Soils After a 4-Week Laboratory Incubation*

Insecticide/ Soil History	Parent Pesticide	Extractable Metabolites	Soil-Bound	$^{14}CO_2$
	^{14}C Recovered in % of Applied			
Carbofuran				
Untreated Soil	56.0	6.5	29.5	5.8
Carbofuran Soil	0.8	1.9	71.8	14.0
Cloethocarb Soil	1.6	2.2	76.9	11.0
Isofenphos Soil	46.4	3.0	39.6	9.3
Cloethocarb				
Untreated Soil	67.0	2.0	22.8	8.1
Cloethocarb Soil	6.4	0.1	73.9	18.8
Carbofuran Soil	24.0	2.1	43.9	15.3
Isofenphos Soil	45.6	1.2	35.8	10.2
Isofenphos				
Untreated Soil	62.8	16.1	9.2	10.0
Isofenphos Soil	12.9	3.6	23.6	52.4
Fonofos Soil	76.3	13.5	8.0	4.7
Carbofuran Soil	74.9	12.4	7.3	4.5
Fonofos				
Untreated Soil	50.3	1.8	27.2	19.3
Fonofos Soil	2.1	1.2	33.7	51.2
Isofenphos Soil	35.3	3.0	37.3	25.6
Carbofuran Soil	60.4	1.7	20.8	15.0
Chlorpyrifos				
Untreated Soil	57.1	6.0	17.1	15.2
Chlorpyrifos Soil	55.2	5.2	8.1	26.7
Fonofos Soil	33.9	32.6	18.6	9.1
Carbofuran Soil	55.8	13.4	15.4	10.9

*Adapted in part from (22,27,28)

studies in which the persistence of chlorpyrifos has not been affected by repeated field applications (21,32).

It is at this point that too many studies of enhanced insecticide degradation conclude. Evidence has been generated that in soils, plots, or fields selected for study the pesticide exhibits decreased persistence when it has been repeatedly applied. These companion soils have often been selected as a result of observed field failures, with 'control' soil collected from an adjacent fencerow, plot, or field. But does the comparison of rates of insecticide degradation in history and non-history soils represent sufficient evidence for proving that an insecticide has undergone enhanced degradation? We believe not for several reasons. First, it is recognized that there is a tremendous amount of natural variation in the rate of pesticide degradation between soils due to such environmental factors as organic matter content, pH, texture, and moisture. For example, Laskowski et al. (33) reported that there was between a 2- and 36-fold difference in the degradation rate of 12 pesticides in different soil types. Merely because soil from one field displays a much more rapid rate of degradation than an adjacent one does not mean that the microbial adaptation for enhanced degradation has occurred. Second, by definition, enhanced microbial degradation is a microbial process, and unless it is demonstrated that the microbial community is involved in the rapid degradation observed, it is impossible to distinguish it from abiotic mechanisms of pesticided degradation. Third, current understanding of enhanced degradation recognizes that the mechanism of enhanced microbial degradation involves an inducible microbial adaptation for pesticide catabolism (18,34). It has been demonstrated that the level of total microbial biomass in soil can sometimes greatly influence the rate of pesticide degradation (35). Should the term 'enhanced degradation' be used to describe the increased rate of degradation seen in soils from fields that have greater levels of microbial biomass (e.g. manured fields)? It is evident from these considerations that for enhanced degradation to be truly characterized, the microbiological aspects of the phenomenon must be investigated.

Microbiology of Enhanced Insecticide Degradation

A first step in the laboratory methodology regarding the microbiology of enhanced degradation involved determination of microbial involvement in the rapid insecticide degradation noted in some history soils. The strategy employed was to expose soils with suspected enhanced degradation to various antimicrobial treatments and monitor any inhibition of insecticide degradation. The rapid insecticide degradation assay in which evolution of $^{14}CO_2$ was used as an indicator of degradation proved ideal for such an investigation. Samples of soil for the mineralization assay were either autoclaved to destroy total microbial activity, treated with 100 µg/g of chloramphenicol to inhibit soil bacteria, or treated with 100 µg/g of cycloheximide to inhibit soil fungi prior to ^{14}C-insecticide application. To investigate reports that the insecticide ethoprop interfered with enhanced carbofuran degradation and that soil microorganisms might be utilizing carbofuran as a nitrogen source, soil samples pretreated with 25 µg/g of ethoprop or sodium nitrate were also incubated.

Results of these assays are presented in Figure 1. With carbofuran, isofenphos, and fonofos (not shown), sterilization by autoclaving totally inhibited rapid pesticide degradation in history soils. By itself this seems to indicate that the rapid degradation is microbially mediated. However, autoclaving also destroys immobilized enzymes present in soil, and it has been shown that these enzyme systems are responsible for the extremely rapid organo-phosphorus insecticide degradation observed in some soils (36). In order to better demonstrate the involvement of the soil microbial community in pesticide degradation, it is desirable to use antimicrobial treatments that are more selective than autoclaving. Although the antibacterial compound chloramphenical greatly inhibited the degradation of these insecticides, the antifungal compound cycloheximide did not. Thus, this type of assay can provide information not only on the microbial nature of the phenomenon but also on the microbial groups involved. The primary role of the soil bacteria in the enhanced degradative process has been well documented (15,27,34). It is worth noting that an excess supply of nitrate did not affect carbofuran degradation, but the insecticide ethoprop apparently inhibited the rapid degradation of carbofuran through some unknown mechanism. Use of sterilized soil assays has commonly been used to distinguish enhanced degradation from abiotic degradation (15,29).

A second step in the study of the microbiology of enhanced degradation involved an attempt to enumerate the microbial population in soil capable of beneficially catabolizing these insecticides. This is a critical step because it distinguishes between mere cometabolism and microbially beneficial catabolism. A most-probable-number (MPN) assay, by which microbial numbers are statistically estimated from a series of dilutions, was used to determine levels of microorganisms capable of metabolizing carbofuran, isofenphos, or fonofos in soils displaying rapid pesticide degradation. MPN techniques have been extensively utilized for the study of microorganisms in aquatic systems and the basic methodology described by Somerville et al. (37) was used. The basal salts medium in which microbial growth was assayed contained either isofenphos or fonofos as a sole carbon source, or carbofuran as a sole carbon and nitrogen source.

Results with isofenphos were quite conclusive. In nonhistory soils there were no microorganisms detected that could catabolize isofenphos as a sole carbon source, whereas in soils with an enhanced rate of isofenphos degradation between 6,000 and 12,000 isofenphos-degrading microorganisms were present per gram of soil. A similar population of isofenphos degraders could be induced in untreated soil by pretreatment with 100 ppm of isofenphos. At higher isofenphos pretreatment levels increases in the total number of soil bacteria as measured by standard plate count methods could be detected. Thus, the numbers of total bacteria in isofenphos-history and untreated soils increased 20-fold and 7-fold in response to a 5,000 μg/g isofenphos application, respectively. In the case of fonofos, no population of fonofos-catabolizing microorganisms could be detected in the soils with rapid rates of fonofos degradation. This indicates that the rapid degradation of fonofos in these soils may be unrelated to any actual microbial adaptation and thus may not involve enhanced degradation. In the case of carbofuran, both soils displaying an enhanced rate of carbofuran degradation and untreated soils contained significant populations of carbofuran-metabolizing microorganisms. In general, the numbers of carbofuran degraders was greater in carbofuran-history soils displaying an enhanced rate of carbofuran degradation (150,000-458,000/g soil) than in untreated soils (23,000-183,000/g soil). This implies that there may be qualitative differences in the rate at which soil microorganisms can degrade a given insecticide as well as quantitative differences in populations of degraders. Although work on estimating the populations of phenoxyalkanoic herbicide degraders and carbamothioate herbicide degraders in soil has shown higher population levels in soils displaying enhanced rates of herbicide degradation (38,39), in some cases no significant difference in population levels has been noted (40).

A third step in the study of the microbiology of enhanced degradation involved isolation of pure cultures of insecticide-degrading soil microorganisms. The strategy was to inoculate a minimal microbial medium with either isofenphos supplied as the sole carbon source, or carbofuran supplied as the sole carbon and nitrogen source, and then isolate individual colonies from the medium for further characterization and identification. Two bacterial strains were isolated from soil displaying an enhanced rate of isofenphos degradation, an *Arthrobacter sp.* and a *Pseudomonas sp.* (22,27). The *Arthrobacter* proved to be the most effective degrader of isofenphos and could utilize it or its hydrolysis products, isopropyl salicylate and salicylate, as sole carbon sources for growth while mineralizing the aromatic ring (22). An earlier study of enhanced isofenphos degradation had resulted in the isolation of a *Corynebacterium sp.* from isofenphos-history soils that could mineralize isofenphos (41). Repeated efforts to isolate carbofuran-degraders failed to identify microorganisms that could grow on a minimal medium containing carbofuran. Although an early study resulted in isolation of a *Pseudomonas sp.* that had some capability to degrade carbofuran in culture (15), the only carbofuran-catabolizing microorganism successfully isolated from carbofuran-history soil was an *Achromobacter sp.* that utilized carbofuran as a sole nitrogen source (42).

It seems likely that there are a variety of soil microorganisms, primarily bacteria, that can adapt for insecticide catabolism and thus mediate enhanced degradation. By itself, isolation of an insecticide-catabolizing microorganism from soil is not sufficient evidence of enhanced degradation, but coupled with data on increased rates of degradation in history soils it does complete the evidence necessary to substantiate this phenomenon. One of the additional bonuses that results from isolation of these insecticide-degrading microorganisms is the opportunity to study the enzymology and genetics of the process.

Cross-Adaptations for Enhanced Degradation

A final consideration of enhanced degradation of soil insecticides in the cornbelt is the specificity of the adaptation that leads to enhanced microbial degradation. It has been noted that in some cases an insecticide is degraded rapidly in soil from a field to which it has never been applied before, but a similar insecticide has. The specificity of enhanced degradation can be studied at all levels from field persistence behavior to interactions with microbial enzymes. Cross-adaptations seem to be especially characteristic of enhanced carbamate insecticide degradation. Carbofuran degraded rapidly in a soil previously exposed to cloethocarb, while the rate of cloethocarb degradation was somewhat enhanced in soil with carbofuran history (Table II). A number of related carbamate insecticides were degraded in soil repeatedly treated with carbofuran (17), and the degree of cross-adaptation for carbamate insecticide degradation in soil has been shown to depend on structural similarity. Likewise, the carbofuran-catabolizing *Achromobacter sp.* isolated from carbofuran-history soil could utilize a number of carbamate insecticides as sole nitrogen sources (42). The adaptation for enhanced degradation of organophosphorus insecticides, however, appears to be much more specific. Neither fonofos nor chlorpyrifos was rapidly degraded in isofenphos-history soil, and only fonofos was degraded rapidly in fonofos-history soil (Table II). No cross-adaptations for enhanced degradation of organophosphorus insecticides were noted in soil displaying enhanced isofenphos degradation, and the *Arthrobacter sp.* isolated could metabolize only isofenphos in pure culture and did not metabolize or cometabolize other organophosphorus insecticides (22).

It is likely that the specificity of enhanced degradation depends on the microbial metabolic use of the insecticide. Hydrolysis of any N-methylcarbamate insecticide yields methylamine as one product, and if the bacteria are utilizing this product as a nitrogen source, the ability to hydrolyze carbamate insecticides in general would be advantageous (42). However, microbial degradation of organophosphorus insecticides as carbon sources proceeds via initial hydrolysis and secondary metabolism of the aromatic phenolic metabolites. Because the hydrolytic metabolites of organophosphorus insecticides tend to be somewhat unique, this might explain the high specificity of the enhanced degradation of both isofenphos (27) and diazinon (43).

Towards an Experimental Definition of Enhanced Degradation

In summary, the enhanced degradation of an insecticide involves its rapid degradation by a population of soil microorganisms that has adapted to beneficially catabolize it following exposure to it or a similar insecticide. This enhanced rate of degradation may or may not result in failure of the compound to control the target pest depending on environmental conditions. For enhanced degradation to be thoroughly investigated studies must be carried out to demonstrate an increased rate of degradation in soils with prior insecticide exposure, to identify the rates and products of degradation in similar soils under controlled conditions, and to elucidate the microbiological aspects of the phenomenon including the identification of an adapted, insecticide-catabolizing microbial population. Using these criteria, only carbofuran and isofenphos have been sufficiently experimentally scrutinized so as to characterize their enhanced degradation in soil. Some evidence suggests the possibility that other insecticides may also have undergone enhanced degradation, but further study is required to provide sufficient evidence.

Literature Cited

1. Krysan, J. L.; Miller, T.A. Methods for the Study of Pest Diabrotica; Springer-Verlag; New York, NY, 1986.
2. Hill, R. E.; Hixson, E.; Muma, M. H. J. Econ. Entomol. 1948, 41, 392-401.
3. Lilly, J. H. Ann. Rev. Entomol. 1956, 1, 203-222.
4. Ball, H. J; Weekman, G. T. J. Econ. Entomol. 1962, 55, 439-441.
5. Apple, J. W.; Walgenbach, E. T.; Knee, W. J. J. Econ. Entomol. 1969, 62, 1033-1035.
6. Chio, H.; Chang, C. S.; Metcalf, R. L.; Shaw, J. J. Econ. Entomol. 1978, 71, 389-393.
7. Turpin, F. T.; Dumenil, L. C.; Peters, D. C. J. Econ. Entomol. 1972, 65, 1615-1619.

8. Felsot, A.S.; Steffey, K.L.; Levine, E.; Wilson, J.G. J. Econ. Entomol. 1985, 78, 45-52.
9. Jennings, V.; Stockdale, H. Herbicides and Soil Insecticides Used in Iowa Corn and Soybean Production, 1977; Iowa State University Cooperative Extension Service: Ames, IA, 1978.
10. Ahmad, N.; Walgenbach, D. D.; Sutter, G. R. Bull. Environ. Contam. Toxicol. 1979, 23, 572-574.
11. Gorder, G. W.; Tollefson, J. J.; Dahm, P. A. Iowa State J. Res. 1980, 55. 25-33.
12. Greenhalgh, R.; Belanger, A. J. Agric. Food Chem. 1981, 29, 231-235.
13. Felsot, A. S.; Wilson, J. G.; Kuhlman, D. E.; Steffey, K. L. J. Econ. Entomol. 1982, 75, 1098-1103.
14. Newton, J.P. M.S. Thesis, University of Nebraska, Lincoln, 1978.
15. Felsot, A. S.; Maddox, J. V.; Bruce, W. Bull. Environ. Contam. Toxicol. 1981, 26, 781-788.
16. Read, D. C. Agric. Ecosyst. Environ. 1983, 10, 37-46.
17. Harris, C. R.; Chapman, R. A.; Harris, C.; Tu, C. M. J. Environ. Sci. Health 1984, B19, 1-11.
18. Felsot, A. S. Ann. Rev. Entomol. 1989, 34, 453-476.
19. Annonymous Ag. Consult. Fieldman 1984, 40, 14.
20. Doersch, R. E.; Doll, J. D.; Wedberg, J. L.; Grau, C. R.; Harvey, R. G.; Kenney, J. E. Pest Control in Corn 1983; University of Wisconsin-Extension Service: Madison, WI; p. 36.
21. Harris, C. R.; Chapman, R. A.; Tolman, J. H.; Moy, P.; Henning, K.; Harris, C. J. Environ. Sci. Health 1988, B23, 1-32.
22. Racke, K. D.; Coats, J. R. J. Agric. Food Chem. 1988. 36, 193-199.
23. Wilde, G.; Mize, T. J. Econ. Entomol. 1984, 13, 1079-1082.
24. Chapman, R. A.; Moy, P.; Henning, K. J. Environ. Sci. Health 1985, B20, 313-320.
25. Niemczyk, H. D.; Chapman, R. A. J. Econ. Entomol. 1987, 80, 880-882.
26. Kaufman, D. D.; Katan, Y.; Edwards, D. F.; Jordan, E. G. In Agricultural Chemicals of the Future; Hilton, J. L., Ed.; Beltsville Symposia in Agricultural Research No. 8; U. S. Dept. of Agriculture: Totowa, MD, 1985; pp 437-451.
27. Racke, K. D.; Coats, J. R. J. Agric. Food Chem. 1987, 35, 94-99.
28. Racke, K. D.; Coats, J. R. J. Agric. Food Chem. 1988, 36, 1067-1072.
29. Abou-Assaf, N.; Coats, J. R.; Gray, M. E.; Tollefson, J. J. J. Environ. Sci. Health 1986, B21, 425-446.
30. Chapman, R. A.; Harris, C. R.; Moy, P.; Henning, K. J. Environ. Sci. Health 1986, B21, 269-276.
31. Felsot, A. J. Environ. Sci. Health 1984, B19, 13-27.
32. Racke, K. D.; Coats, J. R.; Titus, K. R. J. Environ. Sci. Health 1988, B23, 527-539.
33. Laskowski, D. A.; Swann, R. L.; McCall, P. J.; Bidlack, H. D. Residue Rev. 1983, 85, 139-147.
34. Kearney, P. C.; Kellogg, S. T. Pure Appl. Chem. 1985, 57, 389-403.
35. Frehse, H.; Anderson, J. P. E. In Pesticide Chemistry: Human Welfare and the Environment; Miyamoto, J., Ed.; Int. Union Pure Appl. Chem.: Pergamon Press, Oxford, England, 1983; pp 23-32.

36. Getzin, L. W.; Rosefield, I. J. Agric. Food Chem. 1966, 16, 598-601.
37. Somerville, C. C.; Monti, C. A.; Spain, J. C. Appl. Environ. Microbiol. 1985, 40, 726-734.
38. Fournier, J. C.; Codaccioni, P.; Soulas, G. Chemosphere 1981, 10, 977-984.
39. Mueller, J. G.; Skipper, H. D.; Lawrence, E. G.; Kline, E. L. Weed Sci. 1989, 37, 424-427.
40. Moorman, T. B. Weed Sci. 1988, 36, 96-101.
41. Murphy, J. J.; Cohick, A. D. Abstr. 40th Nat. Meet. Entomol. Soc. Am., 1985, paper 121.
42. Karns, J. S.; Mulbry, W. W.; Nelson, J. O.; Kearney, P. C. Pestic. Biochem. Physiol. 1986, 25, 211-217.
43. Forrest, M.; Lord, K. A.; Walker, N.; Woodville, H. C. Environ. Pollut. 1981, A24, 93-104.

RECEIVED February 1, 1990

Chapter 7

Enhanced Degradation of Insecticides in Soil

Factors Influencing the Development and Effects of Enhanced Microbial Activity

R. A. Chapman and C. R. Harris

Agriculture Canada Research Centre, 1400 Western Road, London, Ontario N6G 2V4, Canada

Treatment of soils with pesticides at 10 ppm generated enhanced microbial activity within 6 weeks for 14 of 18 materials tested. Increases in degradation rates ranged from 2 - 100x depending on the material and soil type. The aryl methylcarbamates and isofenphos showed the largest increases. Generally, only activity against the sulfoxide and sulfone metabolites resulted from treatment with thioether containing materials (aldicarb, phorate, terbufos, etc.). Enhanced activity to carbofuran was not generated at low temperatures (3°C), low soil moistures or with low insecticide concentrations (<1 ppm), but once generated was relatively insensitive to low temperature, low soil moisture and further carbofuran treatment. Enhanced activity to all materials decreased in the absence of further insecticide treatment. Times to non-detectability of enhanced activity ranged from ca. 52 wk for fensulfothion and chlorethoxyphos to >164 wk for carbofuran and isofenphos. The effectiveness of enhanced microbial populations was dependent on the form of the insecticide encountered. Granular formulations were less affected than more dispersed forms. For Furadan 15G there was a strong soil moisture dependence. The activity of enhanced microbial populations was significantly reduced by thermal or chemical biocide treatment. Cross enhancement occurred generally among the aryl methylcarbamates and among a few structurally similar organophosphorus insecticides.

Felsot et al. (1) demonstrated that the disappearance rate of carbofuran from soil was affected by previous treatments with the insecticide in 1981. Since then, the phenomenon now generally described as enhanced microbial degradation has been intensively examined, particularly for insecticides used to control corn

0097–6156/90/0426–0082$06.00/0
Published 1990 American Chemical Society

rootworms, Diabrotica spp., in North America (2). The existence of enhanced degradation of herbicides in soil has been known for 40 years (3). The effect was first reported for a soil-applied carbamothioate herbicide in 1980 (4,5) and this prompted renewed interest in the field for herbicides (6). Several fungicides are now known to be affected by this phenomenon (7-9).

Laboratory Generated Enhanced Soil Microbial Activity

In 1982, we decided that the conditions required to produce a "history soil" capable of causing enhanced carbofuran degradation should be better defined. A laboratory experiment was designed that incorporated the following features: 1) a soil that had not been previously treated with any pesticide, 2) a method of treatment that introduced only the pure test chemical to the soil (10), 3) frequent mixing of the soil to maintain a homogeneous distribution of the microbial populations and 4) an extensive series of samples incubated under the same conditions that could be treated at intervals and still permit comparison of disappearance rates with a previously untreated control. Under these conditions, enhanced microbial activity developed for carbofuran during the 28 days required for the initial 10 ppm treatment to decline to ca. 0.5 ppm (11). Clearly a long "history" of carbofuran treatment was not a requirement. Using this technique, we have examined a variety of insecticides and herbicides for their ability to rapidly generate enhanced microbial activity in soils.The results of these studies are summarized in Table I. During the course of the work it became clear that a quantitative measure of the degree of enhancement was required

Table I. E-Factors for Laboratory Treated Soils

	E-Factor for Soil Type Indicated			
Material	Sand	Sandy Loam	Clay Loam	Muck
Butylate		1.3		
Carbofuran	35	44	28	100
Furadan 4.8F		47		
Furadan 15G		16*		
Chlorpropham		13		
Cloethocarb		36		
Diazinon		4.8	3.5	
DOWCO 429X			3.6	
EPTC		0.9		
Fensulfothion(p)	14	8.8	4.6	6.9
(t)		24		
Isofenphos			1.0	
Isofenphos(th)			42	
Oftanol 15G			49	
Phorate(p)		0.5		
(t)		1.0		
Tefluthrin			1.2	

p = parent only; t = parent + metabolites (sulfoxide/sulfone)
*based on testing with carbaryl; th = technical

to keep results in their proper perspective. We have found the ratio of the pseudo first order disappearance rate constants to be useful for comparison. (The rate constant for previously treated / the rate constant for not previously treated = the E(nhancement)-factor. For example, the first entry for carbofuran in Table I is derived as follows: 2.46 day^{-1} / 0.07 day^{-1} = 35). For soils in which degradation is very rapid, the factors are based on only a few daily observations and therefore may vary somewhat among experiments. Tests for enhanced degradation were made 28 days after the initial 10 ppm application. Analytical grade materials were used for all testing and for the initial applications except for the carbofuran and isofenphos formulations and technical isofenphos as noted.

Enhanced microbial activity developed quickly for soils treated with analytical carbofuran, chlorpropham, cloethocarb, diazinon, DOWCO 429X and fensulfothion with E-factors ranging from 4 to 100 depending on the chemical and soil type. Commercial flowable and granular formulations of carbofuran also rapidly generated enhanced activity. The behavior of the granular formulation was particularly interesting. The carbofuran concentration after 28 days was still above 5 ppm and too high to reliably test for activity against the usual 10 ppm test treatment. This difficulty was circumvented by taking advantage of the known cross-activity of carbofuran-enhanced microbes to carbaryl (12) and by using 10 ppm of carbaryl for the enhanced degradation test. The soil was indeed active, but the disappearance of the residual carbofuran from the original granular treatment was apparently unaffected. Enhanced activity also developed quickly for soils treated with technical and formulated isofenphos, but those treated with analytical isofenphos only developed activity occasionally. Soils treated with analytical grade EPTC, butylate, tefluthrin and phorate did not develop activity within 28 days. This was somewhat surprising behavior for the carbamothioates which apparently can develop activity within 6 days of a first soil treatment (13). Their behavior and that of isofenphos implies that there are still unidentified factors involved in the generation of enhanced activity and suggests that pure chemicals, formulated products and metabolites need to be examined to fully understand the phenomenon.

Field-Generated Enhanced Soil Microbial Activity

Samples from field experiments carried out in microplots (ca. 2 m^2) (14, Chapman,R.A., Harris,C.R. unpublished data) have proven valuable in determining the rate of development of enhanced degradative activity. These small plots were developed originally to allow economical evaluation of the efficacy and persistence of insecticides in various, previously untreated soil types at one location (15). Their size permits the simultaneous study of a large number of materials, treatments and retreatments under natural conditions in a small area. They also make it possible to prevent the transfer of soil and associated microbial populations among plots by careful cleaning of the simple equipment required for treatment, sampling and plot maintenance. Table II lists the E-factors determined in the laboratory for soil samples removed from the microplots 6 weeks after the 1st and 3rd annual subsurface band applications. The method used was that described previously for the laboratory-treated soil (11) (ie. field sample vs. not previously treated control). Also included in Table II is the corresponding factor obtained from the ratio of the 0 - 8 week disappearance rates for the 3rd and 1st applications of the commercial formulations from the field microplots.

Enhanced microbial activity was present for carbofuran, chlorethoxyphos, DOWCO 429X, isofenphos, fensulfothion and trimethacarb and the sulfoxide/sulfone metabolites of aldicarb, disulfoton, fensulfothion and terbufos within 6 weeks of the initial application. E-factors ranged from 2 to 31. Chlorpyrifos and

Table II. E-Factors for Microplot Treated Clay Loam

Material (Formulation)	E-Factor: Laboratory Test 1st/Control	E-Factor: Laboratory Test 3rd/Control	E-Factor: Field 3rd/1st
Aldicarb (Temik 15G)(p)	1.0	-	-
(t)	5.1	-	-
Carbofuran (Furadan 15G)	31	Y	1.6
Chlorethoxyphos (Fortress 15G)	2.0	2.1	1.3
Chlorpyrifos (Lorsban 15G)	-	0.8	1.1
Disulfoton (Di-Syston 15G)(p)	0.8	0.9	1.1
(t)	7.6	7.9	4.3
DOWCO 429X (5G)	5.4	5.4	1.6
Fensulfothion (Dasanit 15G)(p)	6.0	6.8	1.1
(t)	12	14	1.4
Fonofos (Dyfonate 20G)	0.8	5.7	1.6
Isofenphos (Amaze 20G)	6.8	27	1.6
Tefluthrin (Force 1.5G)	-	0.8	1.4
Terbufos (Counter 15G)(p)	1.0	1.3	1.4
(t)	4.4	5.0	2.0
Trimethacarb (Broot 15G)	Y	Y	1.7
Phorate (Thimet 15G)(p)	-	1.1	1.0
(t)	-	5.3	2.1

p = parent only; t = parent + metabolites (sulfoxide/sulfone)
Y = active in aqueous incubation test

tefluthrin treated soils were not tested for year 1 because of the negative results observed for the 3 year samples. A factor could not be calculated for trimethacarb because the test for enhanced degradation was done by an alternate aqueous incubation procedure (16).

E-factors for the soils receiving the 3rd annual treatment were similar to those receiving the 1st except for those treated with fonofos and isofenphos indicating that repeated annual treatments with most materials did not increase the enhanced degradative activity of the soil in the field. Of all the materials tested which generated enhanced activity, fonofos was the only one requiring more than 6 weeks after the 1st treatment to generate it. The 3-year sample from phorate-treated soil rapidly degraded the sulfoxide/

sulfone metabolites at rates similar to those observed for the corresponding metabolites in the aldicarb and terbufos-treated soils.

The "enhancement factors" calculated from the 0 - 8 week disappearance rates of the commercial formulations from the original microplot treatments (14, Chapman,R.A., Harris,C.R. unpublished data) (year 3 vs. year 1) were much smaller than those derived from the shorter term laboratory tests of the same soils using pure chemicals and controlled temperature and soil moisture. This demonstrates the difficulty one faces in establishing the presence of enhanced degradative activity based on studies of the persistence of field applications even when a previously untreated plot is available for comparison. There are at least two reasons for these factors being smaller. First, as we have demonstrated, enhanced activity develops rapidly (within 2-3 weeks in some cases (14)). This results in increased 0 - 8 week disappearance rates even in the 1st year treatment plots and thus reduces the enhancement factor. Second, soil treated with a granular formulation of insecticide is a heterogeneous system with respect to the location of the insecticide. The concentration of insecticide in "soil" measured on samples taken from this system represents a combination of the insecticide remaining on the original granules and that which has become dispersed through the soil. If the insecticide remaining on the granule is not as susceptible to enhanced degradation as that which is dispersed through the soil, the total concentration in "soil" will decrease less rapidly than expected for an active soil. This is reflected in a reduced enhancement factor.

Factors Affecting the Generation of Enhanced Activity by Carbofuran in Soil

The laboratory and microplot experiments have shown that enhanced degradative capability is generated quickly following a single treatment in a variety of soil types by at least 14 insecticides or their metabolites. They also demonstrate that various factors can influence the activity of the enhanced microbial population following its development. To better understand these two aspects of the phenomenon, we have studied carbofuran intensively both with regard to the generation of enhanced degradative activity and the subsequent behavior of the active microbial population generated in a sandy loam soil (17). For these tests 28°C, 11.3% moisture, 10 ppm carbofuran treatments and 10 ppm carbofuran retreatments after 28 days were considered "standard". Soils maintained at 3°C following an initial treatment demonstrated no enhancement, but those at 15 and 28°C showed E-factors of 16 and 41, respectively (Table III).

Table III. Effect of Temperature on E-Factors Generated by Carbofuran

Temperature (°C)	E-Factor
3	1.0
15	16
28	41

The effect of different soil moistures on the activity developed is shown in Table IV. There was a critical lower moisture level between 9 and 4.5% for this soil below which enhanced activity failed to develop. The enhanced activity developed at 36% (9x) was significantly lower than that developed at intermediate moistures.

To examine the effect of initial concentration on the development of activity, four soil types were initially treated with five levels of carbofuran ranging from 0.01 to 100 ppm. The results of the subsequent activity tests are summarized in Table V. Concentrations of 1 ppm were sufficient to cause activity to develop in the three mineral soils. Concentrations in excess of 1 ppm were required for the organic soil. Activity appears to develop quickly over a broad range of conditions that one would encounter naturally during use of the insecticide.

Table IV. Effect of Soil Moisture on E-Factors Generated by Carbofuran

Moisture (%)	E-Factor
1.3	0.2
4.5	0.4
9	46
11.3	41
18	50
27	54
36	9

Table V. Effect of Insecticide Concentration on E-Factors Generated by Carbofuran

	E-Factor			
Concentration (ppm)	Sand	Sandy Loam	Clay Loam	Muck
0.01	1.1	0.8	1.3	-
0.1	3.1	0.7	1.0	1.8
1.0	35	14	22	1.0
10	35	13	28	100
100	-	-	-	110

Factors Affecting the Activity of Soil Microbial Populations Enhanced by Carbofuran Treatment

Enhanced microbial populations, once developed, were found to be relatively insensitive to changes in temperature and moisture. Carbofuran continued to disappear from an active soil held at 3°C with an E-factor of 12 (Table VI). The E-factors declined with lower soil moistures, but remained significant even in the 4.5-2.3% range (Table VII).

The effect of additional 10 ppm carbofuran treatments made at monthly intervals on the E-factors observed for a sandy loam soil collected from the same location in three consecutive years are shown in Table VIII. Because the second and subsequent applications disappeared rapidly from the soil, it was possible to calculate an enhancement factor 1 week following these applications. Tests were

Table VI. Effect of Temperature on Enhanced Activity for Carbofuran

Temperature (°C)	E-Factor
3	12
15	40
28	44

Table VII. Effect of Moisture on Enhanced Activity for Carbofuran

Moisture (%)	E-Factor
1.3	<0.1
2.3	1.6
4.5	8.6
6.8	21
9	22
11.3	36

also made at 8 and 12 weeks to determine if changes in activity in the interval between the last treatment and the test were observable.The enhanced activity generated in the soil collected in 1982 and 1983 was unaffected by further treatments, but four treatments of the soil collected in 1984 did increase the E-factor about 5x over that observed after the 1st and 2nd treatments. The E-factors observed for the initial treatments of the 1983 and 1984 soils were about 1/4 those observed in the 1982 soil. Apparently there is another critical factor involved. E-factors were relatively constant over the 12 weeks following the last application.

Table VIII. Effect of Treatment Intensity on Enhanced Activity for Carbofuran

	Pretest	E-Factor		
Treatment No.	Interval (wk)	1982	1983	1984
1	4	45	10	10
	8	40		
	12	39		
2	1	23-36	10	7.4
	4	21-36		
	8	49		
	12	45		
4	1	26-54	10	54
	4	83		
	8	42		
	12	33		

Longevity of Enhanced Activity

The soils in our 3 year microplot persistence studies (14, Chapman,R.A., Harris, C.R. unpublished results) provided an opportunity to observe the persistence of enhanced activity in the absence of further treatment. E-factors observed at times ranging from 52 to 164 weeks following the third annual treatment are listed in Table IX. Soils previously treated with carbofuran, isofenphos and trimethacarb were tested by the aqueous incubation procedure mentioned previously (16) and are assessed as either active or slightly active.

Enhanced activity toward chlorethoxyphos had disappeared prior to the first test at 52 weeks. Similar behavior was observed for fensulfothion for which activity had decreased to such a low level by

Table IX. Longevity of Enhanced Activity in Microplot Treated Clay Loam

	E-Factor at Time(wk) Indicated						
Material	20	52	60	72	144	152	164
Carbofuran		Y	Y	Y	S	S	S
Chlorethoxyphos		1.0	1.1	1.2			
Disulfoton(t)		19	8.0	13			
DOWCO 429X		3.1	2.9	3.3			
Fensulfothion	S						
Fonofos		3.4	2.4	0.9			
Isofenphos		Y	S	S	12	2.0	3.9
Terbufos(t)		2.8	1.7	2.0	1.4	1.0	1.1
Trimethacarb		Y	Y	Y			
Phorate(t)		4.7	1.3	1.3			

Y = active; S = slightly active; t = parent + metabolites

20 weeks following the 3rd treatment that the plots were not sampled further. Enhanced activity toward fonofos and phorate (total) disappeared during the 52-72 week period. Soils treated with the other materials listed remained active during this period. Carbofuran and isofenphos treated soils retained measurable enhanced activity at 164 weeks following the last treatment, but terbufos (total) did not. DOWCO 429X, disulfoton and trimethacarb plots could not be retained for the 144-164 weeks.

Interaction Between Commercial Formulations and Enhanced Soil Microbial Populations

The low "enhancement factors" derived from the disappearance rates of commercial formulations from microplots listed in Table II suggest that differences may exist in the effectiveness of an enhanced microbial population depending on the form of the target chemical.

To examine the interaction between formulation and the enhanced biological activity of the soil, the disappearance rates of analytical and granular formulations of carbofuran, fensulfothion, isofenphos and trimethacarb were determined in samples of clay loam possessing enhanced activity for each insecticide, respectively. Soil moisture had been observed to be a significant factor in the disappearance of granular carbofuran from a sandy loam (18) so the experimental design incorporated two soil moistures, representative of usual early spring conditions (20%) and representative of the upper 1-2 cm of soil in mid-summer (10%). The results are summarized in Table X.

The E-factors for analytical carbofuran, isofenphos and trimethacarb were not markedly moisture dependent. Those for fensulfothion (parent and total) were 4-5x at the higher moisture. Except for carbofuran at 20% moisture, the E-factors for the granular formulations were much smaller than for the corresponding analytical materials. E-factors for all the granular formulations increased at the higher moisture. The interaction of granular carbofuran and

Table X. Effect of Insecticide Form on the Effectiveness of Enhanced Microbial Activity

	E-Factor for Form and Moisture Indicated			
	Analytical		Granular	
Material	10	20	10	20
Carbofuran	9.8	12	0.6	19
Fensulfothion(p)	2.3	11	1.5	3.5
(t)	5.3	22	2.1	5.4
Isofenphos	57	128	12	16
Trimethacarb	10	9.2	2.2	4.3

p = parent; t = parent + sulfone metabolite

enhanced microbial populations appeared to be particularly sensitive to soil moisture with no enhancement observable at 10% and a 19x enhancement at 20%. Although the E-factors for granular isofenphos were much lower than for the analytical material, the values at both moistures were high compared to most values for granular materials. The results in Table X show differences in behavior that may help to explain two of the major insecticide efficacy problems that appear to be associated with enhanced degradation, 1) the enhancement of granular carbofuran degradation under high moisture conditions and 2) the enhancement of granular isofenphos degradation over a broad range of soil moisture. Formulation and moisture have also been found to affect the rate of development of enhanced activity (14), with granular materials being less effective than more dispersed materials particularly at lower moistures.

Effect of Thermal and Chemical Biocide Treatments of Soil on Enhanced Microbial Activity

Thermal Treatment. Yarden *et al.* (8) reported that a 4 wk solarization (solar heating of soil under plastic (19)) of soil displaying enhanced degradation of the fungicide MBC greatly reduced the enhanced activity. To determine if the microbial populations responsible for the enhanced degradation of insecticides were thermally labile, soils with enhanced activity to aldicarb, carbofuran, DOWCO 429X and isofenphos were incubated at 45°C for 4 weeks. This soil temperature has been shown to be readily attainable under plastic at our geographical location (Tomlin,A.D, Lazarovitz,G. unpublished data). The effect of this thermal treatment on the E-Factors for these soils is shown in Table XI. The enhanced activity was destroyed in all cases. In fact, the degradative activity of the thermally treated soils was reduced to below that of the untreated control soils in some cases.

Table XI. Effect of Thermal Treatment on Enhanced Activity

	E-Factor	
Material	Before	After
Aldicarb (p)	1	1
(t)	13	0.3
Carbofuran	13	0.7
DOWCO 429X	1.5	0.1
Isofenphos	26	1.5

To test the effect of solarization in the field on both the generation of enhanced activity and the activity of enhanced microbial populations, microplots containing clay loam were treated with granular carbofuran. Some plots were covered with plastic immediately (July 6) and the remainder were left uncovered. Soil samples were taken at intervals from all plots and tested for enhanced activity using our previously developed aqueous incubation procedure(16). Table XII shows the results of some of these tests. The times required for 50% of the initial 10 ppm of carbofuran to disappear from the test solution (DT/50) are a measure of the enhanced activity in the soil (DT/50 > 8 days, not active; < 5 days, active). Enhanced activity was generated in the uncovered plots

Table XII. Effect of Solarization on Enhanced Activity

		DT/50(days) for Time(wk) Indicated							
Treatment	Period(wk)	2	4	5	7	10	12	15	36
Natural	0 - 10	4.8	4.5	4.4	4.7	5.1	4.9	5.1	5.4
Solarized	0 - 10	>14	>14	>14	11.6	>14	4.5	4.9	9.6
Solarized	3 - 10 1)	4.1	4.8	5.1	7.4	5.7	5.5	5.0	5.5
Solarized	3 - 10 2)	4.6	4.9	5.2	7.2	8.9	5.4	8.1	12.7

within 2 weeks, but not in the solarized plots. At 3 weeks, more plots were covered and the sampling was continued. The plastic covers were removed from all plots after 10 weeks (August 13) and the plots were further sampled at 12, 15 and 36 weeks. The non-solarized plots became active by 2 weeks and the activity remained the following spring as expected from previous work. The plots solarized for 10 weeks immediately following carbofuran treatment did not develop enhanced activity during the solarization period, but did in the subsequent 2 weeks. The activity that developed was unusual in that, 1) the phenolic hydrolysis product of carbofuran, which usually accumulated in the test solution, was only present in trace amounts, and 2) enhanced activity did not persist to the following spring. Solarization of plots in which enhanced activity was allowed to develop naturally for 3 weeks did not cause a rapid decrease in this activity with the soils being only slightly less active at 7 weeks. Following this, different and somewhat erratic results were observed between the two plots with one remaining active and the other becoming inactive. The differences may be due to the depth to which the enhanced microbial population moved into the soil prior to solarization. The farther an enhanced population exists below the surface the less likely it would be affected by the solarization because of the insulating effect of the intervening soil. This aspect of the experiment is currently being investigated more fully. As pointed out by Yarden et al.(8), it is unlikely that solarization would be economical if used solely to affect microbial populations enhanced to degrade insecticides. In those cases where the cumulative benefits merit its use, it would appear best, based on our limited results with carbofuran, to design the treatment to coincide with insecticide application.

Chemical Biocide Treatment. Chlorpyrifos is one insecticide that does not appear to undergo enhanced microbial degradation in soil. Since toxicity to microbes has been reported for both chlorpyrifos (Somasundaram,L., Coats,J.R., Racke,K.D. unpublished results) and its degradation product, 3,5,6-trichloro-2-pyridinol, which can accumulate in soil(20,21), this is not surprising. The role of the trichloropyridinol in the behavior of chlorpyrifos in soil is difficult to establish, but Racke et al.(20) have shown that concentrations of 100 ppm of the pyridinol severely inhibit the microbial degradation of both isofenphos and carbofuran in soil. We have examined the effects of treatment levels of 1, 10, 100 and 1000 ppm of the trichloropyridinol on the activity of soil microbial populations enhanced to degrade carbofuran, DOWCO 429X and

Table XIII. Effect of Trichloropyridinol on Enhanced Activity

Soil Enhanced by	E-Factor for Level(ppm) Indicated -	1	10	100	1000
Carbofuran	27	29	31	0.8	0.6
DOWCO429X	2.5	2.8	1.9	0.5	0.3
Isofenphos	91	46	73	80	0.3

isofenphos. The results are shown in Table XIII. Activity was tested 28 days after the trichloropyridinol treatment. Treatment levels of 1 and 10 ppm had no effect. At 100 ppm, the degradation of carbofuran and DOWCO 429X was inhibited, but this concentration had no effect on isofenphos degradation. The levels required to affect the microbes are such that it would not be feasible to treat large quantities of soil. The treatment of formulated granules with biocides to provide a protected sphere around the granule appears to be possible, but the likelihood of eventual development of enhanced degradation toward the biocide should not be overlooked.

Cross Enhancement

Cross enhancement is the enhanced degradation of a chemical in a soil in which enhanced microbial activity was initially generated by a different chemical. Numerous examples for pesticides are found in the reviews by Felsot (2) and Roeth (6). This first appeared to be an important area of investigation in this field because it provided information on materials that might be used with some degree of assurance of performance, even in soils enhanced to degrade other chemicals, and allow their use in a chemical rotation scheme to control the pest of interest. The discovery that the generation

Table XIV. Cross Enhancement Factors for Carbamate Pesticides

	E- or CE-Factor for Soil Enhanced to					
	Aldicarb	Carbofuran			Cloethocarb	Chlorpropham
Test Material		1	2	3		
Aldicarb (p)	1.0	1.2	4.7	2.3		
(t)	5.1	1.1	10	6.4		
Bendiocarb		Y				
Bufencarb		3.4	5.2			
Butylate		0.9	0.6		0.8	1.0
Carbaryl		7.0	16	7.2	Y	
Carbofuran	3.8	10	22	12	71	1.0
Cloethocarb		6.6	8.4	17	36	
Chlorpropham		0.9	0.5	0.7		13
EPTC		0.9	0.5			
Furathiocarb		1.0	0.5			
Methomyl		1.4	1.1	1.2		
Oxamyl	2.4	0.6	3.4	3.4	Y	
Phorate		1.5	2.3	1.0	0.7	0.7
Propoxur		9.2	14			
Trimethacarb		13	27			

of enhanced activity can be a rapid process and that it can persist for extended periods indicates that chemical rotation will have limited usefulness for soil-applied insecticides where efficacy 8-10 wk after application is often required. Studies of cross enhancement may still provide information on the nature of the basic processes involved in this phenomenon. Our observations on cross enhancement are summarized in Table XIV for carbamate enhanced soils towards various carbamates and phorate and in Table XV for carbofuran and various organophosphorus insecticide enhanced soils towards organophosphorus insecticides and tefluthrin. Only a small fraction of the possible combinations have been examined. Entries for three different carbofuran enhanced soils are included to provide a sampling to the variation that can be observed. Entries 1 and 2 are for sandy loam collected in 1983. They received 8 treatments at 10 ppm and 5 treatments at 100 ppm, respectively. Entry 3 is for sandy loam collected in 1984 which received 2 treatments at 10 ppm. Cross enhancement factors (CE-Factors) were calculated in a manner analogous to the enhancement factors. In four cases, the quality of the data did not allow quantitative treatment and the presence or absence of cross enhancement is indicated by Y or N, respectively.

Cross enhancement was observed between carbofuran and all the aryl methylcarbamates examined except furathiocarb. The behaviour of the oximino methylcarbamates (aldicarb, methomyl, oxamyl) was more variable. Both aldicarb and cloethocarb enhanced soils showed cross enhancement to carbofuran and oxamyl. The chlorpropham enhanced soil was not cross enhanced to carbofuran. The carbamate enhanced soils were not cross enhanced to butylate, EPTC or phorate.

No cross enhancement was observed between carbofuran enhanced soil and tefluthrin or any of the 11 organophosphorus insecticides listed in Table XV. Diazinon and DOWCO 429X showed mutual cross

Table XV. Cross Enhancement Factors for Organophosphorus Insecticides and Carbofuran

	E- and CE-Factors for Soil Enhanced to							
	Carbofuran			Diaz.	Disulf.	DOWCO	Fensulf.	Phorate
Test Material	1	2	3					
Chlorfenvinphos	0.8	0.5	1.2					
Chlorpyrifos	0.8	0.3	1.0					
Diazinon	0.8	0.5	0.9	3.5		2.4		
Disulfoton(p)	1.1	1.4			0.9		0.9	
(t)	0.9				7.9			
DOWCO 429X				2.0		3.6		
Fensulfothion(p)	0.8	0.3	1.0				8.8	
(t)	0.5		0.9				24	
(sulfide)							2.0	
Fenthion							0.7	
Fonofos	0.8	0.9						
Isofenphos	N	N						
Tefluthrin	0.9	1.3						
Terbufos(p)	0.4	1.2	0.9		2.1		0.9	1.9
(t)		0.6	0.9		3.9		0.9	3.4
Phorate(p)	1.5	2.3	1.0				0.9	1.1
(t)	1.2		0.9				1.5	5.3

enhancement. Soils enhanced to the sulfoxide and sulfone metabolites of disulfoton and phorate were cross enhanced to the corresponding metabolites of terbufos. Fensulfothion (a sulfoxide) enhanced soil did not show cross enhancement to the sulfoxide/sulfone metabolites of disulfoton, terbufos or phorate. It did show slight enhancement to its parent sulfide, but not to the structurally similar material fenthion.

The results of a study of carbofuran treatment intensity on the cross enhancement factors for carbaryl, cloethocarb and oxamyl are shown in Table XVI for the sandy loam soil collected in 1982 and 1984. Little change in the CE-factors for carbaryl was observed for 1, 2 and 4 treatments of the 1982 soil. The response of carbaryl was similar to that of carbofuran (see Table VIII). The factors increased significantly for cloethocarb and oxamyl between the 2nd and 4th treatments. Higher CE-factors for cloethocarb and oxamyl and a lower

Table XVI. Effect of Treatment Intensity on Cross Enhancement to Carbofuran

	Pretest		CE-Factors		
Treatment No.	Interval(wk)	Year	Carbaryl	Cloethocarb	Oxamyl
1	4	1982	13	1.1	0.7
		1984	7.5	13	2.2
2	1	1982	9.0	1.2	0.8
		1984	5.7	17	3.6
4	1	1982	7.4	7.9	1.9
		1984	18	145	2.7

factor for carbaryl were observed for the 1984 soil after the 1st treatment. After the 4th treatment, the factor for oxamyl had changed little, but those for carbaryl and cloethocarb had increased markedly reflecting a similar increase in the E-factor for carbofuran that occurred with this soil (see Table VIII). It is clear that the effect of treatment intensity on cross enhancement is not simple.

Summary

Our observations show: 1) some enhanced microbial activity is generated quickly (6 weeks or less) for most of the soil applied insecticides (or their insecticidal metabolites) currently in use, 2) the soil temperatures, soil moistures, soil types and chemical concentrations required, both to generate and maintain the enhanced activity of the soil microbes, are those normally encountered in agriculture, 3) enhanced degradative activity, once generated, can persist for variable periods exceeding 164 weeks for some materials and 4) insecticides applied as granular formulations are often much less susceptible to the effect of enhanced microbial populations than more finely dispersed forms. Clearly, agrochemical scientists face a difficult problem if they wish to exert some control over soil microbial populations and their adaptability to degrade insecticides.

It is important to remember that, for insecticides, enhanced degradation in soil is not a new phenomenon, but only a newly observed one. It is unreasonable to think that the effect of the first encounter between an insecticide and a soil microbial population in 1983 or 1989 would differ from that same encounter occurring in 1965 or 1970. It follows that, some soil applied insecticides were subjected to "enhanced degradation" within a few weeks of their introduction. Therefore, it is unscientific to describe a particular chemical as "once having been a good material but now is of questionable value because enhanced degradative activity developed after a history of use". A more scientific approach would be to ask, which variables involved have changed to permit the natural and rapid adaptation of the soil microbial population to the degradation of this material to become a factor controlling its efficacy?

Acknowledgment

The authors wish to thank our laboratory staff, Colleen Cole, Carol Harris, Karin Henning and Pat Moy, and our field crew, Glen McFadden and Murray Cates, under the supervision of Dr. Jeff Tolman, for their untiring efforts and innumerable suggestions during the course of the work described.

Literature Cited

1. Felsot, A.; Maddox, J. V.; Bruce, W. Bull. Environ. Contam. Toxicol. 1981, 26, 781-8.
2. Felsot, A. S. Ann. Rev. Entomol. 1989, 34, 453-76.
3. Audus, L. J. Plant Soil 1949, 2, 31-6.
4. Obrigawitch, T.; Martin, A. R.; Roeth, F. W. Proc. North Cent. Weed Control Conf. 1980, 35, 20.
5. Schuman, D. B.; Harvey, R. G. Proc. North Cent. Weed Control Conf. 1980, 35, 19-20.
6. Roeth, F. W. Rev. Weed Sci. 1986, 2, 45-65.
7. Bailey, A. M.; Coffey, M. D. Phytopathology 1985, 75, 135-7.
8. Yarden, O.; Katan, J.; Aharonson, N.; Ben-Yephet, Y. Phytopathology 1985, 75, 763-7.
9. Walker, A.; Brown, P. A.; Entwistle, A. R. Pestic. Sci. 1986, 17, 183-93.
10. Miles, J. R. W.; Tu, C. M.; Harris, C. R. Bull. Environ. Contam. Toxicol. 1979, 22, 312-8.
11. Harris, C. R.; Chapman, R. A.; Harris, C.; Tu, C. M. J. Environ. Sci. Health 1984, B19, 1-11.
12. Chapman, R. A.; Harris, C. R.; Harris, C. J. Environ. Sci. Health 1986, B21, 125-41.
13. McCusker, V. W.; Skipper, H. D.; Zublena, J. P.; Dewitt, T. G. Weed Sci. 1988, 36, 818-23.
14. Harris, C. R.; Chapman, R. A.; Tolman, J. H.; Moy, P.; Henning, K.; Harris, C. J. Environ. Sci. Health 1988,B23, 1-32.
15. Harris, C. R.; Svec, H. J.; Sans, W. W. J. Econ. Entomol. 1971, 64, 493-6.
16. Chapman, R. A.; Moy, P.; Henning, K. J. Environ. Sci. Health 1985, B20, 313-9.
17. Chapman, R. A.; Harris, C. R.; Harris, C. J. Environ. Sci. Health 1986, B21, 57-66.
18. Chapman, R. A.; Harris, C. R.; Harris, C. J. Environ. Sci. Health 1986, B21, 125-41.
19. Katan, J. Annual Review of Phytopathology 1981, 19, 211-36.
20. Racke, K. D.; Coats, J. R.; Titus, K. R. J. Environ. Sci. Health 1988, B23, 527-39.
21. Chapman, R. A.; Harris, C. R. J. Environ. Sci. Health 1980, B15, 39-46.

RECEIVED February 21, 1990

PESTICIDES AND MICROBIAL ADAPTATION

Chapter 8

Enhanced Degradation of S-Ethyl *N,N*-Dipropylcarbamothioate in Soil and by an Isolated Soil Microorganism

W. A. Dick, R. O. Ankumah[1], G. McClung[2], and N. Abou-Assaf

The Ohio Agricultural Research and Development Center, The Ohio State University, Wooster, OH 44691

Enhanced degradation of EPTC occurs in soil after repeat applications of EPTC. Studies were conducted to evaluate the mechanisms of enhanced degradation in soil and by an isolated soil microorganism (*Rhodococcus* sp.). Inoculation of a soil without a history of EPTC treatment with 1.0% (w/w) of a soil with enhanced EPTC degradation capabilities increased the rate of EPTC degradation in the previously untreated soil. Degradation of ^{14}C-labelled (1-propyl position) and unlabelled EPTC by the microbial isolate yielded *N*-depropyl EPTC (a product of α-propyl hydroxylation) and EPTC-sulfoxide (sulfur oxidation). It is proposed that initial reactions of soil microorganisms involve both hydroxylation and sulfoxidation, resulting in products that are further metabolized to CO_2. The hydroxylation reaction is thought to be dominant when degradation of EPTC occurs at enhanced rates.

Enhanced pesticide degradation refers to the phenomenon whereby a pesticide is degraded at an increased rate in soil previously treated with the pesticide, or a compound of similar structure, as compared to its rate of degradation in a comparable untreated soil. To date, more than 25 different agricultural pesticides have been reported to have the potential for enhanced degradation in soil including pesticides classified as herbicides, insecticides, and fungicides (Racke, this volume). The earliest report of enhanced degradation of a carbamothioate pesticide was that by Rahman et al. (1) in New Zealand where the herbicide EPTC (*s*-ethyl *N,N*-dipropyl carbamothioate) failed to give adequate weed control in fields after repeat applications. Since that initial report, one of the most

[1]Current address: Department of Agricultural Science, Tuskegee University, Tuskegee, AL 36088

[2]Current address: U.S. Department of Agriculture, Agricultural Research Station, Room 100, Building 050, Beltsville Agricultural Research Center—West, Beltsville, MD 20705

0097–6156/90/0426–0098$06.00/0

widely studied enhanced degradation systems has been that of EPTC in soil.

Enhanced EPTC degradation in soil is supposedly caused by soil microorganisms which appear to acquire additional degradative capabilities (2-4). Although microbial involvement in enhanced pesticide degradation has been recognized, the mechanisms responsible for this phenomenon remain unresolved. Two theories have been proposed to explain microbial involvement in enhanced degradation (5). The chance mutation theory postulates the presence of mutant microorganisms which become dominant with the addition of a substrate in absence of competition. The lag period normally observed before degradation begins is the time required for the mutant population to build to an appreciable level. The adaptive enzyme theory postulates adaptive enzymes, induced by pesticide additions, are produced by microorganisms. The lag phase is the period required to fully develop the adaptive enzyme potential.

While many soils contain the necessary ingredients (i.e. a susceptible pesticide, an adaptable microorganism, and a suitable environmental condition) (6), enhanced degradation is not always observed. This observation has been difficult to explain, although one possible reason is the inability to obtain similar but untreated (control) soils to establish baseline degradation rates. Equally puzzling is how to account for the persistence of EPTC degradative ability in a soil that no longer contains the pesticide to serve as substrate for the microorganisms. Possibly the enzymes induced by EPTC may also act on other substrates thus maintaining the microorganisms' ability to rapidly exploit EPTC as a substrate (7). Transfer of degradative genes on plasmids among microorganisms is another mechanism whereby the survival of the EPTC degradative system may continue from one year to another (8). A microorganism, isolated from soil that had been repeatedly exposed to EPTC, contained four plasmids, one of which mediated the degradation of EPTC (9). Extracellular enzymes may also result in the maintenance of enhanced pesticide degradation in soil. Many extracellular enzymes are active against functional groups commonly contained within soil pesticide molecules. Some extracellular enzymes maintain their activity in soil for long periods, yielding a "steady-state" soil enzyme component unrelated to current microbial proliferation.

To sort out the relative importance of these, or additional hypotheses for explanation of enhanced EPTC degradation in soil, it is important that the microorganisms involved be identified and that the biochemical pathway(s) by which they degrade EPTC be determined. Although several different pathways have been proposed for carbamothioate degradation in soil, current information is based on studies conducted in animal and plant systems (10-12). Carbamothioate degradation in mammals involves conversion to a sulfoxide which further undergoes cleavage at the carbamoyl bond (13). In plants, EPTC was hypothesized to undergo hydroxylation at the carbon alpha to the sulfur, followed by carbamate cleavage (14). Plant studies have also determined that EPTC degradation involves hydrolysis at the ester linkage to form mercaptan, amine and CO_2 (15).

Lack of information concerning the microorganisms and the mechanisms involved in enhanced EPTC degradation has seriously limited attempts to control the rapid breakdown of this herbicide, and other carbamothioate pesticides, in soil. This study was initiated to 1) evaluate enhanced EPTC degradation in field and laboratory soils, 2) isolate soil microorganism(s) active in degrading EPTC and 3) determine the biochemical pathway(s) of EPTC degradation by the isolated microbes.

Materials and Methods

Soils. Characteristics of the three surface (0 to 15 cm sample depth) soils used in this study are given in Table I. The Brookston soil was collected from a field located near Canal Winchester, Ohio that had previously been treated with Eradicane for 1, 2, 3, and 4 consecutive years or had remained untreated. The Plano and Dothan soils were collected from sites in Wisconsin and South Carolina, respectively, which had histories of EPTC and butylate (*s*-ethyl-*N*,*N*-diisobutyl carbamothioate) use.

Table I. Soil Characteristics

Soil			Organic
Series (texture)	Subgroup	pH	carbon (%)
Brookston (clay loam)	Typic Argiaquoll	7.5	4.5
Plano (silt loam)	Typic Argiudoll	5.9	3.8
Dothan (sandy loam)	Plinthic Paleudult	5.5	3.0

Chemicals. Technical EPTC and butylate and ^{14}C-1-propyl EPTC (specific activity, 35 mCi/mole) were supplied by the Stauffer Chemical Company. The radiopurity of the ^{14}C-EPTC was greater than 97% as determined by thin layer chromatography (TLC). Millimole quantities of unlabelled and ^{14}C-labelled EPTC-sulfoxide and EPTC-sulfone were synthesized by the method of Casida et al. (16). Purification was accomplished by TLC with ultraviolet radiation (UV) detection.

Analytical Procedures. Gas chromatographic (GC) analyses of EPTC and metabolites were performed using glass columns containing 5% carbowax 20M on Chrom WHP and 3% OV-17 on Supelcoport (100/200 mesh). A Varian 3700 gas chromatograph with a nitrogen specific thermoionic detector was used.

Gas chromatographic/mass spectrometry (GC/MS) results were obtained at the Ohio State University Chemical Instrumentation Center using a Finnigan 4021 GC/MS instrument. Both electron impact and chemical ionization were performed on samples following separation of compounds by a gas chromatograph equipped with capillary columns containing 3% OV-17 or 5% carbowax 20M.

Thin-layer chromatography (TLC) for separation of EPTC and metabolites was accomplished using silica gel 60 F_{254} chromatoplates developed with hexane-acetone (6:1), hexane-ether (4:1) or hexane-ethyl acetate (3:2). Detection of unlabelled and or radiolabelled pesticide and metabolites was by UV adsorption and autoradiography, respectively. Quantitation of ^{14}C-activity was accomplished by liquid scintillation counting (LSC) of scraped sections of gel.

Non-protein thiol content in the culture media, after precipitation of protein using 0.2M trichloroacetic acid, was assayed using Ellman's reagent (17), with 2-mercaptomethanol used as a standard.

Microbial Isolate. A microbial isolate capable of growing on EPTC as a sole carbon and energy source was isolated from Jimtown loam soil (Typic Argiaquoll; organic carbon, 4.5%; pH, 7.5) by an enrichment batch culture technique. One gram field-moist soil was added to 50 ml basal salt medium (BSM) (18) supplemented with 100 or 200 mg L^{-1} EPTC (BSME). At 10-day intervals, 1 ml of inoculum was transferred into a flask containing fresh media. The flasks were plugged with cotton and incubated at 27°C on a rotary action shaker at 250 rpm. After four transfers, 0.1 ml dilutions were plated out on nutrient agar (NA) plates. Individual isolates were purified and screened for their ability to grow on basal salt agar plates with EPTC as the carbon source (BSAE). Isolates went through three transfers on BSAE plates and were then tested for their ability to grow in BSME. Isolate JE1 was retained for all subsequent studies based on its ability to rapidly degrade EPTC and to utilize EPTC as a sole carbon source. JE1 has been tentatively identified as *Rhodococcus* sp. on the basis of its morphology and cell wall composition.

Metabolic Studies. To measure EPTC and butylate degradation in soil, aliquots of air-dried Brookston soil were treated with 4 mg kg^{-1} of EPTC or butylate and incubated at 25°C in the dark. At various intervals, samples were removed from the incubator and EPTC or butylate remaining in the soil was extracted with 10:3 toluene:water. The toluene layer was analyzed for EPTC or butylate by GC. The minimum detectable level of EPTC or butylate was 0.05 mg kg^{-1} soil and the average recovery exceeded 90%.

The minimum inoculum level of an EPTC history soil which results in enhanced degradation was evaluated by adding 0.1, 1.0, 10 and 100% (w/w) of a 3-yr history Brookston soil to a non-history Brookston soil. EPTC was added at a rate of 4 mg kg^{-1} soil and the amount remaining after various time intervals was measured as previously described.

The effect of various antibiotics on enhanced EPTC degradation was investigated by treating aliquots of the 3-yr history Brookston soil with kanamycin, streptomycin, chloramphenicol, or cycloheximide at a rate of 100 mg kg^{-1} soil. The samples were then treated with EPTC (10 mg kg^{-1} soil) and the remaining EPTC was extracted and measured after 0, 3, and 7 days as previously described.

Rhodanese assays were conducted on EPTC and butylate history soils as described by Tabatabai and Singh (19).

The experimental design for all experiments conducted on the soils was a randomized block design with three replicates. Separate analyses of variance were performed at each sampling time. Mean separations were determined using the Least Significant Difference

(LSD) test only when the F-test was significant at the 0.05 level of significance.

Microbial isolate JE1 in BSME (100 mg EPTC L^{-1}) was grown on a rotary shaker at 27°C to midlog phase (2-3 days; $O.D._{600}$, 0.06-0.08). An aliquot of the culture was treated with EPTC (50 mg L^{-1}), and EPTC remaining in solution was measured at 2 h intervals. An aliquot containing a similar amount of EPTC but no microbiol cells was used as a control. A midlog culture was also treated with ^{14}C-1-propyl EPTC (100 mg L^{-1}) and, at 8 h intervals, the headspace above the culture was swept into NaOH and the ^{14}C-activity was measured by LSC.

For identification of metabolites formed during EPTC degradation, the same procedure was used with the exception being the final EPTC concentration was adjusted to 100 mg L^{-1} after the culture had reached the midlog phase. At 8 h intervals, for 32 h, samples were asceptically removed and centrifuged. The supernatants were extracted with toluene, and then analyzed for undegraded EPTC remaining by GC and TLC. The remainder of the toluene extract was concentrated to approximately 0.5 ml under nitrogen and analyzed for EPTC and metabolites using GC or TLC and liquid scintillation counting. The aqueous fraction was lyophilized, dissolved in methanol, and analyzed by GC, TLC (hexane-ether, 9:1 or hexane-acetone, 6:1) and LSC.

Results

Metabolic Studies in Soil. Enhanced degradation of EPTC in a Brookston soil was evident after only a single field application of EPTC (Table II). The rate of EPTC degradation in soil after one year of EPTC use was not significantly different ($P<0.05$) from rates observed in history soils treated for 2, 3, or 4 consecutive years. Butylate, a carbamothioate with structure similar to EPTC, was degraded more rapidly in a 4-year history Brookston soil than in the same soil with no prior history of EPTC use (Figure 1). The degradation of butylate, however, was slower than that of EPTC applied at the same rate.

Table II. Degradation of 4 mg EPTC kg^{-1} in Brookston soil treated with Eradicane for 1, 2, 3 or 4 consecutive years

Previous Eradicane use	Days after application				
	0	3	7	10	14
Years	- - -EPTC remaining, mg kg^{-1} soil- - -				
0	3.86	3.63	2.76	2.41	1.31
1	3.87	1.82	0.64	0.44	0.34
2	3.84	1.73	0.78	0.33	0.30
3	3.82	1.87	0.55	0.27	0.22
4	3.86	1.78	0.74	0.36	0.29
LSD(0.05)†	NS	0.49	0.44	0.12	0.45

† LSD = least significant difference values calculated when the F-test was significant at $P \leq 0.05$.

An experiment was conducted to determine the minimum amount of inoculum of an enhanced soil required to bring about the development of enhanced degradation in a non-history soil. At all sampling times, an inoculum level of 1.0% or higher resulted in significantly ($P<0.05$) higher degradation rates than the uninoculated non-history soil (Figure 2).

To examine the class of microorganism responsible for enhanced EPTC degradation in the Brookston soil, various antibiotics were applied at a rate of 100 mg kg^{-1} soil. Chloramphenicol, an antibacterial compound, greatly inhibited EPTC degradation. The amount of EPTC remaining after 7 days was greater with this treatment than in a soil not previously exposed to EPTC (Table III). Cycloheximide, streptomycin, and kanamycin had little effect in arresting the enhanced degradation of EPTC in the Brookston soil.

Table III. Effect of various antibiotics on degradation of EPTC (10 mg kg^{-1} soil) in a 3-year Eradicane history Brookston soil

Treatment	Days after application		
	0	3	7
	EPTC remaining, mg kg^{-1} soil		
None (non-history soil)	9.47	8.56	7.47
None (3-year history soil)	9.44	5.86	1.80
Cycloheximide	9.88	5.24	1.56
Streptomycin	9.46	4.32	1.75
Kanamycin	9.32	4.93	1.86
Chloramphenicol	9.79	8.99	8.75
LSD (0.05)†	NS	0.79	0.86

† LSD = least significant difference values calculated when the F-test was significant at $P \leq 0.05$.

The correlation between rhodanese activity and rates of EPTC degradation in enhanced and non-history soils was investigated because of observations by Reed et al. (20) that bacterial isolates capable of growing on EPTC and butylate also exhibited high rhodanese activity. Mixed results were obtained with two out of the four EPTC history samples having significantly ($P<0.05$) higher rhodanese activities than the non-history soil (Table IV). Of the three soils tested, none had significantly higher rhodanese activity in the butylate history soils as compared to the non-history soil. In fact, with the Dothan A and B soil samples, significantly lower rhodanese activity levels were observed in the butylate history soils.

Metabolic Studies with Microbial Isolate JE1. EPTC was found to be efficiently metabolized by JE1 (Figure 3). Growth of JE1 was associated with the degradation of EPTC over an 8 h period. In contrast, EPTC levels remained constant over the same 8 hour period in the uninoculated control. Degradation of ^{14}C-labelled EPTC and appearance of metabolites into an aqueous or organic soluble fraction was also measured. An initial rise in ^{14}C-activity in the

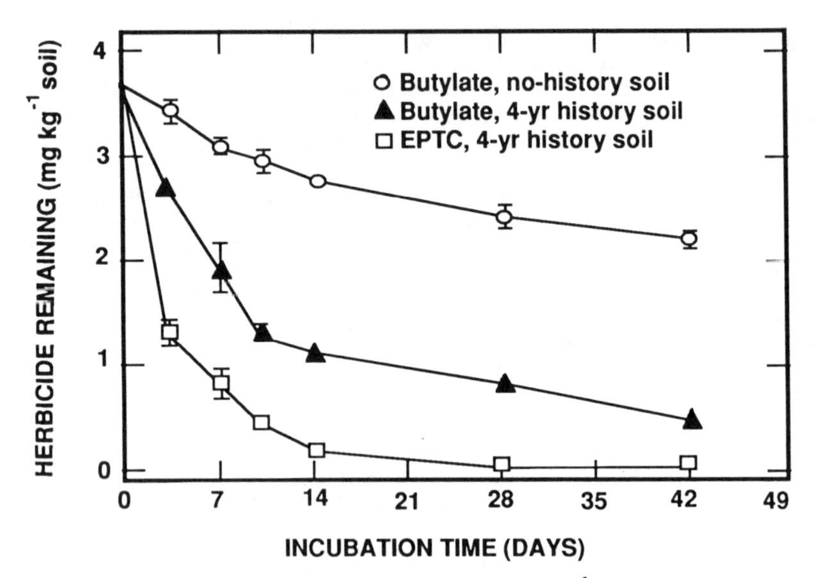

Figure 1. Degradation of butylate (4 mg kg^{-1}) and EPTC (4 mg kg^{-1}) in a Brookston soil treated with Eradicane for four consecutive years. Bars denote standard deviations.

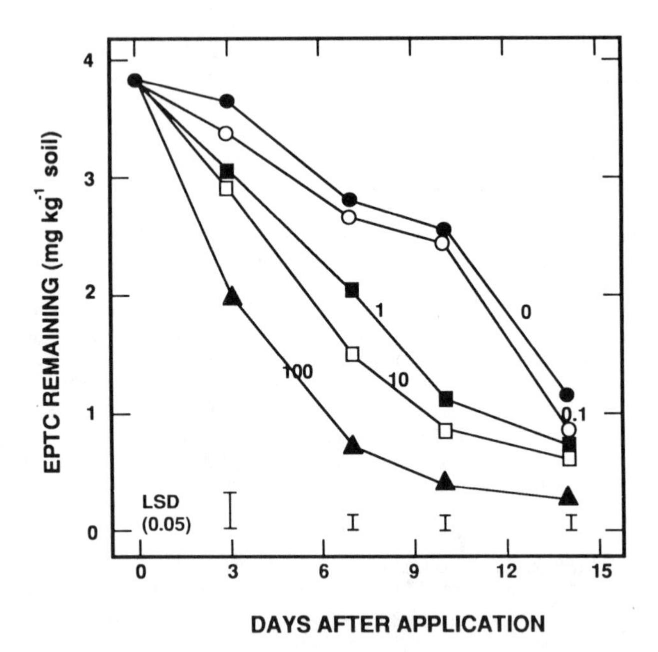

Figure 2. Degradation of EPTC (4 mg kg^{-1}) in a non-history Brookston soil inoculated with various percentages (w/w) of the same soil which had been treated with Eradicane for three consecutive years.

Table IV. Rhodanese activity of soils with enhanced degradation rates of EPTC and Butylate

Soil history	Rhodanese activity in soil specified, nmoles g^{-1} h^{-1}			
	Plano	Dothan A	Dothan B	Brookston
Non-history	564	330	730	400
EPTC	688	397	940	355
Butylate	682	292	520	-
LSD (0.05)[†]	140	8	207	10

† LSD = least significant difference values calculated when the F-test was significant at $P \leq 0.05$.

aqueous fraction was observed followed by a decline to initial levels (Figure 4). The decrease in ^{14}C-activity of the organic fraction was coupled with rapid evolution of $^{14}CO_2$.

Examination of the organic fraction during EPTC degradation using TLC and autoradiography revealed four metabolites (Metabolites 1, 2, 3 and 5; Figure 5) distinct from the parent molecule. Three of the four metabolites did not accumulate appreciably as incubation proceeded suggesting they were rapidly converted to other compounds by JE1. Metabolite 5, however, gradually accumulated with time as the concentration of EPTC declined. After 24 h, 24% of the initial ^{14}C-activity in EPTC was present as Metabolite 5.

Metabolite 1 (Figure 5) was tentatively identified as EPTC-sulfoxide, based on similar chromatographic characteristics with standards. Additional confirmation using GC/MS was not possible because this compound is thermally unstable. Metabolites 2 and 3, which were less polar than EPTC sulfoxide (Metabolite 1), had chromatographic characteristics similar to the hydroxylated EPTC at the β- and α-propyl position (14) although further attempts to confirm this were not made.

Metabolite 5 was the least polar of the metabolites contained in the organic fraction of JE1 cultures actively degrading EPTC. It was tentatively identified as *N*-depropyl EPTC on the basis of its chromatographic properties on thin-layer plates (Figure 5). When GC analysis (3% OV-17 column) was applied for characterization of the organic fraction, Peak #1 was observed to substantially increase in area with incubation time (Figure 6). GC/MS analysis of the organic fraction using a 3% OV-17 capillary column confirmed the identity of this compound as being N-depropyl EPTC. Three other metabolites in the organic fraction were identified by GC/MS when a 5% carbowax 20M column was used. These compounds were propionaldehyde, *s*-methyl formic acid and *s*-ethyl formic acid.

Additional GC analyses of the aqueous fraction of JE1 cultures actively degrading EPTC indicated the presence of traces of propylamine, but no dipropylamine. Also, a strong mercaptan-like odor developed in JE1 cultures actively degrading EPTC suggesting the formation of a mercaptan. Non-protein thiol content (17) was found to accumulate in the media as the incubation progressed.

Evidence that EPTC may be hydroxylated at the a-carbon of the ethyl group to yield α-hydroxy-ethyl EPTC was also observed. However, α-hydroxy-ethyl EPTC is unstable and yields carbonyl

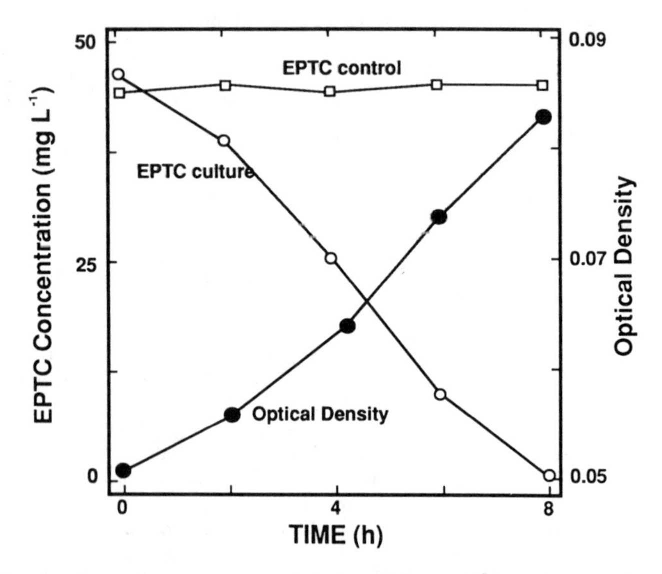

Figure 3. Degradation of EPTC (50 mg L^{-1}) and growth of microbial isolate JE1 in minimal salt media.

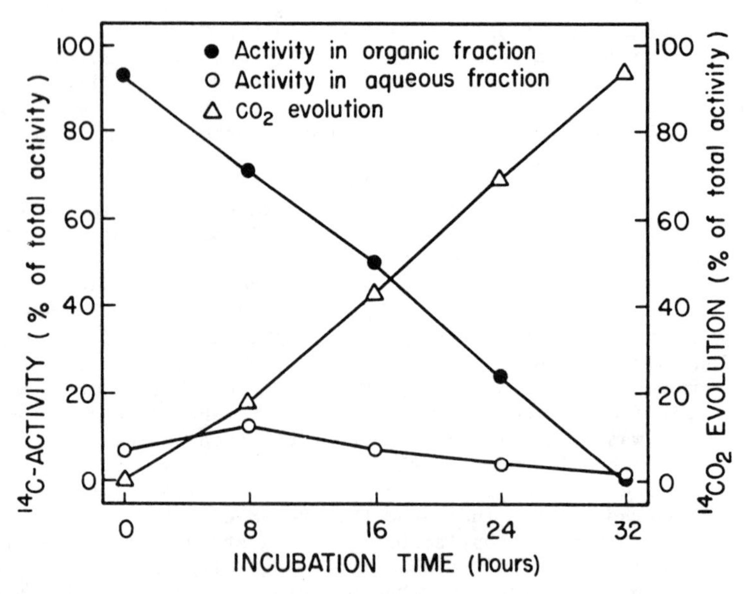

Figure 4. Fractionation of ^{14}C-label during metabolism of ^{14}C-1-propyl EPTC (100 mg L^{-1}) by microbial isolate JE1 in minimal salt media.

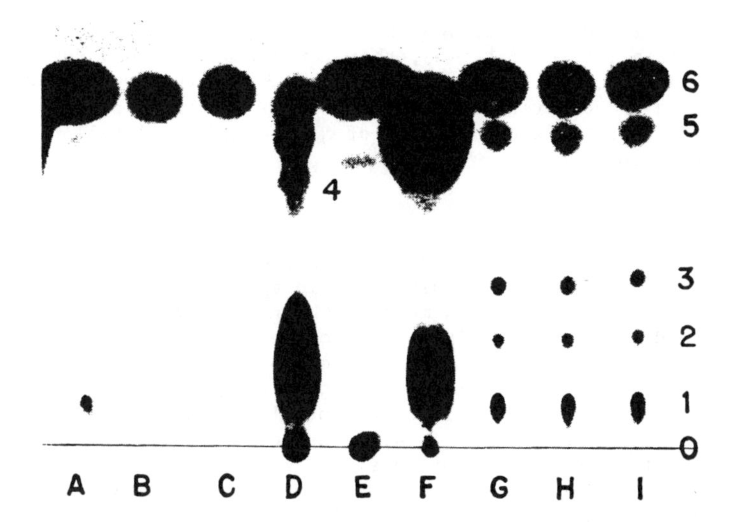

Figure 5. Autoradiogram of metabolic products of EPTC produced by microbial isolate JE1. Thin-layer chromatograph was developed with hexane/ethyl acetate (3:2). A, 8 h incubation; B, 0 h incubation; C, sterilized control; D, E, F, m-chloroperoxy benzoic acid + EPTC; G, H, I, 24 h incubation. Metabolites: 1, EPTC-sulfoxide; 2, 3, unknowns; 4, EPTC-sulfone; 5, *N*-depropyl EPTC; 6, EPTC.

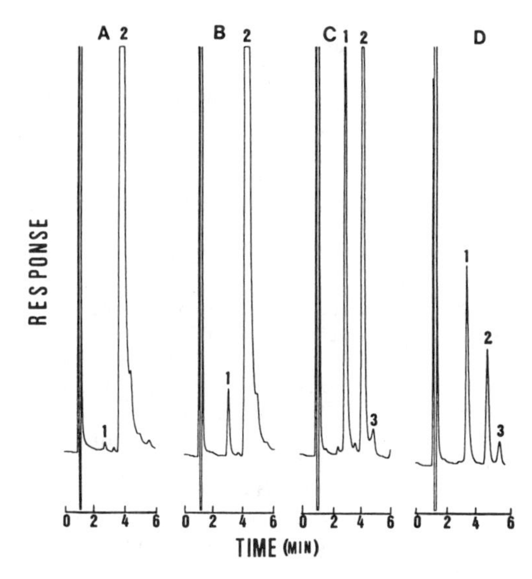

Figure 6. Gas chromatograms of JE1 metabolic products after A, 0 h; B, 8 h; C, 16 h; D, 24 h incubations. Separation on 3% OV-17 glass colums. Peaks: 1, *N*-depropyl EPTC; 2, EPTC; 3, unknown.

sulfide (COS) and acetaldehyde when it decomposes. The presence of this unstable intermediate was, therefore, implicated by detection of acetaldehyde (16). The extent of this hydrolysis during EPTC degradation as compared to the other reactions taking place could not be assessed, however, because of a lack of EPTC labelled in the appropriate position.

Discussion

Metabolic Studies in Soil. Enhanced EPTC degradation in the Brookston soil after only a single field application (Table II) is consistent with observations made by Obrigawitch et al. (21) and Schuman and Harvey (22). Cross enhancement between EPTC and butylate also occurred in the Brookston soil (Figure 1). However, butylate degradation was slower than that of EPTC. Butylate has been found to give acceptable weed control in EPTC history fields where EPTC has failed (21, 23-24). One possible explanation may be that butylate degrades to form a stable metabolite that has herbicidal activity and can thus provide control even when the parent molecule has disappeared (12-13).

The inoculation experiment results (Figure 2) indicate that only small amounts of soil with enhanced degradation capabilities for a specific pesticide have the potential to induce enhanced degradation in non-history soils. Similar results were reported by Engvild and Jensen (25) and more recently by Yarden et al. (26). Yarden et al. (26) demonstrated that 2% of a soil with enhanced degradation of benzimidazol-2-yl carbamate (MBC) was sufficient to cause a rapid increase in the rate of degradation of this same compound in a non-history soil. Development of enhanced degradative capabilities in non-history soils through inoculation with enhanced soils may be due to a direct amplification of the pesticide degrading population. Thus, a previous application of a pesticide may not be the only means of causing a development of enhanced degradation. A possible consequence is that any movement of an enhanced soil by wind or water may lead to misleading conclusions when comparing degradation in a field repeatedly treated with a pesticide against a supposedly non-enhanced edge or fence row soil sample.

Microorganisms active in degrading EPTC in the Brookston soil were very sensitive to chloramphenicol (Table III). Chloramphenicol exhibits broad spectrum activity against both gram positive and gram negative bacteria. Cycloheximide, an effective inhibitor of fungi, did not affect the rate of EPTC degradation. Kanamycin and streptomycin, which are more effective against gram positive than gram negative bacteria, also did not alter the rate of EPTC degradation. This may lead to the conclusion that the active degraders in the Brookston soil belong to the gram negative class of bacteria. These results are inconsistent with the types of microorganisms we have isolated from soil and which are capable of growing on EPTC as the sole carbon/energy source. In almost all cases, gram positive microorganisms were obtained, and attempts by Tam et al. (9) to isolate an EPTC-degrading microorganism also yielded a gram positive bacterium. Further research to characterize the microorganisms responsible for enhanced EPTC degradation in soils is obviously needed.

Rhodanese assays were conducted on EPTC and butylate-history soils and non-history soils (Table IV). The methodology to measure rhodanese activity in soil is thoroughly documented and, if correlated with the rate of EPTC or butylate in soil, would provide a rapid means of identifying soils with enhanced degradative capabilities. However, the results were inconclusive and limit the use of the rhodanese assay for such a purpose. The difficulty in developing a simple enzyme assay for identification of soils with enhanced EPTC degradation, as opposed to several of the organophosphate herbicides, is due to the lack of information on the mechanisms involved in EPTC degradation.

Metabolic Studies with Microbial Isolate JE1. EPTC is rapidly metabolized by microbial isolate JE1 which utilizes it as its sole carbon/energy source (Figure 3). The degradation products were identified either by TLC or GC/MS analysis. Hydroxylation and sulfoxidation were found to be important reactions leading to the mineralization of EPTC by soil microorganisms. The rapid evolution of $^{14}CO_2$ from ^{14}C-1-propyl EPTC suggests that mineralization of this portion of the herbicide molecule is very rapid.

The relative amounts of the *N*-depropyl EPTC both early in the incubation period and later suggest that hydroxylation of the α-propyl carbon of the *N,N*-dialkyl moiety is a major route in the microbial metabolism of EPTC. Hydroxylation of the other carbons of the *N,N*-dialkyl portion of the carbamate was found to be a less preferred route compared to the hydroxylation of the α-propyl carbon. Sulfoxidation of the carbamate may be second in the importance to the hydroxylation reactions observed.

On the basis of the metabolic products identified and their relative amounts formed during incubation, a degradative pathway involving hydroxylation and sulfoxidation mechanisms is proposed (Figure 7). EPTC is first hydroxylated at the α-propyl carbon to form α-hydroxy-propyl EPTC. This compound is unstable and breaks down to form *N*-depropyl EPTC and propionaldehyde. The *N*-depropyl EPTC is further metabolized to *s*-ethyl formic acid and propylamine. The *s*-ethylcarboxylic acid is then demethylated to *s*-methyl formic acid. This product is hypothesized to degrade to form CO_2 and methyl mercaptan.

Although no conclusive identifications were made, the chromatographic characteristics of the other minor metabolites found in the organic fraction suggest that hydroxylations of carbons other than the α-propyl carbon may occur during degradation of EPTC.

The availability of different pathways for the metabolism of EPTC by these microbes may account for the very efficient degradation of this carbamothioate by soil microorganisms. It may also explain why dietholate slows down degradation of EPTC in enhanced soils without having any effect on the normal EPTC degradation (27-29). A possible explanation for enhanced EPTC degradation may be that in soil, where EPTC has been used repeatedly, all available pathways become operative. The addition of dietholate, which has been theorized to inhibit the hydroxylation pathway (30-31), may lessen the importance of this pathway,

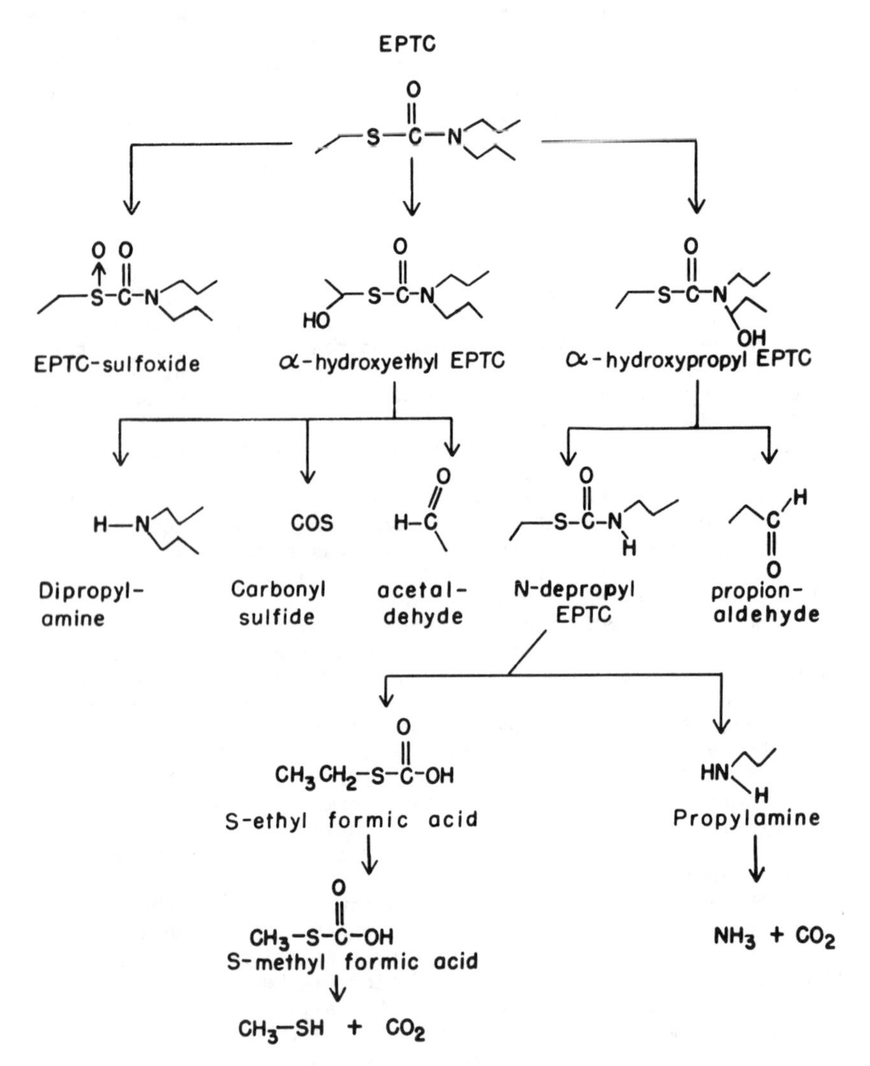

Figure 7. Proposed pathways for the metabolism of EPTC by microbial isolate JE1.

resulting in a decrease in the rate of degradation of EPTC in the enhanced soil.

Conclusions

Development of enhanced degradation of EPTC in a Brookston soil was rapid, being evident after only a single year treatment. The application of the soil rhodanese assay to identify soils with enhanced EPTC capabilities was not successful. Studies with Brookston soil also indicated that addition of only a small amount of an enhanced soil (0.1%) to a non-enhanced soil was capable of increasing EPTC degradation rates in the non-enhanced soil.

A microorganism (tentatively identified as *Rhodococcus* sp.) capable of utilizing EPTC as a sole carbon/energy source was isolated from soil. Metabolism of labelled (1-propyl position) and unlabelled EPTC by this microorganism yielded two major metabolites, *N*-depropyl EPTC and EPTC-sulfoxide. It is proposed that initial reactions in EPTC metabolism involve hydroxylation of the α-hydroxy-propyl EPTC. Decomposition of this unstable compound yields *N*-depropyl EPTC which is subsequently metabolized to CO_2. Sulfoxidation of EPTC is a less important initial reaction. A possible explanation of enhanced EPTC degradation in soil is that the pathway involving hydroxylation of the α-propyl carbon becomes the major route of degradation. In a non-enhanced soil, however, this pathway is active only at low levels and the sulfoxidation reaction is the major route of EPTC degradation.

Acknowledgments

Salaries and research support provided in part by the North Central Region Pesticide Impact Assessment Program (OH000363-SS) and by state and federal funds appropriated to The Ohio State University and the Ohio Agric. Res. & Develop. Ctr., Wooster, OH. EPTC (including ^{14}C-1-propyl EPTC) and butylate were supplied by Stauffer Chemical Company.

Literature Cited

1. Rahman, A.; Atkinson, G. C.; Douglas, J. A.; Sinclair, D. P. New Zealand J. Agric. 1979, 137, 47-49.
2. Lee, A. Soil Biol. Biochem. 1984, 16, 529-31.
3. Skipper, H. D.; Murdock, E. C.; Gooden, D. T.; Zublena, J. P.; Amakiri, M. A. Weed Sci. 1986, 34, 558-63.
4. Wilson, R. G. Weed Sci. 1984, 32, 264-68.
5. Audus, L. J. In The Physiology and Biochemistry of Herbicides; Audus, L. J., Ed.; Academic Press: New York, 1964; p 163-206.
6. Hutzinger, O.; Veerkamp, W. In Microbial Degradation of Xenobiotic and Recalcitrant Compounds; Leisinger, T.; Hutter, R.; Cook, A. M.; Neusch, J., Eds.; Academic Press: New York, 1981; pp 3-45.
7. Greaves, M. P.; Davies, H. A.; Marsh, J. A. P.; Wingfield, G. I. CRC Critical Rev. Microbiol. 1976, 5, 1-35.
8. Waid, J. S. Residue Rev. 1972, 44, 65-71.

9. Tam, A. C.; Behki, R. M.; Khan, S. U. Appl. Environ. Microbiol. 1987, 53, 1088-93.
10. Gray, R. A. Proc. California Weed Control Conf., 1971, pp 128-34.
11. Kaufman, D. D. J. Agric. Food Chem. 1967, 15, 582-91.
12. Tuxhorn, G. L.; Roeth, F. W.; Martin, A. M.; Wilson, R. G. Weed Sci. 1986, 34, 961-65.
13. Casida, J. E.; Gray, R. A.; Tilles, H. Science 1974, 184, 573-74.
14. Chen, Y. S.; Casida, J. E. J. Agric. Food Chem. 1978, 26, 263-66.
15. Fang, S. C. In Degradation of Herbicides; P. C. Kearney; Kaufman, D. D., Eds.; Marcel Dekker: New York, 1969; pp 147-64.
16. Casida, J. E.; Kimmel, E. C.; Ohkawa, H.; Ohkawa, R. Pestic. Biochem. Physiol. 1975; 5, 1-11.
17. Ellman, G. L.; Courtney, K. D.; Andres, V., Jr.; Featherstone, R. M. Biochem. Pharmacol. 1961, 7, 88-95.
18. Kearney, P. C.; Kaufman, D. D. Science 1965, 147, 740-41.
19. Tabatabai, M. A.; Singh, B. B. Soil Sci. Soc. Am. J. 1976, 40, 381-85.
20. Reed, J. P.; Kermer, R. J.; Keaster, A. J. Bull. Environ. Contam. Toxicol. 1987, 39, 776-82.
21. Obrigawitch, T.; Martin, A. R.; Roeth, F. W. Weed Sci. 1983, 31, 417-22.
22. Schuman, D. B.; Harvey, R. G. Proc. North Central Weed Control Conf., 1980, pp 19-20.
23. Gunsolos, J. L.; Fawcett, R. S. Proc. North Central Weed Control Conf., 1980, p 18.
24. Rahman, A.; James, T. K. Weed Sci. 1983, 31, 783-89.
25. Engvild, K. C.; Jensen, H. L. Soil Biol. Biochem. 1969, 1, 295-300.
26. Yarden, O.; Aharonson, N.; Katan, J. Soil Biol. Biochem. 1987, 19, 735-39.
27. Obrigawitch, T.; Wilson, R. G.; Martin, A. R.; Roeth, F. W. Weed Sci. 1982, 30, 175-81.
28. Prochnow, C. L., Proc. West. Soc. Weed Sci., 1981, 34, 55-56.
29. Roeth, F. W. Rev. Weed Sci. 1986, 2, 45-65.
30. Miaullis, B.; Nohynek, G. J.; Periero, F. Proc. British Crop Protection Conf., 1982, pp 205-10.
31. Wilson, R. G.; Rodebush, J. E. Weed Sci. 1987, 35, 289-94.

RECEIVED January 22, 1990

Chapter 9

The Role of Fungi and Bacteria in the Enhanced Degradation of the Fungicide Carbendazim and the Herbicide Diphenamid

N. Aharonson[1], J. Katan[2], E. Avidov[1], and O. Yarden[2]

[1]Department of Chemistry of Pesticides and Natural Products, Agricultural Research Organization, The Volcani Center, Bet Dagan, 50–250, Israel
[2]Department of Plant Pathology and Microbiology, Faculty of Agriculture, The Hebrew University of Jerusalem, Rehovot, 76–100, Israel

Single application of carbendazim (MBC) or repeated applications of diphenamid conditioned the soil for enhanced degradation. Low initial doses of the pesticide or mixing a small volume of carbendazim-history soil with untreated soil sufficed to condition the entire soil volume for enhanced degradation. Soil disinfestation or treatments with fungicides such as triphenyltin acetate or TMTD were generally effective in decreasing enhanced degradation of the two pesticides. Fungi capable of degrading MBC or diphenamid were isolated, enumerated and identified from soils with and without enhanced degradation. There were no differences in the degradation between fungi isolated from previously treated or nontreated soils. Differences were found only between mixed bacterial cultures. These results, together with data about cross enhancement between some metabolites and their respective parent compounds, and the similarity in the metabolism between soil and bacterial cultures suggests that for these pesticides bacteria are the major component responsible for enhanced degradation, though fungi degraders are also present in the soil. The work suggests possible enzymatic reactions related to enhanced degradation.

Enhanced degradation of soil-applied pesticides has been reported in recent years for a growing number of insecticides, fungicides, nematicides and herbicides. The diversity of pesticides, belonging to different classes of chemicals and which were found associated with enhanced degradation, was well documented in several review articles (1-4). Enhanced degradation is usually linked with repeated soil application of the same or a structurally related pesticide. Enhanced degradation and the accompanying failure of pest control

0097–6156/90/0426–0113$06.00/0

have created an increasingly serious agricultural problem for farmers and have raised some fundamental questions with regard to shifts in microbial populations in soil (5-8).

It is generally accepted that microbial metabolism as expressed by population shifts and/or activity of degraders are the major factors contributing to enhanced degradation and conversion of pesticides to biologically non-active compounds. This has been demonstrated in experiments in which the process of enhanced degradation was stopped or delayed by soil sterilization or by antimicrobial treatments (9,10). The microbial population exists in a dynamic equilibrium that can be altered by modifying environmental conditions (e.g. addition of pesticides or other biologically active substances). Microorganisms may respond to pesticides as substrates and thereby derive energy or utilizable nutrients for metabolism. The pesticide may also undergo degradation by analog-induced or constitutive cometabolism, whereby the pesticide itself does not serve as an energy source (1,3).

The exact mechanisms for microbial adaptation to the pesticide molecule in soils that develop enhanced degradation capacity are not completely understood. These processes could be viewed from the ecological and population aspects, from their biochemical and enzymatic reactions, or from the genetic aspects, in which extrachromosomal elements may be involved as part of the process.

This summary presents data on two pesticides with different molecular structures and biological activities, with the emphasis on the microorganisms associated with the development of enhanced degradation and the possible enzymatic reactions that are taken place in these processes.

Enhanced Degradation

Previous application of the same pesticide or of a structurally related compound is the major factor in inducing enhanced degradation. In certain cases enhanced degradation might develop in the soil following one pretreatment. The fungicide carbendazim (methyl benzimidazol-2-ylcarbamate, MBC) represents such a group of compounds. After one treatment the half-life of MBC was shortened from 17 days to 3-4 days (11 and Figure 1). For some pesticides even the small amount that reaches the soil through foliar spray applications is sufficient to induce enhanced degradation, as has been shown by Walker *et al*. with the fungicide iprodione (12) and by Yarden *et al*. with the fungicide MBC (11). On the other hand, the maximal rate of enhanced degradation was obtained with compounds such as the herbicide diphenamid (N,N-dimethyl 2,2-diphenyl-acetamide), only after several repeated treatments. Four soil treatments with diphenamid were needed before its half-life was decreased from 50-60 days to 3-4 days (9). It is not surprising that enhanced degradation has developed to varying degrees in different soils and under different agricultural regimes. This could be the result of the intrinsic chemical properties of the pesticide, but could also derive from the fact that, among other processes, certain soils are "suppressive" while others are "conducive" to enhanced degradation, similar to the well established phenomenon with soilborne pathogens (3).

Although MBC and diphenamid differ considerably from each other in the ease with which they induce enhanced degradation, their

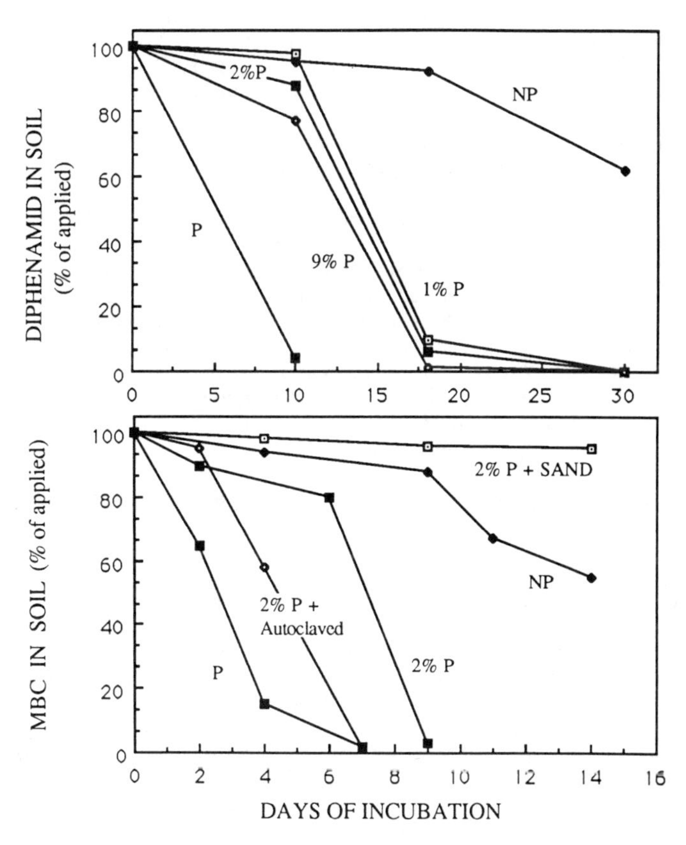

Figure 1. Diphenamid and MBC degradation in mixtures composed of 1-9 percent of soil that has acquired enhanced degradation capacity (P) with non-history soil (NP) and, for MBC soil, also mixed with autoclaved soil and sterile sand. MBC data was adapted from Ref. 11.

behavior in problem soils (i.e., soils with enhanced degradation) was found in many other respects to be quite similar. The behavior of a third pesticide, the herbicide EPTC (S-ethyl dipropyl thiocarbamate), was also found to be very similar to that of MBC and diphenamid. Some of these similarities, which are discussed below, suggest a certain generalization of the process which may be true also for other classes of chemicals.

An additional mechanism by which a soil may acquire enhanced degradation is by mixing a small volume of soil that has developed enhanced degradation, with soil that has not been treated in the past with the pesticide (i.e., non-problem soil). Such mixing induced enhanced degradation even when the previously treated soil formed only 2% of the mixture for MBC or 1% for diphenamid (Figure 1). The major noticeable change in such soil mixtures was the sharp reduction in the lag period before degradation started. Mixing the soil with previously nontreated but autoclaved soil enhanced the induction effect. These findings could be explained by the fact that in autoclaved soil, with a relative abundance of nutrients and absence of antagonists, the amplification of the pesticide - degrading microorganisms became much more pronounced. Similar results were described by Walker et al., for the degradation of the fungicides iprodione and vinclozolin (12). The implications of these findings might be serious. In fact, under field conditions, this mechanism may well be responsible for the spread of enhanced degradation to adjacent soils, similar to the phenomenon of soil contamination with soilborne pathogens.

Effect of Soil Disinfestation and Antimicrobial Agents on Enhanced Degradation

Soil Disinfestation. Soil fumigation with methyl bromide or soil solarization strongly reduced MBC and diphenamid degradation in nontreated soils (Figure 2) and in soils that had developed enhanced degradation (Figure 3). Field disinfestation of previously nontreated (non-problem) soils resulted in slower degradation of MBC and diphenamid in comparison with the degradation in the non-disinfested control soils. The disinfestation of soils that had acquired enhanced degradation for either MBC or diphenamid slowed the rate of degradation of both pesticides. The degradation in the enhanced degradation soil, after disinfestation, was similar to that in the nontreated control soil, and in some cases degradation in the former was even slower than in the nontreated soil.

Soil disinfestation can be regarded as a general broad-spectrum antimicrobial treatment which results in a drastic, but incomplete reduction in microbial activity, and can be compared to soil sterilization by autoclaving or gamma irradiation. Therefore, the effectiveness of disinfestation in suppressing enhanced degradation clearly demonstrates the important role of microorganisms in the degradation of these pesticides in regular soil, as well as in soil showing enhanced degradation. Moreover, the data point towards means that can be developed for the purpose of controlling enhanced degradation under field conditions. This is especially relevant under greenhouse and intensive cropping conditions, where soil disinfestaion constitutes a regular practice.

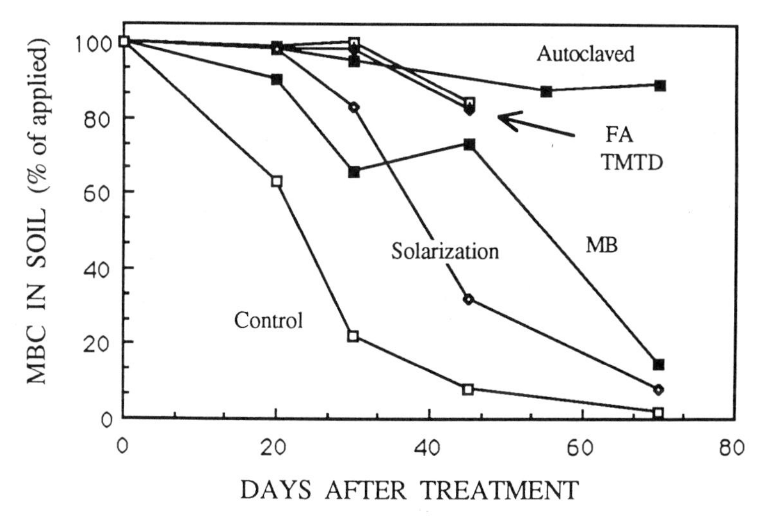

Figure 2. Effect of soil disinfestation with methyl bromide (MB), solarization, autoclaving and treatment with TMTD and fentin acetate (FA), on the degradation of MBC that was added at 10 µg/g soil. The fungicides TMTD and FA were added at 20 µg/g soil. Data from Phytopathology, 1985, 75, 763.

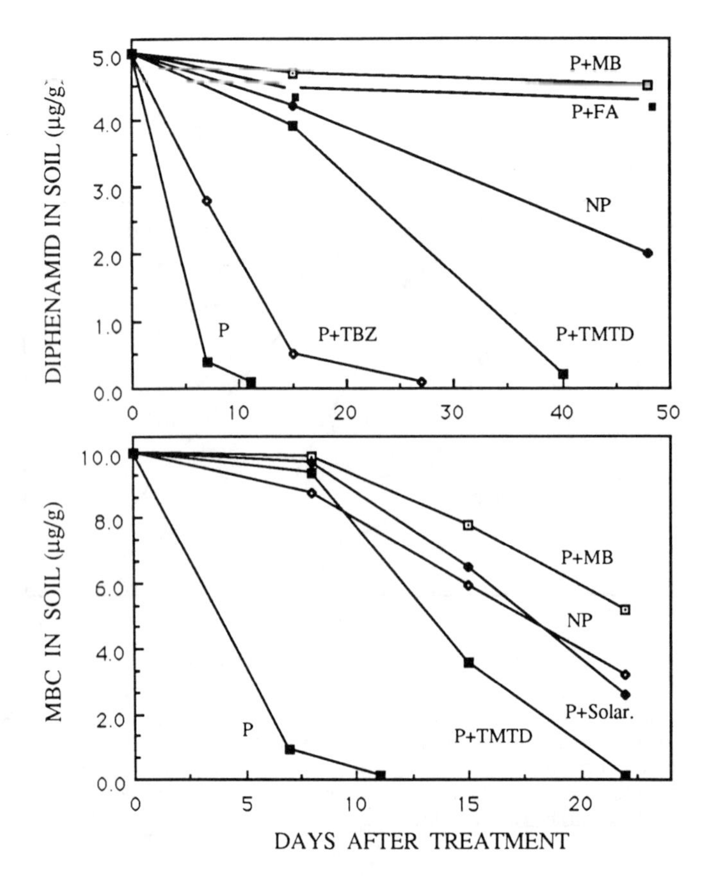

Figure 3. Effect of soil disinfestation and some fungicides on the degradation of diphenamid and MBC in soils that have acquired enhanced degradation capacity (P), as compared with a nontreated soil (NP). MB=methyl bromide, FA=fentin acetate. Data adapted from Ref.9 and Ref.11.

Soil disinfestation with methyl bromide, vapam, or by solarization controlled degradation of certain pesticides such as EPTC to various degrees (13). Apparently some of the degraders were not affected by these biocides, a fact that was reflected by partial reduction in degradation of EPTC, when compared with the stronger inhibitory effect in sterile soil.

Effect of Some Fungicides and Other Antimicrobial Agents. It is expected that the use of more specific biocides, which have a narrower spectrum of antimicrobial activity as compared with disinfestation, will also result in slowing down degradation of certain pesticides in regular soils or in soils showing enhanced degradation. However, the reduction in rates of degradation would depend strongly on the biocide in question and on its concentration (14).

Certain biocides exhibit a relatively narrow range of activity and therefore might be useful for identification of the degraders associated with enhanced degradation. Classical parameters, such as total microbial biomass, enzymatic activity, microbial respiration and enumeration, are very rough measurements of microbial activity. Thus, when dealing with the degradation of pesticides, which are applied to the soil in very small amounts, such parameters might not be sensitive enough to correlate with rates of degradation (15). Application of certain biocides, such as cycloheximide, PCNB, oxytetracycline, captan, chloramphenicol or TMTD, which demonstrate a certain degree of specificity, may serve as an additional useful tool for the identification of microbial degraders in soils with enhanced degradation (16). However, it should be pointed out that employing biocides for identification should be considered with caution, since these biocides are often not very specific and may affect a larger group of microorganisms. The concentration of the biocide is very critical in determining its degree of specificity.

The results for three fungicides - biocides in controlling degradation were as follows: The fungicides tetramethylthiuram disulfide (TMTD) and triphenyltin acetate (fentin acetate) were very effective in delaying degradation of several soil-applied pesticides (Figure 3). Fentin acetate was found as effective as the broad-spectrum biocide methyl bromide in inhibiting degradation of diphenamid in a soil with enhanced degradation capacity, whereas the fungicides TMTD and TBZ were less active. These findings also demonstrate the potential of specific chemicals to suppress enhanced degradation. The effectiveness of various fungicides in suppressing enhanced degradation may suggest that fungi are involved in the enhanced degradation of MBC or diphenamid. Again, this might be questionable since the effectiveness of many fungicides extends beyond fungi, especially when the fungicides are applied at high concentrations. For example, the fungicide PCNB is also very toxic to actinomycetes (17).

Microbial Studies

Enhanced degradation is the result of microbial processes in which an enrichment in the population or activity of microorganisms capable of degrading the pesticide, or induction of enzymatic reactions already present in the soil microflora, are expected. General counts of the total populations of bacteria, fungi and actinomycetes did not reveal significant quantitative differences regarding diphenamid and MBC-history and nonhistory soils, as will be described. This is not surprising, since the degraders constitute only a small proportion of the total soil microflora. Therefore, even if a change occurs in segments of the population, it might not be detected through a general count.

Fungal Degraders. In order to assess the degradative capabilities of fungi, individual colonies from each tested soil were randomly transferred to potato dextrose agar amended with 10 µg/g MBC and assessed for their capability to degrade the fungicide (19). Irrespective of the origin of the isolates, nearly 80% were identified as *Alternaria alternata*, *Bipolaris* or *Ulocladium* sp. An MBC-degrading fungus isolated from MBC-history soil was identified as *Acremonium falciforme*. The results showed that there was no significant difference in the number of degraders or in the rate of degradation by the fungi populations that were isolated from a soil with or without enhanced degradation.

A similar experiment was carried out with fungi capable of degrading diphenamid, isolated from diphenamid history or nonhistory soils (18). The capacity of the various fungi to degrade diphenamid varied, and the distribution of the degraders was about the same in both soils. Most of the fungi that were isolated from the two soils were capable of degrading diphenamid to a certain extent. The most efficient fungal degraders were identified as *Fusarium*, *Aspergillus* and *Penicillium* spp.

From this study it was concluded that fungi probably were not the major organisms responsible for the enhanced degradation of diphenamid and MBC, even though many of them might be important in "normal" degradation of these pesticides in regular soils before the latter developed enhanced degradation. However, in such studies it is still an open question whether the behavior of these fungi in culture reflects their capacity to degrade the pesticide in the soil.

Bacterial Degraders. The main difference between soils with and without enhanced degradation, with regard to microbial degraders, was observed in mixed bacterial cultures originating from the two soils. The degradation of MBC (19), diphenamid (18) and EPTC (20) was found to be much faster in mixed bacterial cultures originating from soils that developed enhanced degradation capacity than from nonhistory soils (Figure 4). The role of these organisms in enhanced degradation was demonstratead by inoculating them into nonhistory soils, which resulted in an enhancement of degradation in the latter soils. Fungi were not present in these bacterial cultures. The high pH (7.5), continuous shaking of the medium, and the early predomination in the medium by the bacteria, provided an unfavorable environment for fungi growth.

The possibility that actinomycetes might be involved in the enhanced degradation of diphenamid was examined by applying to the culture PCNB, which is inhibitory to actinomycetes. The medium used for estimating the number of soil bacteria also permits the

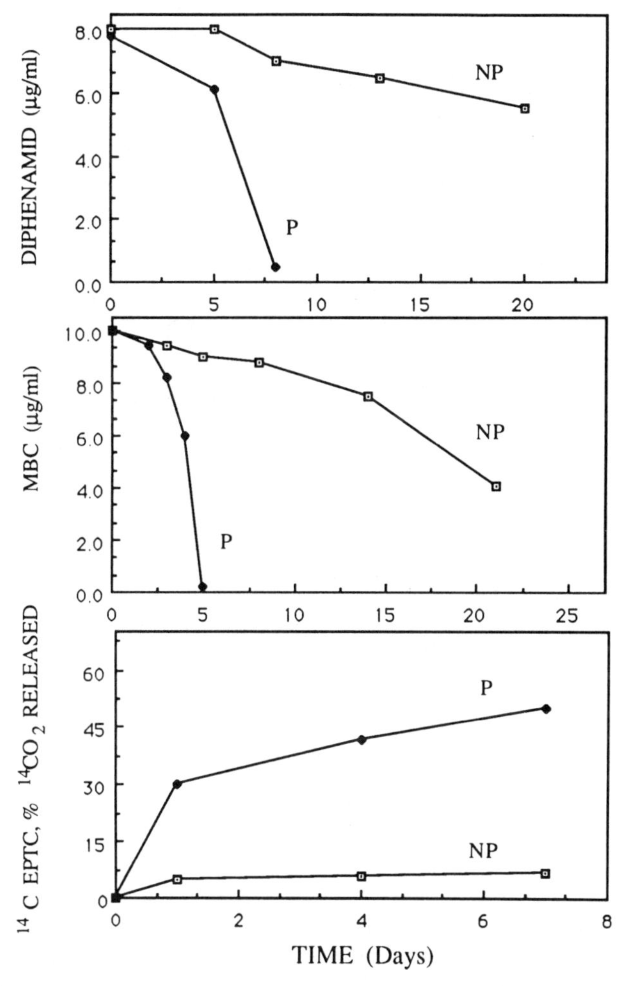

Figure 4. Diphenamid, MBC and EPTC degradation by mixed bacterial cultures obtained from history (P) and non-history (NP) soils, respectively. EPTC degradation was assessed by measuring $^{14}CO_2$ evolution.

development of actinomycete colonies. PCNB, as described by Farley and Lockwood (17), was found to inhibit growth of actinomycetes in culture media whereas bacteria were unaffected. Actinomycetes numbers were reduced 90% by 10 µg/ml PCNB, and 99% by 25-200 µg/ml. Numbers of bacteria were not reduced in soil extracts, even at PCNB concentrations as high as 200 µg/ml.

In this experiment, at a concentration of 10 µg/ml of PCNB in the mixed bacterial culture, 99% of the actinomycetes propagules in the culture were eliminated without any effect on diphenamid degradation. On the other hand, the addition of the bactericide chloramphenicol to the bacterial cultures obtained from diphenamid - history soils strongly inhibited diphenamid degradation (Table I).

The fact that some fungicides such as TMTD and fentin acetate were very effective in curbing enhanced degradation of several pesticides, led us to consider the possibility that fungi could be involved in the enhanced degradation. However, it has been shown that these fungicides, at the tested concentrations, were also very effective in controlling the bacterial degraders in the mixed bacterial cultures. The addition of 10 µg/ml of each of the fungicides to the bacterial culture slowed down the degradation to the same rate as that of cultures from non-history soils, whereas the antifungal agent cycloheximide (50 µg/ml) did not affect the degradation (Table I).

Table I. Degradation of Diphenamid (10 µg/ml) in Mixed Bacterial Culture, Derived from Soil with Enhanced Degradation Capacity, and which was Amended with Thiram (TMTD), Fentin Acetate (FA), Chloramphenicol (Chloram), Cycloheximide (Cyclo) or PCNB

Days after treatment	Biocide added						
	TMTD	FA	Chloram.	Cyclo.	PCNB	NP[z]	P[y]
	(% of applied remaining)						
0	100	100	100	100	100	100	100
2	103	100	100	-	-	96	100
6	105	102	103	43	3	90	5
10	107	103	100	11	0	94	-
20	73	72	100	-	-	79	-

[z]NP=non-history soil, [y]P=soil that acquired enhanced degradation.

Data adapted from Ref. 18

Persistence of Enhanced Degradation

The persistence of enhanced degradation capacity in the soil under laboratory conditions and in the field has been explored to a certain extent. It has been shown that enhanced degradation of MBC lasted in the field for at least 2 years and in the laboratory for 3 years (11). Similar results were obtained with diphenamid. Soil that was treated with diphenamid in 1982 and again in 1984, was found one year later to have developed enhanced degradation under practical field conditions and preserved it for at least one year.

In the laboratory, some of the diphenamid-treated soils lost their activity after 1 or 2 years, but the enhanced degradation could be restored in these soils much faster than it took to develop enhanced degradation in the first place in the previously non-treated soil. Handling and storage conditions of the soil in the laboratory are critical for preserving the activity.

In mixed bacterial cultures the activity of the diphenamid degraders was rapidly lost. Within 2 weeks of the first diphenamid treatment, the microbial degraders in the culture lost their capability to degrade diphenamid. The addition of various nutrients, or of nontreated soil as a source of nutrients, did not restore degradation. With MBC the results were somewhat different. The activity of the degraders was maintained for a longer period of time, but there were fluctuations in the activity, upon subsequent transfers, ranging from almost no degradation to very rapid degradation (19).

Failure to preserve the capability for degradation of diphenamid in the mixed bacterial culture might be one of the reasons for the failure to isolate a pure culture of individual strains of bacteria capable of degrading diphenamid. Several other recent studies reported similar observations, in which the bacterial culture from soil with enhanced degradation was very active but individual active strains could not be isolated. Such reports included the herbicide linuron (21), the fungicides iprodione (22) and metalaxyl (23), and the insecticide aldicarb (24). It is possible that failure to isolate bacterial degraders may be associated with pesticides in which the bacterial degraders fail to grow on the respective pesticide as the sole carbon source.

Pathways of Degradation and Enzymatic Reactions Associated with Enhanced Degradation of Carbendazim and Diphenamid

Data on the pathways and the enzymatic reactions involved in enhanced degradation and the specificity of the enzymes are important for better understanding of the development of enhanced degradation. Studies with MBC and diphenamid and with some of their degradative analogs have revealed the following information (Table II). MBC and its metabolite 2-AB were rapidly degraded in soil that developed enhanced degradation of MBC. The degradation of MBC and 2-AB in the non-history soil, during the time of the experiment, as shown in Table II was negligible. On the other hand, the rate of degradation of the derivative benzimidazole or of the fungicide thiabendazole was not enhanced in the MBC-history soil. Cross enhancement was observed only between MBC and 2-AB. These findings suggested that the reaction associated with enhanced degradation of MBC does not involve ring cleavage, ring substitution or hydroxylation. The reaction seems to involve the amino moiety that is attached to the imidazole ring.

The data on diphenamid showed cross enhancement only between diphenamid and its monodemethylated degradation product (diphen M-1). The degradation of the bidemethylated derivative (diphen M-2) was not enhanced when diphen M-2 was applied to the

Table II. Degradation of MBC, Diphenamid and their Derivatives in Soil and by Mixed Bacterial Cultures, Derived from Soils with Enhanced Degradation Capacity to Either Pesticide, Respectively

Chemical name	Structure	Degradation in mixed bacterial culture		
		Days of incubation		
		0	4	8
		(% of applied)		
Diphenamid	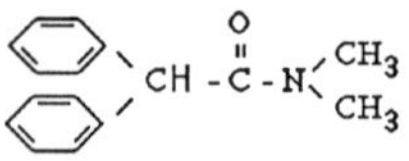 $(C_6H_5)_2CH-C(=O)-N(CH_3)_2$	100	72	5
N-Methyl-2,2-diphenyl acetamide (diphen M-1)	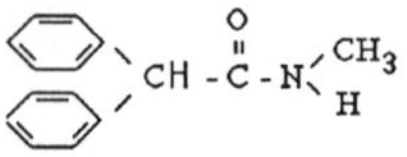 $(C_6H_5)_2CH-C(=O)-N(CH_3)H$	100	65	2
2,2-Diphenylacetamide (diphen M-2)	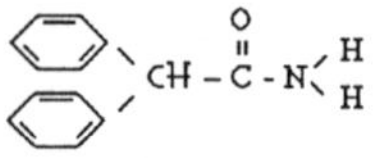 $(C_6H_5)_2CH-C(=O)-NH_2$	100	100	103

Chemical name	Structure	Degradation in soil		
		Days of incubation		
		0	4	12
		(% of applied)		
Carbendazim (MBC)	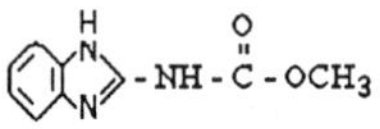 benzimidazol-2-yl $-NH-C(=O)-OCH_3$	100	20	0
2-Aminobenzimidazole	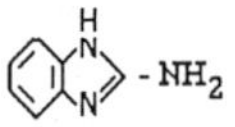 benzimidazol-2-yl $-NH_2$	100	6	0
Benzimidazole	benzimidazole	100	92	---

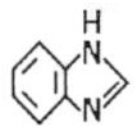

Data was adapted from Ref 18 and 19

diphenamid-history soil. Under these experimental conditions (Table II) neither diphenamid nor its derivatives were degraded in non-history soil during the first 2 weeks. The monodemethylated derivative was identified as a degradation product of diphenamid in a mixed bacterial culture that was obtained from soil with enhanced degradation, and in Fusarium culture, in which its concentration reached after 6 days 34% of the total diphenamid applied (Figure 5). The bidemethylated derivative (diphen M-2) was not detected in the soil or in culture extracts, and since it was degraded more slowly than diphen M-1, its absence from the extract suggested that diphen M-2 was not produced by the microorganism that degrades diphenamid in soil with enhanced degradation capacity (18).

Previous reports on the metabolism of diphenamid in mammals and in the soil have shown that the degradation proceeded via demethylation of the molecule. In mammals, demethylation is known as an oxidative process, mediated by the mixed function oxidases. It is therefore suggested that enhanced degradation of diphenamid in the soil might be associated with induction of mono-oxidases in soil bacteria. Such a mechanism could be considered as analogous to the development of resistance to pesticides by insects or weeds, by the induction of oxidative processes. One might consider enhanced degradation as the buildup of resistance to the pesticide by the soil bacterial population analogues to the usual well defined resistance by the target pest.

Fungi probably degrade diphenamid by the same oxidative enzymes, but their rate of degradation in the soil is much slower: several months instead of several days, as observed with bacterial cultures and with soils that developed enhanced degradation. Slower degradation by fungi in comparison with bacterial degradation has been shown for several other pesticides in soils with enhanced degradation, such as carbendazim (19), isofenphos (25), lindane (26), linuron (21) and metazachlor (16).

Conclusions

Enhanced degradation of pesticides and the rapid loss of their effectiveness is an additional example of the undesirable consequences of abandoning crop rotation. In such cases monoculture is often accompanied by repeated use of the same pesticides. This is analogous to the buildup, in soil, of populations of soilborne pests and pathogens.

A shift in microbial degradation in soils that acquired enhanced degradation capacity toward diphenamid, MBC and EPTC seems to be associated with bacterial degraders, rather than with fungi. Moreover, the results for diphenamid have shown that enhanced degradation is apparently linked with induction of an oxidative reaction in soil bacteria, which might be analogous to the development of resistance in pests. Such a possibility raises several questions regarding the persistence of such changes and the generalization of the process.

Because of the many uncertainties about this phenomenon, the simple analogy from the buildup of populations of soilborne pests and pathogens, and their control, should always be kept in mind. Therefore, the approach for dealing with enhanced degradation should

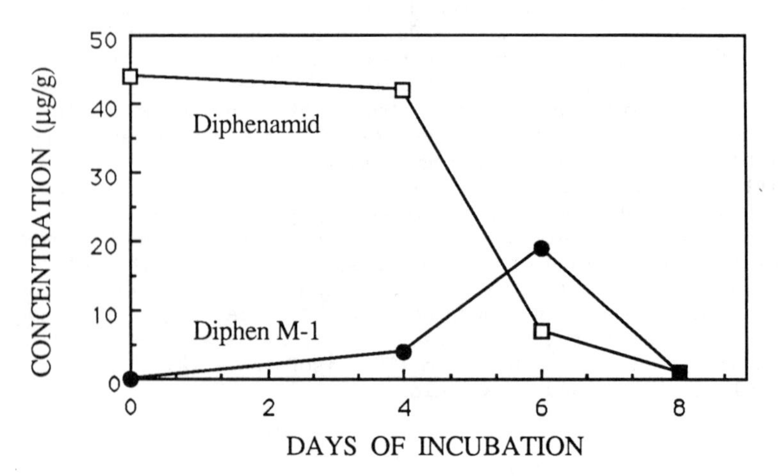

Figure 5. Disappearance of diphenamid and the formation of its demethylated metabolite (desmethyl diphenamid, diphen M-1) in *Fusarium* culture.

be similar to that taken for controlling soilborne pathogens, i.e., to detect the problem well in advance, to develop preventative means, and only as the last step to treat the soil which has developed enhanced degradation.

Literature Cited

1. Roeth, F.W. Rev. Weed Res. 1986, 2, 45-65.
2. Kaufman, D. D.; Katan, J.; Edwards, D.F.; Jordan, E.J. in: Agricultural Chemicals of the Future, Hilton, J.L. Ed.; Rowman & Alanheld: New Jersey, 1985, pp. 437-451.
3. Katan, J.; Aharonson, N. in: Ecological Studies; Vol 73, Toxic Organic Chemicals in Porous Media, Gerstel, Z.; Chen, Y.; Mingelgrin, U.; Yaron, B. Eds. Springer Verlag, 1989,pp.193-207.
4. Felsot, A.S. Ann. Rev. Entomol. 1989, 34, 453-476.
5. Suett, D.L; Walker, A. Aspects of Appl. Biol. 1988, 17, 213-222.
6. Tollefson, J. Proc. Brit. Crop Prot. Conf.-Pests & Diseases. 1986, 3, p. 1131-1136.
7. Read, D.C. Agric. Ecosystems Environ. 1986, 15, 51-61.
8. Obrigawitch, T.; Martin, A.R.; Roeth, F.W. Weed Sci. 1983, 31, 187-192.
9. Avidov, E.; Aharonson, N.; Katan, J. Weed Sci. 1988, 36, 519-523.
10. Racke, K.D.; Coats, J.R. J. Agric. Food Chem. 1988, 36, 193-199.
11. Yarden, O.; Aharonson, N.; Katan, J. Soil Biol. Biochem. 1987, 19, 735-739.
12. Walker, A.; Brown, P.A.; Entwistle, A.R. Pestic. Sci. 1986, 17, 183-193.
13. Tal, A.; Rubin, B.; Katan, J.; Aharonson, N. J. Weed Sci., 1989, 37, 434-439.
14. Avidov, E.; Aharonson, N.; Katan, J.; Rubin, B.; Yarden, O. Weed Sci. 1985, 33, 457-461.
15. Ingham, E.R. Crop Protection. 1985, 4, 3-32.
16. Allen, R.; Walker, A. Pestic. Sci. 1988, 22, 297-305.
17. Farley, J.D.; Lockwood, J.L. Phytopathology, 1968, 58, 714-715.
18. Avidov, E.; Aharonson, N.; Katan, J. Weed Sci. 1990, 38, (in press).
19. Yarden, O.; Salomon, R.; Katan, J.; Aharonson, N. Can. J. Microbiol. 36, 1990 (in press).
20. Tal, A. Ph.D Tesis, The Hebrew University of Jerusalem, Israel 1988.
21. Glad, G.; Goransson, B.; Popoff, T.; Theander, O.; Torstensson, N.T.L. Swedish J. Agric. Res. 1981, 11, 127-134.
22. Head, I.M.; Cain, R.B.; Suett, D.L.; Walker, A. Brighton Crop Protection Conf.-Pests & Diseases. 1988, p. 699-704.
23. Ana Maria Bailey; Coffey, M.D. Can. J. Microbiol. 1986, 32, 562-569.
24. Read, D.C. J. Econ. Entomol. 1987, 80, 156-163.
25. Racke, K.D.; Coats, J.R. J. Agric. Food Chem. 1987, 35, 94-99.
26. Wada, H.; Senoo, K.; Takai, Y. Soil Sci. Plant Nutr. 1989, 35, 71-77.

RECEIVED January 22, 1990

Chapter 10

Influence of Pesticide Metabolites on the Development of Enhanced Biodegradation

L. Somasundaram and Joel R. Coats

Department of Entomology, Iowa State University, Ames, IA 50011

Adaptation of soil microorganisms for rapid degradation of soil-applied pesticides can occur as a result of the complex interactions between the soil, the pesticide, the microbes, and environmental conditions. The current research addresses the role of breakdown products from the pesticide in the development of the condition. Several factors that influence enhanced microbial degradation include: nutrient value of the metabolite molecule, toxicity of the metabolite to soil microorganisms, and the availability of the metabolite to soil microbes. Comparisons of several pesticides, and their respective degradation products provide insight into the question of why soil microbial populations can develop rapid degradation capabilities for some pesticides but not others.

Enhanced microbial degradation is a soil-pesticide-microbe interaction, influenced by all three factors. In addition to these primary factors, environmental factors and management practices influencing these factors may also affect the degradation process. Because enhanced biodegradation has been reported to occur in several ecosystems, including rice (1), corn (2), sorghum (3), vegetables (4,5), turf grass (6), and cattle dips (7), it is essential to understand the factors influencing this phenomenon.

Many researchers have investigated the role of soil (8,9), microbes (10,11), management practices (12, 13), and environment (14) on the enhanced degradation of pesticides. The properties of the pesticides that influence enhanced degradation have not been studied in detail. Because pesticides are metabolized after application to the soil, some of them within a short time, the properties of pesticide metabolites may also influence the degradation process significantly.

NOTE: This chapter is Journal Paper No. J–13794 of the Iowa Agriculture and Home Economics Experiment Station, Ames, IA. Project No. 2306.

0097–6156/90/0426–0128$06.00/0

Pesticide Properties Influencing Enhanced Biodegradation

One of the properties of pesticides that may influence the induction or inhibition of enhanced biodegradation is their toxicity to the soil microbes responsible for degradation. The resistance of some pesticides to microbial adaptation may be attributed to their microbial toxicity. Soil microorganisms involved in pesticide catabolism utilize the pesticide as a carbon or nutrient source (15-17), reflecting that the nutritive value is an important property in rendering the pesticide susceptible to enhanced biodegradation. The ability of microbes to utilize a pesticide is largely determined by the presence of utilizable carbon and organic nitrogen moieties. The presence of more recalcitrant saturated ring structures (e.g., cycloate and molinate) or an increase in the number of halogens (e.g., 2,4,5-T and tefluthrin) could result in resistance to microbial adaptation.

The availability of pesticides in soil is a critical factor in the induction or inhibition of enhanced biodegradation. The pesticides, whether toxic to the microbes or serving as a suitable substrate, should be available to the microorganisms to exert their toxicity or provide nutrient value. Thus, availability, low microbial toxicity, and high nutritive value seem to be the properties that could favor enhanced degradation of a pesticide.

Pesticide Metabolites. Pesticides are degraded through different mechanisms, and some of them, such as terbufos and phorate, are almost completely metabolized within a short period after their application to soil (18). In general, hydrolytic reactions are the most significant ones in the microbial metabolism of pesticides, although oxidations are often important as well. Pesticide catabolism by adapted soil microbes usually involves an initial hydrolysis, followed by further metabolism and utilization of hydrolysis products as nutrient or energy sources. The effect of pesticide degradation products on the fate of subsequently applied pesticides needs to be better understood. The properties of pesticide metabolites such as nutritive value, microbial toxicity, and availability to soil microbes may be critical in the induction of enhanced degradation of their parent compounds.

Potential of Metabolites to Condition Soils for Enhanced Biodegradation

The presence of some metabolites in flooded rice soils (19,20) and in cranberry bogs (21) has resulted in rapid degradation of their parent compounds. To assess the role of metabolites in conditioning the soil for their respective parent compounds (in nonflooded situations), we pretreated a soil up to 4 times with a hydrolysis product and studied the fate of the subsequently applied ^{14}C-labeled parent compound for several insecticides and herbicides.

A clay loam soil (pH 6.5, organic matter 3.4%) not treated with any pesticide in the last 5 years was used in this study. This soil was pretreated 1, 3, and 4 times, at weekly intervals, with 5 ppm of a hydrolysis product. A week after the last pretreatment, all the treatments were surface-treated with the ^{14}C-labeled parent pesticide and incubated for 3 weeks. The details of soil incubation and metabolism studies have been given elsewhere (22).

2,4-Dichlorophenoxyacetic Acid. Pretreating soil with 2,4-dichlorophenol, the hydrolysis product of 2,4-D, resulted in 76% of applied ^{14}C-2,4-D being mineralized to $^{14}CO_2$ within 3 days, as compared with 3% in soil that received no 2,4-dichlorophenol (Figure 1). At the end of the 3-week incubation, there was a proportionate increase in the amount of $^{14}CO_2$ evolved and decrease in the soil-bound residues formed as the number of 2,4-dichlorophenol pretreatments increased (Table I). The potential of structurally similar compounds such as protocatechuic acid and vanillic acid in conditioning soils for rapid degradation of 2,4-D has been documented (23).

2,4,5-Trichlorophenoxyacetic Acid. 2,4,5-Trichlorophenol, the hydrolysis product of 2,4,5-T, did not condition the soil for enhanced biodegradation of ^{14}C-2,4,5-T (Table I). The 2,4,5-trichlorophenol, unlike 2,4-dichlorophenol, did not serve as a suitable microbial substrate.

Parathion. Prior exposure of soils to p-nitrophenol, the hydrolysis product of parathion, resulted in increased mineralization of parathion to $^{14}CO_2$. The amount of $^{14}CO_2$ evolved increased in proportion to the number of pretreatments with p-nitrophenol (Table II). In soils pretreated up to 4 times with p-nitrophenol, nearly two-thirds of applied parathion was converted to soil-bound residues and $^{14}CO_2$ as compared with 39% in soils with no p-nitrophenol pretreatment. Sudhakar-Barik et al. (19) observed enrichment of parathion-hydrolyzing microbes in soils treated with p-nitrophenol.

Diazinon. 2-Isopropyl-6-methyl-4-hydroxypyrimidine, the hydrolysis product of diazinon, did not condition the soils for enhanced degradation of diazinon (Table II). Despite the low microbial toxicity and high availability (discussed elsewhere in this chapter), the hydroxypyrimidine metabolite did not predispose soils for rapid degradation of diazinon. Enhanced biodegradation of diazinon in rice soils has been previously reported (1). Evidently, the soil we studied did not contain microbes capable of adapting for diazinon enhanced degradation.

Carbofuran. The degradation of ^{14}C-ring carbofuran was not affected in soils pretreated with its hydrolysis product, carbofuran phenol (Table III). In the flooded rice soils of India, carbofuran phenol (without serving as an energy source) enhanced the degradation of carbofuran (20). It is possible that carbofuran phenol may serve as an enzyme inducer, but we found no evidence of enhanced degradation of carbofuran in soils pretreated up to 4 times with carbofuran phenol. A bacterium utilizing the nitrogen of the N-methyl carbamate side chain has been isolated (17). Methyl amine, a secondary hydrolysis product of carbofuran, may be the preferred microbial substrate.

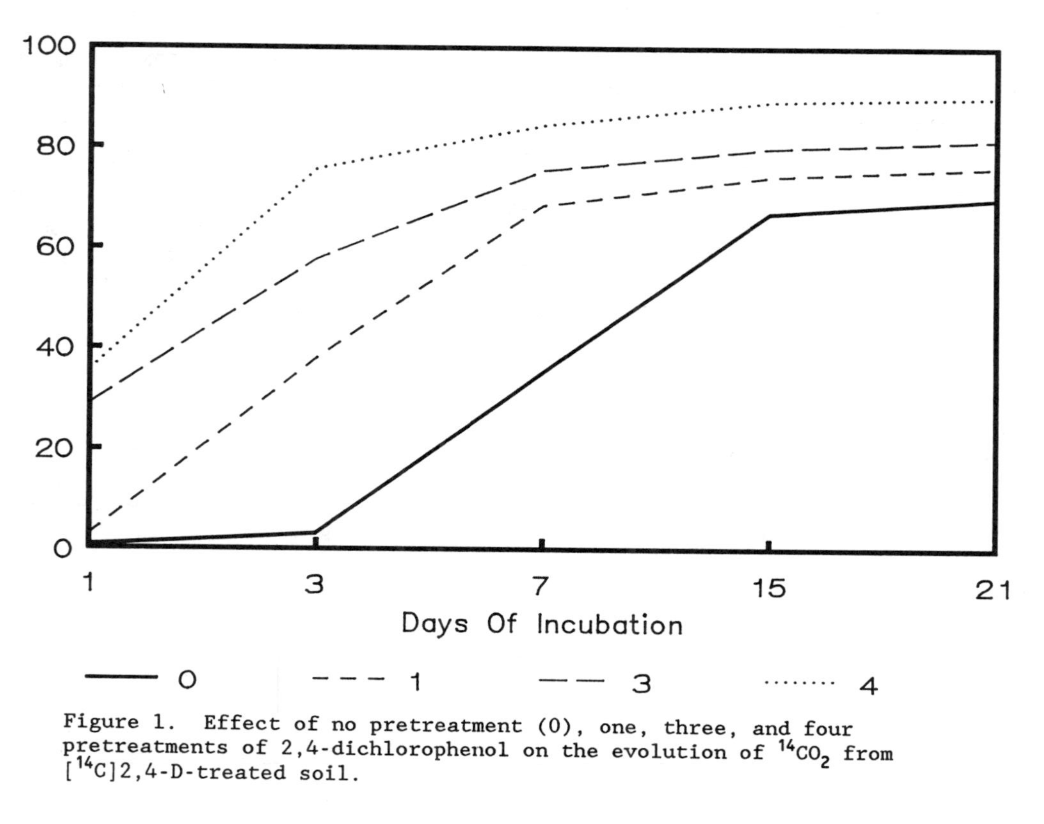

Figure 1. Effect of no pretreatment (0), one, three, and four pretreatments of 2,4-dichlorophenol on the evolution of $^{14}CO_2$ from [^{14}C]2,4-D-treated soil.

Table I. Degradation of pesticides in soil as influenced by the pretreatment of hydrolysis products

Fractions	^{14}C recovered, % of applied ^{14}C pesticide			
	Number of pretreatments with hydrolysis products			
	0	1	3	4
	2,4-Dichlorophenol (from 2,4-D)			
2,4-D	0.21^{a}	0.22^{a}	0.18^{a}	0.26^{a}
2,4-Dichlorophenol	0.63^{a}	0.55^{a}	0.40^{a}	0.53^{a}
Soil-bound	29.46^{a}	21.75^{b}	18.52^{c}	13.18^{d}
$^{14}CO_2$	69.88^{a}	76.02^{b}	81.47^{c}	90.01^{d}
Others^{e}	0.87^{a}	0.62^{b}	0.47^{bc}	0.40^{c}
Total	101.05^{a}	99.17^{a}	101.05^{a}	103.06^{a}
	2,4,5-Trichlorophenol (from 2,4,5-T)			
2,4,5-T	17.84^{a}	15.37^{b}	15.16^{b}	15.75^{b}
2,4,5-Trichlorophenol	4.41^{a}	4.85^{a}	5.30^{ab}	6.30^{b}
2,4,5-Trichloroanisole	35.91^{a}	30.62^{a}	36.52^{a}	36.21^{a}
Soil-bound	30.94^{a}	30.21^{a}	30.47^{a}	29.47^{a}
$^{14}CO_2$	9.26^{a}	5.58^{b}	5.19^{b}	5.29^{b}
Others^{e}	0.74^{a}	0.96^{a}	0.81^{a}	0.63^{a}
Total	99.10^{a}	87.60^{c}	93.47^{b}	93.46^{b}

[a-d]Means in each row with the same letter are not significantly different at 5% level (Student-Newman-Keuls Test).

[e]Includes volatile products other than $^{14}CO_2$ as well as polar, water-soluble products.

Table II. Degradation of pesticides in soil as influenced by the pretreatment of hydrolysis products

Fractions	^{14}C recovered, % of applied ^{14}C pesticide			
	Number of pretreatments with hydrolysis products			
	0	1	3	4
	p-Nitrophenol (from Parathion)			
Parathion	55.01[a]	52.14[b]	40.87[c]	33.84[d]
p-Nitrophenol	0.75[a]	1.78[b]	2.84[c]	3.29[d]
Soil-bound	26.96[a]	27.27[a]	29.85[b]	30.80[b]
$^{14}CO_2$	12.28[a]	13.13[a]	25.57[b]	30.65[c]
Others[e]	1.69[a]	1.91[a]	2.68[a]	3.06[a]
Total	96.70[a]	96.26[a]	101.10[b]	101.65[b]
	Hydroxypyrimidine (from Diazinon)			
Diazinon	37.90[a]	37.58[a]	39.85[a]	38.59[a]
Hydroxypyrimidine	18.43[a]	18.73[a]	22.00[b]	20.52[ab]
Soil-bound	19.59[a]	18.18[a]	18.16[a]	19.85[a]
$^{14}CO_2$	0.95[a]	1.04[a]	0.77[a]	0.95[a]
Others[e]	12.85[a]	13.74[a]	13.03[a]	14.17[a]
Total	89.73[a]	89.29[a]	93.83[a]	94.08[a]

[a-d]Means in each row with the same letter are not significantly different at 5% level (Student-Newman-Keuls Test).

[e]Includes volatile products other than $^{14}CO_2$ as well as polar, water-soluble products.

Table III. Degradation of pesticides as influenced by the pretreatment of hydrolysis products

Fractions	^{14}C recovered, % of applied ^{14}C pesticide			
	Number of pretreatments with hydrolysis products			
	0	1	3	4
	Carbofuran phenol (from Carbofuran)			
Carbofuran	77.01[a]	74.27[b]	77.48[a]	75.20[b]
3-Ketocarbofuran	1.38[a]	1.75[a]	1.67[a]	1.41[a]
Soil-bound	10.84[a]	12.52[b]	13.25[b]	13.06[b]
$^{14}CO_2$	2.75[a]	3.38[a]	3.27[a]	3.08[a]
Others[e]	0.58[a]	1.14[a]	0.58[a]	0.78[a]
Total	92.56[a]	93.07[a]	96.27[a]	93.54[a]
	3,5,6-Tricholoro-2-pyridinol (from Chlorpyrifos)			
Chlorpyrifos	73.78[a]	79.08[b]	79.61[b]	85.08[c]
3,5,6-T-2-pyridinol	0.78[a]	0.73[a]	0.88[a]	0.73[a]
Soil-bound	6.19[a]	4.72[b]	3.69[c]	2.83[d]
$^{14}CO_2$	5.72[a]	0.76[b]	0.59[b]	0.60[b]
Others[e]	11.79[a]	10.71[a]	13.09[a]	10.40[a]
Total	98.27[a]	96.00[a]	97.87[a]	99.65[a]

[a-d]Means in each row with the same letter are not significantly different at 5% level (Student-Newman-Keuls Test).

[e]Includes volatile products other than $^{14}CO_2$ as well as polar, water-soluble products.

Chlorpyrifos. Prior applications of 3,5,6-trichloro-2-pyridinol, the hydrolysis product of chlorpyrifos, increased the persistence of chlorpyrifos (Table III). At the end of a 3-week incubation, there was about 5 to 10% increase in chlorpyrifos recovered in pyridinol-pretreated soils as compared with control soils. Increased persistence of a parent compound in soil pretreated with its degradation product is new to the literature.

Isofenphos. Exposure of soils to salicylic acid, the secondary hydrolysis product of isofenphos, resulted in enhanced degradation of isofenphos (Table IV). Nearly two-thirds of the applied isofenphos was converted to soil-bound residues in soil pretreated 3 and 4 times with salicylic acid. Seventy-eight percent of the applied isofenphos was recovered at the end of the 3-week incubation in the control treatment as compared with 34 to 65% in soils pretreated with salicylic acid. The ability of microbes to metabolize structurally similar compounds such as 3,5-dichlorosalicylate, 3,6-dichlorosalicylic acid (24), and 5-chlorosalicylate (25) to their benefit has been reported. The low microbial toxicity, relative availability (as discussed later in this chapter), and nutritive value of salicylic acid may contribute to its potential to condition soils for enhanced degradation of isofenphos.

Table IV. Degradation of pesticides as influenced by the pretreatment of hydrolysis products

Fractions	^{14}C recovered, % of applied ^{14}C pesticide			
	Number of pretreatments with hydrolysis products			
	0	1	3	4
	Salicylic acid (from Isofenphos)			
Isofenphos	78.04[a]	65.30[b]	36.81[c]	33.69[c]
Isofenphos oxon	8.95[a]	7.81[a]	5.35[b]	4.43[b]
Soil-bound	8.75[a]	12.34[b]	21.32[c]	24.90[d]
$^{14}CO_2$	3.91[a]	9.93[b]	34.47[c]	38.72[d]
Others[e]	0.36[a]	0.42[a]	0.26[a]	0.33[a]
Total	100.02[a]	95.81[a]	98.22[a]	102.07[a]

[a-d]Means in each row with the same letter are not significantly different at 5% level (Student-Newman-Keuls Test).

[e]Includes volatile products other than $^{14}CO_2$ as well as polar, water-soluble products.

Microbial Toxicity

The Beckman Microtox system was employed to assess the relative toxicity of pesticides and their hydrolysis products to bacteria. This system utilizes *Photobacterium phosphoreum*, a marine bioluminescent bacterium phylogenetically related to several genera of bacteria important in soil. The Microtox system measures the light emitted from *P. phosphoreum* that have been exposed to a chemical dissolved in the diluent. The details of theory and operation of Microtox analyzer and experimental conditions used have been described (26-28).

The hydrolysis products of some pesticides susceptible to enhanced degradation (isofenphos, diazinon, carbofuran) yielded higher EC_{50} values, reflecting their low toxicity to bacteria (Table V). Conversely, some pesticides with little or no propensity for enhanced biodegradation yield hydrolysis products that show considerable antibacterial activity (chlorpyrifos, 2,4,5-T). However 2,4-dichlorophenol and p-nitrophenol, which are reported to serve as energy sources to soil microbes, also recorded a low EC_{50} value. The toxicity of hydrolysis products to *P. phosphoreum* may not correspond to toxicity to soil bacteria in all instances, but data indicate that susceptibility to enhanced degradation may be partly influenced by the toxicity or lack of toxicity of the hydrolysis products to bacteria.

Table V. The toxicity of pesticides and their hydrolysis products as determined with the Microtox system

Pesticide	EC_{50}	Hydrolysis product	EC_{50}
2,4-D	100.67	2,4-dichlorophenol	5.0
2,4,5-T	51.63	2,4,5-trichlorophenol	1.77
parathion	8.51	p-nitrophenol	13.74
chlorpyrifos	46.25	3,5,6-trichloro-2-pyridinol	18.58
carbofuran	20.52	carbofuran phenol	60.92
		methylamine	34.64
isofenphos	97.81	salicylic acid	213.92
		isopropyl salicylate	5.62
diazinon	10.30	2-isopropyl-4-methyl-6-pyrimidine	886.42

Availability of Pesticides/Hydrolysis Products in Soils

Another important variable that determines the microbial metabolism of soil-applied pesticides is the availability of the chemical to the microbial systems degrading it. The hydrolysis product and parent pesticide should be available to microbes so as to exert their toxicity or provide nutrient value. The lack of availability of some chemicals may result in resistance to microbial adaptation.

One laboratory approach to the study of availability is the mobility of the pesticides on soil thin-layer chromatography plates as an index of a compound's adsorption/desorption behavior.

Soil Thin-Layer Chromatography. The soil thin-layer chromatography technique (STLC) was employed to assess the mobility of pesticides and their hydrolysis products in soil. Because soil properties such as organic matter, pH, and clay influence the behavior of chemicals in soil, six texturally different soils with a wide range in organic matter content (0.7 to 6.1%) and pH (5.5 to 8.5) were used in this study. In the STLC technique, thin layers of soil serve as the adsorbent phase and are developed with water by using techniques analogous to conventional thin-layer chromatography (29). ^{14}C-Labeled compounds were applied as spots on STLC plates and developed with distilled water by ascending chromatography. The developed plates were exposed to Kodak Royal Blue X-ray film for 2 to 3 weeks. The R_f value (relative frontal movement) for each compound was measured as the front of the spot or streak in the resultant autoradiogram (30).

Mobility in Soils. Chlorpyrifos was not mobile in any of the soils studied, but its hydrolysis product trichloropyridinol was mobile, especially in loamy sand and silt loam soils (Table VI). Parathion, diazinon, and isofenphos were slightly mobile ($R_f < 0.25$), and their hydrolysis products were significantly more mobile than the respective parent compounds ($p < 0.01$). Carbofuran phenol was more mobile in all soils studied (R_f 0.33 to 0.68). 2,4-D was mobile in all six soils studied (R_f 0.56 to 1.00), whereas its hydrolysis product, 2,4-dichlorophenol, was low to intermediate in mobility. 2,4,5-T and 2,4,5-trichlorophenol were nearly immobile in any of the soils studied ($R_f < 0.03$).

Implications of Mobility on the Availability and Degradation of Pesticides in Soil. Repeated application of 2,4-dichlorophenol, p-nitrophenol, and salicylic acid (as observed in current studies) and carbofuran phenol (20) has induced enhanced microbial degradation of their parent compounds. R_f values of these hydrolysis products indicate intermediate to high mobility in soils. The p-nitrophenol, 2,4-dichlorophenol, and salicylic acid were utilized as energy sources by microbes, and their availability in soil may contribute to the induction of rapid microbial metabolism. Carbofuran phenol did not serve as a microbial substrate but also enhanced the degradation of its parent compound, carbofuran (20). Carbofuran phenol is freely available in anaerobic soils, but the significance of its availability is yet to be understood.

The mobility data for isofenphos indicates that, for a pesticide to be susceptible to enhanced degradation, the pesticide need not necessarily be very mobile. Although isofenphos has a low mobility pattern, its salicylic acid metabolite is more readily available to microorganisms. Salicylic acid is a benzoic acid analog, and its high availability is similar to that of benzoic acid and amiben (31). Racke and Coats (11) suggested that the formation of salicylic acid during isofenphos metabolism in soil may represent a key factor in the susceptibility of isofenphos to enhanced degradation. Our studies with soils exposed to salicylic acid have confirmed this view. Some pseudomonads have been reported to carry the salicylic acid degradative plasmid (32). Observations from the mobility and metabolism studies indicate that the less mobile isofenphos may be hydrolytically metabolized by surface-soil microorganisms and that the resultant metabolites are more available to degrading microorganisms.

Table VI. Mobility of pesticides and their hydrolysis products in six soils, as determined with use of soil TLC plates

Pesticide/ hydrolysis product	R_f value in					
	clay loam	loam	silty clay loam	sandy loam	loamy sand	silt loam
chlorpyrifos	0.02	0.02	0.02	0.02	0.02	0.00
t-pyridinol	0.26	0.26	0.58	0.36	0.94	1.00
parathion	0.10	0.10	0.14	0.12	0.18	0.13
p-nitrophenol	0.15	0.23	0.50	0.46	0.89	1.00
diazinon	0.10	0.12	0.12	0.17	0.24	0.21
hydroxypyrimidine	0.80	0.78	0.71	0.74	0.96	0.81
isofenphos	0.09	0.09	0.13	0.16	0.10	0.16
salicylic acid	0.09	0.39	0.59	0.35	0.75	0.96
carbofuran	0.57	0.76	0.69	0.77	0.81	0.75
carbofuran phenol	0.33	0.42	0.33	0.63	0.68	0.48
2,4-D	0.57	0.56	0.67	0.68	1.00	1.00
2,4-dichlorophenol	0.12	0.11	0.20	0.14	0.58	0.46
2,4,5-T	0.00	0.02	0.03	0.03	0.03	0.00
2,4,5-T-phenol	0.00	0.00	0.00	0.00	0.01	0.00

The hydroxypyrimidine hydrolysis product of diazinon is more readily available in all soils tested and is mineralized by microbes (33). Our Microtox studies have demonstrated its low toxicity to bacteria. Availability, low microbial toxicity, and susceptibility to microbial metabolism of this hydrolysis product may favor enhanced degradation of its parent compound in soils with populations of degrading microorganisms, but no adaptation was noted in our laboratory studies.

Chlorpyrifos is immobile in soil and is not available to microbes. However, its pyridinol hydrolysis product is relatively mobile; its microbial toxicity and availability in soil may contribute to the increased persistence of chlorpyrifos observed in pyridinol-treated soils.

Both 2,4,5-T and 2,4,5-trichlorophenol are relatively unavailable in soil, indicating their low availability for microbial degradation.

Influence of Soil Characteristics on Mobility. Most of the pesticides and their hydrolysis products were more mobile in the loamy sand and silt loam soils than in the other soils used. These two soils had relatively high pH's of 8.3 and 8.5, respectively. The pH has an indirect influence on movement of some pesticides by effecting stronger adsorption at low pH values (34,35), resulting in lower mobility. In general, adsorption of chemicals is weak at neutral pH's and above, leading to increased movement (36) and faster degradation in alkaline soils (9).

Simple regression analysis of mobility data versus organic matter content and clay content indicated that greater organic matter and clay content retarded mobility. Silt loam and loamy sand soils had a low organic matter (0.7 and 1.2%) and clay content (13.0 and 8.0). The increased availability of chemicals in these two soils could be attributed to the high pH and low organic matter and clay contents. The susceptibility of pesticides to enhanced degradation only in specific parts of a state or country may be because of the influence of soil characteristics such as pH and organic matter on the availability of chemicals to microorganisms.

Conclusion

Pesticide degradation products are capable of conditioning soils for enhanced degradation of their parent compounds and could play an important role in the induction or inhibition of enhanced microbial degradation of some pesticides. The properties of degradation products such as greater availability, low microbial toxicity, and nutritive value may favor enhanced degradation of their parent compounds in soils with populations of degrading microorganisms. On the basis of these findings and our related research, we define enhanced microbial degradation as a phenomenon in which adapted soil microorganisms make use of the pesticide or pesticide degradation products as an energy or nutrient source, resulting in decreased persistence of the pesticide. Future research should also be directed toward understanding the interaction between microbes and metabolites (primary, as well as secondary) of pesticides.

Acknowledgments

This research was supported by grants from the USDA North Central Region Pesticide Impact Assessment Program and Dow Chemical Company. Any opinions, findings, and conclusions expressed are those of the authors and do not necessarily reflect the views of the granting agencies. Journal Paper No.J-13794 of the Iowa Agriculture and Home Economics Experiment Station, Ames, IA, Project 2306.

Literature Cited

1. Sethunathan, N. Proc. Natl. Acad. Sci.(USA) 1971, 17(1), 18-19.
2. Rahman, A.; Atkison, G.C.; Doughlas, J.A,; Sinclair, D.P. N. Z. J. Agric. 1979, 139(3), 47-49.
3. Wilde, G.; Mize, T. Environ. Entomol. 1984, 13, 1079-1082.
4. Walker, A.; Brown, P.A.; Entwistle, A.R. Pestic. Sci. 1986, 17, 183-193.
5. Harris, C.R.; Chapman, R.A.; Morris, R.F.; Stevenson, A.B. J. Environ. Sci. Health 1988, B23, 301-316.

6. Niemczyk, H.D.; Chapman, R.A. J. Econ. Entomol. 1987, 80, 880-882.
7. McDougall, K.W.; Machin, M.V. Pestic. Sci. 1988, 22, 307-315.
8. Abou-Assaf, N.; Coats, J.R. J. Environ. Sci. Health 1987, B22, 285-301.
9. Walker, A. Pestic. Sci. 1987, 21, 219-231.
10. Fournier, J.C.; Codaccioni, P.; Soulas, G.; Chemosphere 1981, 10, 977-984.
11. Racke, K.D.; Coats, J.R. J. Agric. Food Chem. 1987, 35, 94-99.
12. Abou-Assaf, N.; Coats, J.R.; Gray, M.E.; Tollefson J.J. J. Environ. Sci. Health 1987, B21, 425-446.
13. Somasundaram, L.; Racke, K.D.; Coats, J.R. Bull. Environ. Contam. Toxicol. 1987, 39, 579-586.
14. Chapman, R.A.; Harris, C.R.; Harris, C. J. Environ. Sci. Health 1986, B21, 125-141.
15. Cook, A.M.; Doughton, C.G.; Alexander, M. Appl. Environ. Microbiol. 1978, 36, 668-672.
16. Nelson, L.M. Soil Biol. Biochem. 1982, 14, 219-222.
17. Karns, J.S.; Mulbry, W.W.; Nelson, J.O.; Kearney, P.C. Pestic. Biochem. Physiol. 1986, 25, 211-217.
18. Harris, C.R.; Chapman, R.A. Can. Entomol. 1980, 112, 641-653.
19. Sudhakar-Barik; Wahid, P.A.; Ramakrishna, C.; Sethunathan, N. J. Agric. Food Chem. 1979, 27, 1391-1392.
20. Rajagopal, B.S.; Panda, S.; Sethunathan, N. Bull. Environ. Contam. Toxicol. 1986, 36, 827-832.
21. Ferris, I.G.; Lichtenstein, E.P. J. Agric. Food Chem. 1980, 28, 1011-1019.
22. Somasundaram, L.; Coats, J.R.; Racke, K.D.; J. Environ. Sci. Health 1989, B24, 457-478.
23. Kunc, F.; Rybarova, J. Folia Microbiol. 1984, 29, 156-161.
24. Kruger, J.P.; Butz, R.G.; Atallah, Y.H.; Cork, D.J. J. Agric. Food Chem. 1989, 37, 534-538.
25. Crawford, R.L.; Olson, P.E.; Frick, T.D. Appl. Environ. Microbiol. 1979, 38, 379-384.
26. Anonymous, Interim Manual 015-555879; Beckman Instruments Inc.; Microbics Operations: Carlsbad, CA, 1979.
27. Bulich, A.A.; Greene, M.W.; Isenberg, D.L.; Aquatic Toxicology and Hazard Assessment: Fourth Conference 1981, ASTM STP 737, 338-347.
28. Somasundaram, L.; Coats, J.R.; Racke, K.D. Bull. Environ. Contam. Toxicol. 1990, 44, (2).
29. Helling, C.S.; Turner, B.C. Science 1968, 162, 562-563.
30. Somasundaram, L. Ph.D. Dissertation, Iowa State University, University Microfilms #90-03565, Ann Arbor, MI, 1989.
31. Bailey, G.W.; White, J.L.; Rotherberg, T. Soil Sci. Soc. Am. Proc. 1968, 32, 222-223.
32. Chakrabarty, A.M. J. Bacteriol. 1972, 112, 815-823.
33. Sethunathan, N.; Pathak, J. Agric. Food Chem. 1972, 20, 586-589.
34. Weber, J.B. Residue Rev. 1970, 32, 93-130.
35. Renner, K.A.; Meggitt, W.F.; Penner, D. Weed Sci. 1968, 36, 78-83.
36. Nicholls, P.H. Pestic. Sci. 1988, 22, 123-137.

RECEIVED January 22, 1990

Chapter 11

Molecular Genetics of Pesticide Degradation by Soil Bacteria

Jeffrey S. Karns

Pesticide Degradation Laboratory, Natural Resources Institute, Agricultural Research Service, U.S. Department of Agriculture, Beltsville, MD 20705

Although it is generally accepted that the degradation of soil applied pesticides by microorganisms is responsible for many pesticide performance failures, little is known about the molecular mechanisms responsible for the evolution of new pesticide degradative capabilities. Mobile genetic elements such as plasmids and transposons have been shown to encode enzymes responsible for the degradation of several pesticides. The isolation of pesticide degrading microorganisms and the characterization of genes encoding pesticide degradation enzymes, combined with new techniques for isolating and examining nucleic acids from soil microorganisms, will yield unique insights into the molecular events that lead to the development of enhanced pesticide degradation phenomenon.

The role of soil microorganisms in the enhanced degradation of several soil-applied pesticides has been well documented (1-3). The enhanced degradation phenomenon is apparently a result of the ability of microbial populations to respond to environmental changes (such as the introduction of a pesticide) by rapidly evolving or otherwise acquiring the genetic material that encodes the biochemical mechanisms required to deal with the new conditions. In the case of pesticides, the introduced compound may be a new source of carbon, energy, nitrogen, phosphorous, or sulfur, which if utilized by one or several members of the microbial community allows them to proliferate and out-compete other members of the community. Alternatively, microorganisms may degrade a pesticide because it is toxic to them and by transforming the pesticide through hydrolysis, oxidation, or reduction that toxicity is relieved. In either case there is a natural selection process that occurs because of the introduction of the xenobiotic pesticide molecule into soil.

There are several theoretical mechanisms by which the adaptation of soil microbial communities to yield populations which are capable of rapidly degrading a pesticide may occur. In some cases there may be enzymes that carry out normal cellular functions which happen to also catalyze the transformation of a pesticide molecule due to a resemblance of the pesticide molecule to the enzyme's natural substrate. This type of co-metabolism (4) is probably responsible in part for the normal, slow rates of biodegradation observed for most of today's pesticides under ordinary conditions. Indeed, this type of biodegradation is desirable because it undoubtedly contributes to the elimination of pesticide molecules from the environment, decreasing the likelihood that they will become pollution problems. However, if the organisms that transform the pesticide molecule are able to derive some form of benefit (either by themselves or in concert with several other members of the population) they might be expected to slowly increase their numbers until they reach some threshold at which the rate of pesticide degradation is fast enough to cause noticeable effects on pesticide performance. If this scenario accurately describes the manner in which enhanced degradation phenomena arise in soils, the one conclusion that seems obvious is that all soil incorporated pesticides will ultimately suffer from performance failures due to rapid degradation or will end up being banned for a lack of degradability of either the pesticide or its degradation products.

Another scenario involves the hypothesis that some genetic event must take place within the soil microbial community in order for enhanced degradation to occur. Such an event might involve the mutation of a gene so that the enzyme it encodes degrades the pesticide at a faster rate and/or has a higher affinity for the pesticide molecule. Mutations that affect the way gene expression is regulated may also influence the fate of pesticides in soils. Usually, the natural substrate of an enzyme or a natural metabolite in the same metabolic pathway acts as an inducer of gene expression. This induction causes changes at regulatory locations on the DNA which increases the rate at which messenger RNA (mRNA) is made from the DNA template via transcription. The increased amount of mRNA for that gene results in greater production of that particular enzyme via translation (the process by which proteins are produced). The higher levels of enzyme produced in each cell result in a higher overall level of enzyme activity within the population. Thus, if a mutation affected the regulation of a potential pesticide degradation gene so that the enzyme it encodes is expressed constantly (constitutive expression) or, is induced to high levels by the pesticide (pesticide acting as inducer rather than, or in addition to the natural substrate) the result would be a population of organisms that degraded the pesticide molecule at a greater rate. Along this line, it seems possible that silent, or cryptic, genes that might be present in a bacterium but are normally not expressed could play a role in the evolution of pesticide degradation capabilities within a bacterial population (5).

Mobile DNA in the form of plasmids, transposons, and bacteriophages may also contribute to the genesis of enhanced degradation in soils. Reanney (6) discussed the possible role of such extra-chromosomal elements [ECE's] in the evolution of organisms. He argues that the occurrence of the types of mutations described in the previous paragraphs within essential genes carried on the chromosome of an organism, while possibly aiding the survival of said organism in the short term, may ultimately doom the organism when the condition that made such a mutation advantageous disappears. ECE's may be part of a mechanism by which a cell can duplicate portions of DNA and modify the duplicated genes while leaving the original copy of the gene intact. In this way existing genes can be modified to perform new tasks that may be of advantage to a bacterium without compromising the function of the original copy of the gene. Because of natural selection, the ECE containing a useful function would survive and might be disseminated to other bacteria, resulting in the rapid adaptation of a microbial population. The role of plasmids in the rapid spread of antibiotic resistance genes among bacteria in hospitals is well known.

What I hope to do in this paper is to give the reader a brief introduction to bacterial genetics and then to describe the research that has demonstrated the involvement of plasmids in carrying genes which encode enzymes that degrade pesticides. I will then discuss the possible role of plasmids in the development of the enhanced degradation phenomenon and the types of research that may lead to the delineation of the molecular events that lead to the development of rapid pesticide degradation in soils.

Bacterial Genetics

Types of DNA Found in Bacterial Cells. Most of the DNA in a bacterial cell is contained on one large DNA molecule known as the chromosome or genophore. In *Escherichia coli* the chromosome is about 3.7×10^6 base pairs in size and can encode about 3000 genes. In general, the bacterial chromosome carries all the genes essential to the replication and maintenance of the cell.

Plasmids are autonomous, circular DNA molecules (from around 1 kilobase pairs (kb) to several hundreds of kb in size) that are capable of self-replication within a bacterial cell (Figure 1). Plasmids contain some genes involved in their own replication and transfer between bacteria, but can also harbor additional genes that can impart such biochemical capabilities as antibiotic resistance, utilization of additional nutrients, production of pathogenic factors, nitrogen fixation, or the production of bacteriocins. These plasmid encoded traits may be useful to the cell but are usually not essential. Some plasmids are very limited in the range of bacteria that can act as host while others may have very broad host ranges, being able to replicate within many different genera of bacteria. Some plasmids contain genes that encode the factors required for their transfer to other bacterial cells through the process of conjugation (see below). Some conjugal plasmids also contain regions of DNA that allow them to be integrated into the bacterial genome so that portions of chromosome can be transferred between mating cells. Plasmids also vary in the number of copies that are maintained within a cell, with some being present in one or two copies per cell while others may be present in 20 or more copies per cell. It follows that a gene encoded on one of these high copy-number plasmids is highly amplified and that very large amounts of the enzyme encoded by such a gene would be produced.

Transposable elements are mobile DNA segments that can insert themselves into many sites in cellular DNA (7). Transposable elements are not capable of autonomous replication but are propagated only when they are inserted into chromosomal or plasmid DNA. Insertion sequence elements (IS elements) are the smallest pieces of DNA (0.7 to 1.5 kb) known to function as transposable elements. IS elements contain only the genes needed for their insertion into cellular DNA. Transposons (Tn elements) are more complex transposable elements that do contain genes unrelated to the insertion function. Transposons frequently have a central piece of DNA containing one or several genes flanked by nearly identical insertion elements at its termini (Figure 2, 8).

Bacteriophages are very small virus-like particles that can attack bacterial cells. The virus consists of a nucleic acid core surrounded by a protein coat. Individual bacteriophages have limited host specificities, attacking only closely related species of bacteria. However, bacteriophages are known to exist for most bacterial species. In general, bacteriophages attack their host bacterium by attaching to a specific receptor on the cell surface and injecting their nucleic acid core into the cell. Once inside the cell, the phage nucleic acid is replicated and transcribed resulting in the production of more bacteriophage particles. Mature phage particles are released through the cell membrane in some cases, or in other cases are released when the cells lyse and die. Of interest to the discussion here is the fact that bacterial DNA sequences are accidentally packaged into phage particles at some finite frequency so that bacteriophage can sometimes act as vehicles for transfer of bacterial genes.

Mechanisms of Gene Exchange in Bacteria. There are several mechanisms by which DNA can be exchanged between bacteria in the laboratory (Figure 3). Whether any of them actually occurs in the soil environment under natural conditions is open to debate (9). **Transformation** is the process by which bacteria take up naked DNA from the environment. Only certain genera of bacteria have been shown to undergo natural transformation, among them are *Acinetobacter*, *Haemophilus*, *Bacillus*, and *Staphylococcus* (9,10). These bacteria have certain periods in their life cycle when they are competent (able to take up DNA). There may be other bacteria that have periods of natural competence but have not had it demonstrated in the laboratory yet. Some bacteria that have not been demonstrated to undergo periods of natural competence can be rendered competent through careful manipulation of their environment (hot-cold pulses, presence of certain ions, etc.).

Transduction is the term for the transfer of genes from one bacterium to another by bacteriophage particles. The amount of DNA that can be transferred is limited by the size of the DNA that can be carried within the phage head. The specificity of the bacteriophage for receptors on the host bacterium limits any role of transduction in the transfer of DNA between species of bacteria. However, the protection of the labile DNA molecule by the phage coat

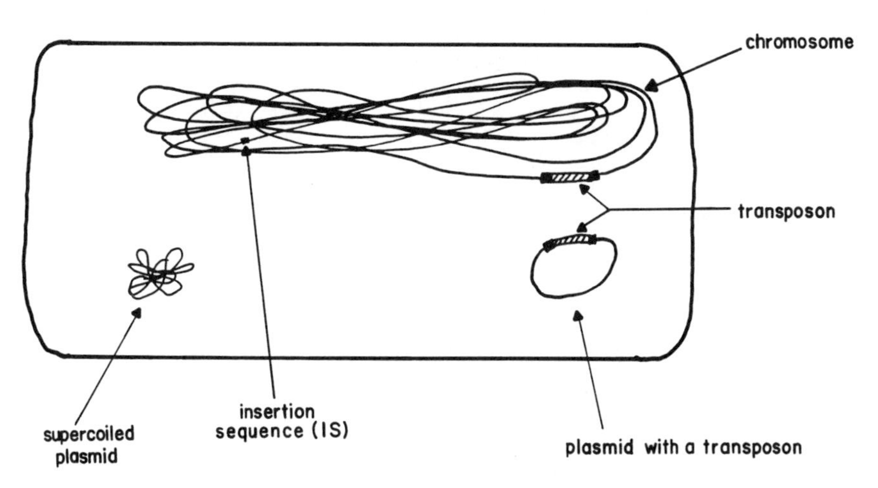

Figure 1. Types of DNA present inside a bacterial cell. All forms of DNA are normally present in the supercoiled form.

IS-L A B C IS-R

Figure 2. Generalized structure of a hypothetical catabolic transposon encoding three enzymes (A, B, and C) involved in the degradation of a pesticide. The clear zones marked IS-L and IS-R represent the nearly identical insertion elements that mark the left and right boundries of the transposon and encode the proteins required for movement of the transposon.

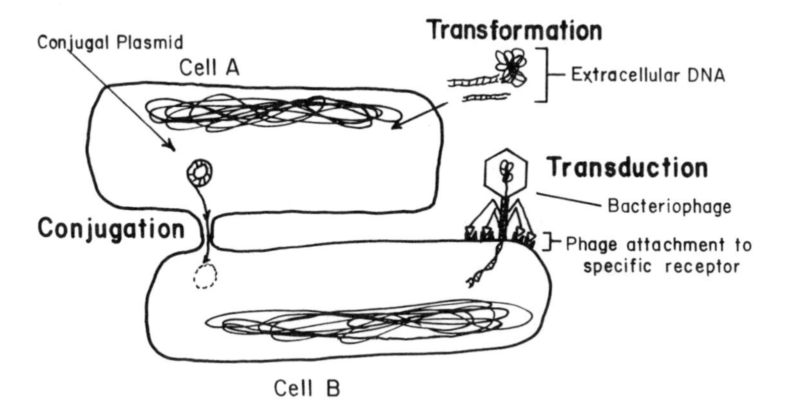

Figure 3. Mechanisms of DNA exchange in bacteria.

might suggest that transduction could play a role in the transfer of genes among soil microorganisms.

Conjugation denotes the transfer of DNA between bacterial cells by cell-to-cell contact. Conjugation is mediated by plasmids (conjugal plasmids or sex factors) which encode factors for the transfer and mobilization of DNA. Not all plasmids are conjugal plasmids. Some plasmids contain only the factors that allow them to be mobilized but do not encode transfer functions. These types of plasmids can only be conjugally transferred when a second plasmid that contains the transfer functions is present. Other types of plasmids exist which contain no transfer or mobilization functions and hence cannot be conjugally transferred under normal conditions. As noted previously, some conjugal plasmids can mediate the transfer of chromosomal DNA between bacteria. Although there can be some debate about whether conjugation plays an active role in gene transfer among members of the soil microbial community (for example are cell densities in soil high enough) this author feels that conjugal transfer of plasmids is likely to be an important means of gene transfer in soil.

Regardless of the means by which DNA is transferred from one cell to another the introduced DNA must be replicated and expressed in order to have any effect on the cell that obtains it. If the transferred DNA is a complete replicon (that is a unit capable of autonomous replication, i.e. a plasmid) that is capable of replication in the recipient cell the obtained genes will be propagated. Since there are differences in the sequences that are recognized as promoters (sequences signaling the start of mRNA synthesis) or ribosome binding sites (sequences signalling the start of translation) by different genera of bacteria, the promoters and ribosome binding sites of the genes of interest must be functional in their new host for the new genes to be expressed. If the introduced DNA is not a complete replicon (such as DNA obtained through transduction or transformation) or is on a plasmid that cannot replicate in the new host cell, a recombination event must occur in order for the new DNA to be replicated and passed on to future generations. Under normal conditions a requirement for this recombination is that the introduced DNA must have a large portion of DNA that is identical to that in the host's DNA (that is have a large degree of homology with the host DNA) in order for homologous recombination, or "crossing over" to occur. However, if the introduced genes are part of a transposon the requirement for regions of DNA homology does not exist and the genes can become a part of the new cell's DNA through the illegitimate recombination mechanism of transposition (11). Thus, the existence of a gene as part of a transposon can radically affect the ability of that gene to spread with the soil microbial community.

Plasmids and Transposons in Pesticide Degradation

Although there have been many reports of pesticide degradation by bacteria, little is known about the genetic aspects of this degradation. There are probably many reasons for this, but two of the most significant are: 1) the types of bacteria that are isolated from soil environments are generally harder to work with than most bacteria that are the subject of genetic and molecular biology studies, and ; 2) in many cases the bacteria transform and inactivate pesticides without deriving any known benefit from the transformation, thus, the pesticide degradation phenotype is not a selectable one making genetic studies very difficult. In spite of the difficulties there have been genetic studies done on several pesticide degrading bacteria which may provide insights into the roles that plasmids and/or transposons may play in the development of the enhanced degradation phenomenon.

Bacterial degradation of the soil incorporated herbicide dalapon (2,2'-dichloropropionic acid, perhaps the first casualty to the enhanced degradation phenomenon) has been shown to proceed through enzymatic dehalogenation to yield propionic acid which can be used as a carbon and energy source (12). The genes encoding two distinct chloroaliphatic acid dehalogenases have been shown to be plasmid encoded in a chloroacetate degrading *Moraxella* sp. (13). Slater and co-workers (14) studied the degradation of dalapon in a microbial community maintained in a chemostat for prolonged periods. The original community contained at least three distinct bacterial members which acted in concert to degrade dalapon but no single organism was capable of completely degrading the herbicide. During the course of the study they witnessed the evolution of *Pseudomonas putida* strain PP3 which was capable of rapid degradation of the herbicide in pure culture. Studies showed that strain PP3 differed from its progenitor (strain PP1, a member of the original community) in its ability to dehalogenate dalapon. Subsequent studies revealed that the dehalogenase genes which had been acquired by strain PP3 were part of a transposable element that could be mobilized from strain PP3 into other bacteria (15).

The bacterial metabolism of phenoxyalkanoic acid herbicides has been studied in some detail. Numerous studies have demonstrated the involvement of plasmids in the degradation of 2,4-dichlorophenoxyacetic acid (2,4-D) (16,17). DNA/DNA hybridization studies comparing 2,4-D degradation plasmids from various organisms have shown substantial homology (16,18), suggesting that these organisms did not evolve this degradative ability independently but rather that a common set of genes has been acquired by many diverse bacterial isolates. DNA sequencing of some of the genes involved in 2,4-D metabolism has shown that they share a common ancestry with plasmid encoded chlorobenzoic acid degradation genes (19), indicating that plasmid transfer and recombination events have played a role in the evolution of the degradative capabilities for these two types of compounds. A 2,4,5-trichlorophenoxyacetic acid (2,4,5-T) degrading *Pseudomonas cepacia* was shown to lose the ability to degrade 2,4,5-T at a rapid rate upon subculture in rich growth medium (20), a trait indicative of plasmid involvement. Although a specific plasmid has yet to be implicated in the degradation of 2,4,5-T by this organism, DNA sequencing studies have shown that there are repeated DNA sequences that bear resemblance to insertion elements located close to genes that have been shown to be involved in 2,4,5-T degradation (21). Since these repeated DNA sequences have not been found in any other strains of *Pseudomonas cepacia* that have been examined, it is likely that they have played a role in the evolution of the 2,4,5-T degradative capability of this organism.

Plasmids have also been shown to be involved in the degradation of the organophosphate insecticide parathion (22,23). The opd (for organophosphate degradation) gene, encoding parathion hydrolase, has been shown to be present in at least three distinct organophosphate pesticide degrading bacteria (22,23,24). It was demonstrated that in Flavobacterium ATCC27551, isolated in the Philippines, and in Pseudomonas diminuta strain MG, isolated in the USA, identical opd genes were carried on plasmids that were otherwise very different (25). DNA sequencing has revealed that the opd genes from these organisms are absolutely identical (26,27). It is interesting to note that cloning studies have shown that the opd gene promoter does not function in Pseudomonas putida or Escherichia coli (26), which like the original hosts are gram negative organisms, but does function in the gram positive organism Streptomyces lividans (28). In addition, a leader peptide encoded by the opd gene functions to insert the processed parathion hydrolase protein in the membrane of the native Flavobacterium and Pseudomonas diminuta strains but causes the protein to be excreted by Streptomyces lividans (26,28). The ability of these regulatory elements to function across such a wide range of microorganisms is strongly suggestive of the important role that plasmid mobility and recombination have played in the dissemination of the opd gene.

While the above mentioned examples are the best studied cases of the genetics of pesticide degradation in bacteria, and strongly implicate plasmids and recombination in the evolution and spread of pesticide degradative abilities among members of the soil microbial community, there are several other cases of plasmid encoded pesticide degradation that, to date, are less well studied. A plasmid has been shown to encode a nitrilase that enables Klebsiella pneumoniae to degrade the herbicide bromoxnil (29). Of special interest to a volume on the enhanced degradation phenomenon are the reports from several laboratories on plasmids that encode the enzymes responsible for the degradation of the carbamothioate herbicides EPTC and butylate (30,31). In my laboratory, we have cloned the gene encoding a carbofuran hydrolase enzyme in an Achromobacter sp. (32). This enzyme has the ability to catalyze the hydrolysis of several N-methylcarbamate insecticides (33). By using the cloned carbofuran hydrolase gene as a specific probe in DNA/DNA hybridization experiments, we were able to show that the gene is encoded on the large plasmid present in this bacterium (32).

Future of Genetic Research on Enhanced Degradation of Pesticides

Much research remains to be done before there can be any true understanding of the genetic factors that play a role in the development and propagation of the enhanced degradation phenomenon in agricultural soils. While it seems very likely that plasmids and transposons will be found to play a role in the evolution of bacterial populations that contribute to the rapid degradation of pesticides, their actual involvement cannot be determined without a thorough understanding of the basic microbial ecology of the phenomenon. To my knowledge, no one has demonstrated that any of the various microorganisms isolated as degraders of any pesticide are actually causative agents of enhanced degradation. Such a demonstration might require the development of an agricultural version of Koch's postulates, where a putative causative organism is isolated from soil, shown to produce the phenomenon when introduced to an inactive soil, and re-isolated from the newly active soil. While such experiments are not simple to perform, it might be possible to conduct them on a greenhouse scale, taking tillage practices, soil types, type of crop, rainfall, etc. into account.

Many of the techniques classically associated with the field of biotechnology can also be used to gain insight into the enhanced degradation phenomenon at both the organismal and molecular levels. Monoclonal antibodies specific for the surface antigens present on pesticide degrading bacteria from soils could be used to study the distribution of those cells in soil (34,35). This technique takes advantage of the ability to isolate antibodies highly specific for antigens on the surface of a particular bacterial cell. By attaching a fluorescent molecule to these antibodies a highly specific probe is obtained that can be detected through the use of fluorescent microscopy techniques (35). Using such probes it is possible to track and enumerate specific microorganisms. In greenhouse simulations this technique could be used to study the growth and spread of these organisms under model field conditions.

The technique of DNA/DNA hybridization offers a very powerful and specific means of detecting the genes encoding pesticide degrading enzymes. This technique takes advantage of the complementary nature of the two strands of the DNA molecule. By fixing denatured, single stranded DNA to solid supports and using tagged (radiolabeled or tagged through attachment of various small reporter molecules) DNA as a specific probe it is possible to detect low levels of DNA sequences of interest in a highly heterologous mixture of DNA. The DNA fixed to the solid support could be from colonies on an agar plate (colony blots, 36), bulk DNA isolated from soil or a population of microbes, or purified DNA cut with restriction enzymes and separated by agarose electrophoresis (Southern blots, 37). Techniques for extracting nucleic acids from soils have been developed recently (38,39,40). Using cloned genes which encode enzymes likely to be involved in the degradation of a given pesticide as probes it would be possible to detect similar genes in organisms from many soils. These highly specific probes would allow the tracking of a given gene even if it were to spread to other organisms within the soil microbial community. The polymerase chain reaction (PCR, 41) can be used to greatly amplify the amount of a particular DNA sequence in a bulk DNA sample (Figure 4). This procedure uses heat stable DNA polymerase and oligonucleotides that are specifically homologous to known nucleotide sequences within the gene of interest. The oligonucleotides act as primers for the in vitro synthesis of new strands of DNA (Figure 4). Since the primers are specific complements to the DNA of interest only that DNA is duplicated. The DNA is remelted to allow another round of synthesis and the procedure is repeated many times. After 20 such cycles there would be over 1 million copies of a DNA sequence that was present in the original sample as only one copy. Thus, using this technology to increase the sensitivity of detection for genes of interest in soils will certainly help in the study of the molecular microbial ecology of soils. PCR has already been shown to be of use in the study of environmental samples (42).

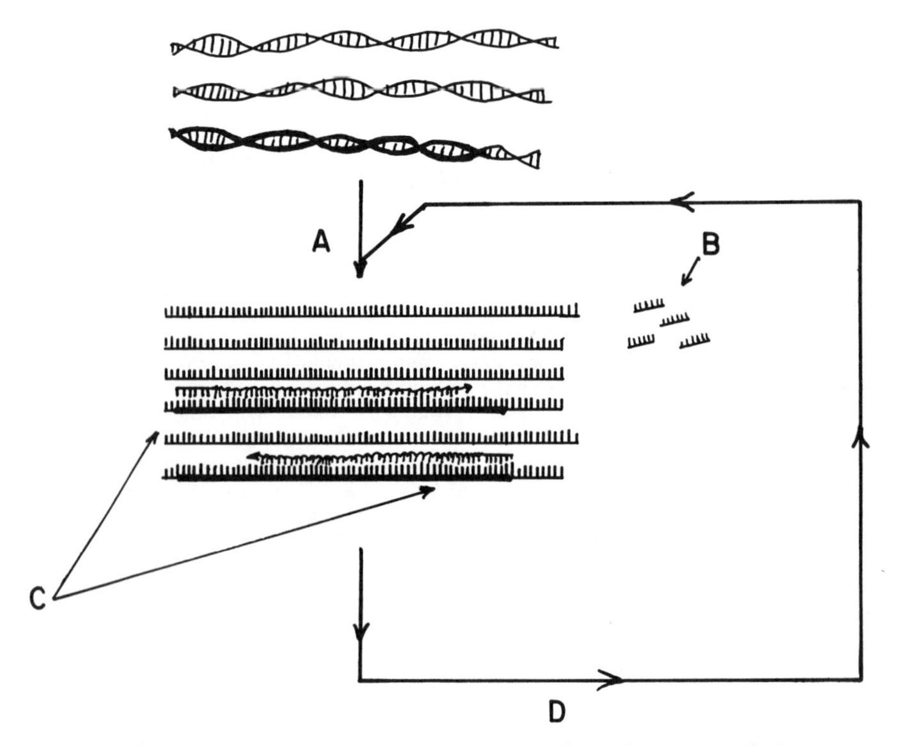

Figure 4. Diagramatic representation of the polymerase chain reaction (PCR) used to amplify particular DNA sequences. At the top are DNA molecules present in a mixture of bulk DNA with the heavy lines representing the gene of interest. In step A the DNA is melted at elevated temperature after which oligonucleotides with a sequence complementary to the DNA of interest (B) are added. The oligonucleotides act to prime the synthesis of new DNA complementary to the DNA of interest by a heat stable DNA polymerase (C). After an adequate period of time the DNA is remelted at elevated temperature and the cycle is repeated (D).

In order to reach this level of sophistication many organisms that degrade a pesticide must be isolated, and the genes encoding pesticide degradation enzymes cloned. These cloned genes would have to be compared in order to determine how much diversity there is in this one biochemical function. In some cases there may be several classes of genes that encode enzymes with identical functions, but that do not cross react in DNA/DNA hybridization experiments. We have demonstrated that there is a degree of diversity among parathion hydrolase genes from different bacterial sources (43). Once the degree of diversity in a given pesticide degradation gene is determined, a set of DNA probes that adequately detects the genes of interest would be developed. In order to use PCR DNA sequences with genes of interest must be known. Thus, such experiments would require unprecedented cooperation between pesticide chemists, soil microbiologists, microbial ecologists, and molecular biologists, but would yield incredible insights into the biological processes that take place in soils.

Literature Cited

1. Wilson, R. G. Weed Sci. 1984, 32, 264-268.
2. Racke, K. D.; Coats, J. R. J. Agric. Food Chem. 1988, 36, 193-199.
3. Harris, R.; Chapman, R. A.; Harris, C.; Tu, C. M. J. Environ. Sci. Health 1984, B(19), 1-11.
4. Horvath, R. S. Bacteriol. Rev. 1972, 36, 146-155.
5. Hall, B. G.; Yokoyama, S.; Calhoun, D. H. Mol. Biol. Evol. 1983, 1, 109-124.
6. Reanney, D. R. Bacteriol. Rev. 1976, 40, 552-590.
7. Campbell, A.; Berg, D.; Botstein, D.; Lederberg, E.; Novick, R.; Starlinger, P.; Szybalski, W. In DNA-Insertion Elements, Plasmids, and Episomes, Bukhari, A.I.; Shapiro, J. A.; Adhya, S. L., Eds.; Cold Spring Harbor Laboratory, Cold spring Harbor, NY. 1977; p 15-22.
8. Grindley, N. D. F.; Reed, R. R. Ann. Rev. Biochem. 1985, 54, 863-896.
9. Trevors, J. T.; Barkay, T.; Bourquin, A. W. Can. J. Microbiol. 1987, 33, 191-198.
10. Smith, H. O.; Danner, D. B.; Deich, R. A. Ann. Rev. Biochem. 1981, 50, 41-68.
11. Cohen, S. N. Nature (London) 1976, 263, 731-738.
12. Kearney, P. C.; Kaufman, D. D.; Beall, M. L. Biochem. Biophys. Res. Commun. 1964, 14:29-33.
13. Kawasaki, H.; Tone, M.; Tonomura, K. Agric. Biol. Chem. 1981, 45, 29-34.
14. Senior, E.; Bull, A. T.; Slater, J. H. Nature (London) 1976, 263, 476-479.
15. Slater, J. H.; Weightman, A. J.; Hall, B. G. Mol. Biol. Evol. 1985, 2, 557-567.
16. Amy, P. S.; Shulke, J. W.; Frazier, L. M.; Seidler, R. J. Appl. Environ. Microbiol. 1985, 49, 1237-1245.
17. Pemberton, J. M.; Fisher, P. R. Nature (London) 1977, 277, 732-733.
18. Don, R. H.; Pemberton, J. M. J. Bacteriol. 1981, 145, 681-686.

19. Ghosal, D.; You, I.-S. Mol. Gen. Genet. 1988, 211, 113-120.
20. Kilbane, J. J.; Chatterjee, D. K.; Karns, J. S.; Kellogg, S. T.; Chakrabarty, A. M. Appl. Environ. Microbiol. 1982, 44, 72-78.
21. Tomasek, P. H.; Frantz, B.; Sangodkar, U. M. X.; Haugland, R. A.; Chakrabarty, A. M. Gene 1989, 76, 227-238.
22. Mulbry, W. W.; Karns, J. S.; Kearney, P. C.; Nelson, J. O.; McDaniel, C. S.; Wild, J. R. Appl. Environ. Microbiol. 1986, 51, 926-930.
23. Serdar, C. M.; Gibson, D. T.; Munnecke, D. M.; Lancaster, J. H. Appl. Environ. Microbiol. 1982, 44, 246-249.
24. Chaudhry, G. R.; Ali, A. N.; Wheeler, W. B. Appl. Environ. Microbiol. 1988, 54, 288-293.
25. Mulbry, W. W.; Kearney, P. C.; Nelson, J. O.; Karns, J. S. Plasmid 1987, 18, 173-177.
26. Mulbry, W. W.; Karns, J. S. J. Bacteriol. 1989, 171, 6740-6746.
27. Serdar, C. M.; Murdock, D. C.; Rohde, M. F. Bio/Technology 1989, 7, 1151-1155.
28. Steiert, J. G.; Pogell, B. G.; Speedie, M. K.; Laredo, J. Bio/Technology 1989, 7, 65-68.
29. Stalker, D. M.; McBride, K. J. Bacteriol. 1987, 109, 955-960.
30. Mueller, J. G.; Skipper, H. D.; Lawrence, E. G.; Kline, E. L. Weed Science 1989, 36, 96-101.
31. Tam, A. C.; Behki, R. M.; Kahn, S. U. Appl. Environ. Microbiol. 1987, 53, 1088-1093.
32. Tomasek, P. H.; J. S. Karns. J. Bacteriol. 1989, 171, 4038-4044.
33. Derbyshire, M. K.; Karns, J. S.; Kearney, P. C.; Nelson, J. O. J. Agric. Food Chem. 1987, 35, 871-877.
34. Macario, A. J. L.; Conway de Macario, E. Monoclonal Antibodies against Bacteria; Academic: Orlando, Fl. 1985.
35. Schmidt, E. L. Bull. Ecol. Res. Comm. (Stockholm) 1973, 17, 67-76.
36. Grunstein, M.; Hogness, D. Proc. Nat. Acad. Sci. USA. 1975, 72, 3961-3965.
37. Southern, E. M. J. Mol. Biol. 1975, 98, 503-517.
38. Holben, W. E.; Jansson, J. K.; Chelm, B. K.; Tiedje, J. M. Appl. Environ. Microbiol. 1988, 54, 703-711.
39. Steffan, R. J.; Goksoyr, J.; Bej, A. K.; Atlas, R. M. Appl. Environ. Microbiol. 1988, 54, 2908-2915.
40. Ogram, A.; Sayler, G. S.; Barkay, T. J. Microbiol. Methods 1987, 7, 57-66.
41. Mullis, K. B.; Faloona, F. A. Methods Enzymol. 1987, 155, 335-350.
42. Chaudhry, G. R.; Toranzos, G. A.; Bhatti, A. R. Appl. Environ. Microbiol. 1989, 55, 1301-1304.
43. Mulbry, W. W.; Karns, J. S. Appl. Environ. Microbiol. 1989, 55, 289-293.

RECEIVED January 30, 1990

Chapter 12

Response of Microbial Populations to Carbofuran in Soils Enhanced for Its Degradation

R. F. Turco[1] and A. E. Konopka[2]

[1]Department of Agronomy, Purdue University, West Lafayette, IN 47907
[2]Department of Biological Sciences, Purdue University, West Lafayette, IN 47907

The size of the microbial populations able to rapidly degrade carbofuran in soils enhanced for its degradation were enumerated by means of substrate addition and fumigation. Use of these techniques followed unsuccessful attempts to enumerate the population using plate or direct counts in the enhanced soils. Overall biomass size declined following application of carbofuran. No biomass suppression was observed in the non-enhanced soils and implies this suppression may be related to the formation of metabolites such as carbofuran-phenol or methylamine. In the enhanced soils, 6% of applied pesticide was initially incorporated into biomass carbon. This contrasts with 0.87% incorporation in the non-enhanced soils. After 15 days there was complete loss of the pesticide; at this time the biomass contained 2% of the applied material.

Enhanced biodegradation of pesticides constitutes an extreme in microbial exploitation of a soil-applied material. We recognize a soil that is enhanced for the degradation of carbofuran as one where pesticide losses are greater than 75% of the applied material within the 10 days following application. This time frame was derived from reports on the rapid loss of carbofuran (1-4). A typical response of an enhanced soil to the application of carbofuran is shown in Fig. 1 (5). The degradation of the applied pesticide is postulated to occur in a multistage process (Fig. 2). Most efforts to understand the impact of carbofuran on the microbial population, or to define the size of the enhanced biomass, have centered on following changes in numbers or types of total or pesticide-degrading microorganisms as estimated by plate counts or MPN techniques. Few efforts have described the population changes that occur during the period of rapid carbofuran degradation, as most work has focused on the kinetics of pesticide degradation in non-

0097-6156/90/0426-0153$06.00/0

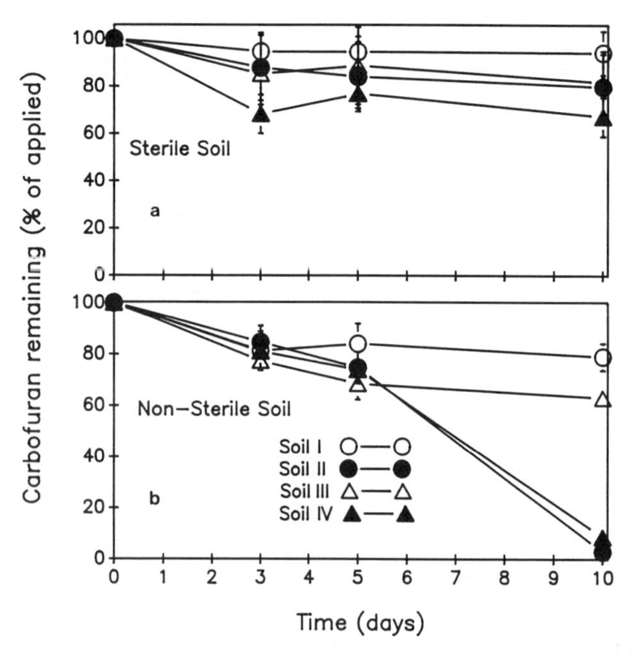

Figure 1. Conversion of carbofuran (100 μg g^{-1} soil) in sterile (^{60}CO irradiated) and non-sterile soils. Each bar is one standard error. (Reprinted with permission from Ref. 5. Copyright 1990 Pergamon Press.)

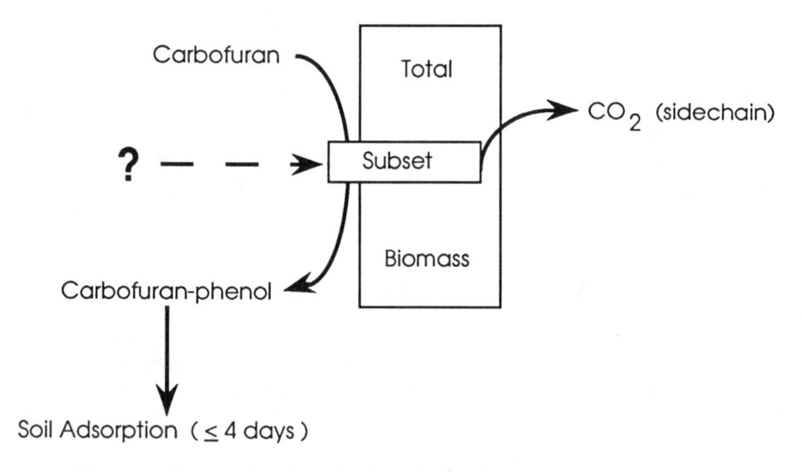

Figure 2. Proposed pathway for the microbial conversion of carbofuran.

enhanced soils (Table I). Duah-Yentumi and Johnson (6) showed carbofuran to have little effect on the distribution of types or total size of microbial populations in soils exposed to single or repeated applications of the pesticide. This work confirmed earlier work (7) that showed carbofuran to have little effect on non-target organisms, but contradicted work of Mathur et al. (8) that showed carbofuran to stimulate the bacterial and actinomycetal population. In all cases cited above, the soils were not considered enhanced and soil microbial population estimates were made from plate counts.

Table I. Attempts to understand microbial population changes resulting soil from application of organics

Compound	Effect	Magnitude	Method	Reference
2,4-D	INC	10-1000 (x)	MPN	31
2,4-D	INC	10-20 (x)	Plate and ^{14}C-CO_2	32
2,4-D & MCPA	INC			33
Captan, Thiram Vedasan	DEC	1.4 (x)	Substrate Addition	18
Carbofuran	INC	Slight	Plate	7
Carbofuran	NE	---	Fumigation	6
Carbofuran	INC	NR	Plate	34
Carbofuran	INC	1.4 (x)	Plate and CO_2	8
Carbofuran	INC*	10-100 (x)	Plate	9
EPTC*				
(6 mg)	NE	---	Plate ^{14}C-MPN	
(60 mg)	INC	10 (x)	^{14}C-MPN	19
HCCH	INC	1-10 (x)	Substrate Addition	17
Phenolic Acids	INC	1-10 (x)	Fumigation, Substrate Addition	16

* = repeat application, INC = increase, NE = no effect, DE = decrease

Microbial Populations and Enhanced Carbofuran Degradation

We are developing an understanding of how microbial populations in 'enhanced' soils respond to carbofuran. Our efforts have centered around two sets of soils in which enhanced carbofuran degradation has been demonstrated (5). We have estimated the microbial population changes in these two systems.

Our initial studies were conducted using plate counts. For plate count media we have used soil extract agar, starch glycerol agar, Henrici and carbofuran-agar (carbofuran as carbon and/or nitrogen source). Little difference in total microbial numbers were evident following treatment of the enhanced or non-enhanced soils with carbofuran. Use of carbofuran as sole carbon source in the plating media failed to improve recovery of degraders from soil. We have isolated over 400 colonies from different solid media containing carbofuran. Following transfer to liquid culture, we have been unable to detect carbofuran degradation within a time period of 10 days or longer. Our findings are similar to the poor recoveries of pesticide degrading bacteria from enhanced soil, reported by (9). However, in recent work we have used a chemostat to select carbofuran-degrading bacteria, and have isolated four carbofuran-degrading strains.

We adapted the soil recovery method of Ramsey (10) and utilized acridine orange direct counts to study the changes in total cell numbers directly in soil. The few consistent changes or shifts in microbial numbers we have observed with this method (Table II), following pesticide application, are consistent with our previous plate count findings. However, this is not completely unexpected as our work has shown that utilization of carbofuran in these soils is limited to side chain hydrolysis that occurs within five days of pesticide application (Fig. 3a,b) (5). Subsequent incubation of soils for 10 weeks with ^{14}C-carbofuran (100 μg carbofuran g^{-1} soil) have shown that total CO_2 evolved from soils was not significantly different for any of the soils (Fig. 4b). Degradation as indicated by ^{14}C-CO_2 evolution from ring labeled carbofuran was minimal during the 10 weeks (Fig. 4a). This extended our earlier conclusion that the enhanced mechanism is limited to removal of the sidechain, occurs in the first 5 days, and is followed by adsorption of the metabolite (carbofuran phenol) onto the soil surface (Fig. 3 and 5). This limits the amount of useable carbon soil microorganisms can easily extract from carbofuran.

Table II. Changes in bacteria numbers in soil as estimated by AODC

	(μg Carbofuran) g^{-1} soil	Time (days) 0	5	10
		---- log cell g^{-1} soil ----		
Soil	0			
I		9.27	9.36	9.41
II		9.28	9.60	9.77
III		9.62	9.75	9.62
IV		9.44	9.42	9.14
	100			
I		9.32	9.53	9.46
II		9.30	9.46	9.82
III		9.40	9.66	9.35
IV		8.97	9.25	8.92

If the entire structure of carbofuran was usable as a nutrient source, a 100 μg g^{-1} application to soil would supply 65 μg of C and 6 μg of N. However, the nutrient contribution of carbofuran appears to be limited to the carbonyl group (11). Our estimates have shown that if carbofuran is serving as a carbon source, taking into

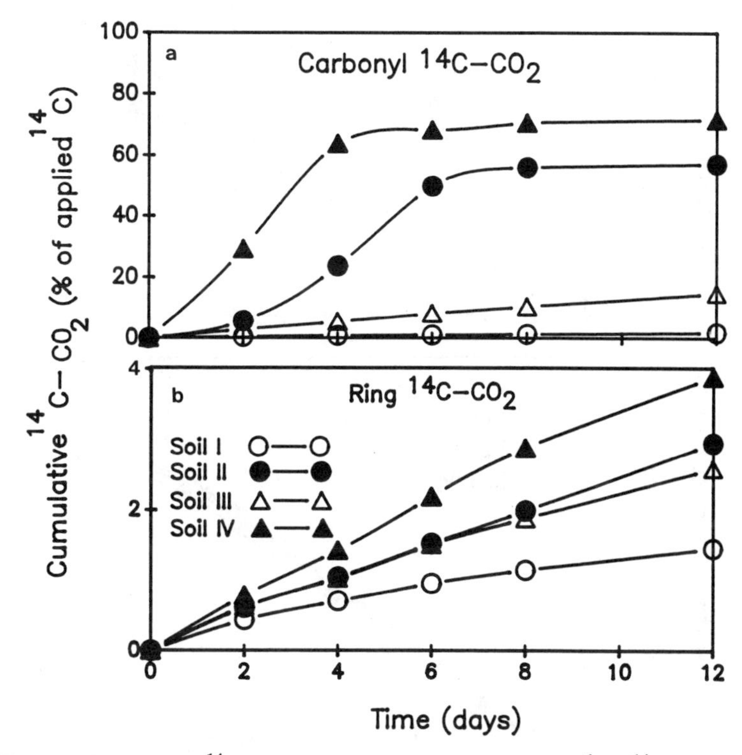

Figure 3. Evolution of $^{14}C\text{-}CO_2$ from soils treated with 100 $\mu g\ g^{-1}$ (a) ^{14}C-carbonyl or (b) ^{14}C-ring carbofuran. (Reprinted with permission from Ref. 5. Copyright 1990 Pergamon Press.)

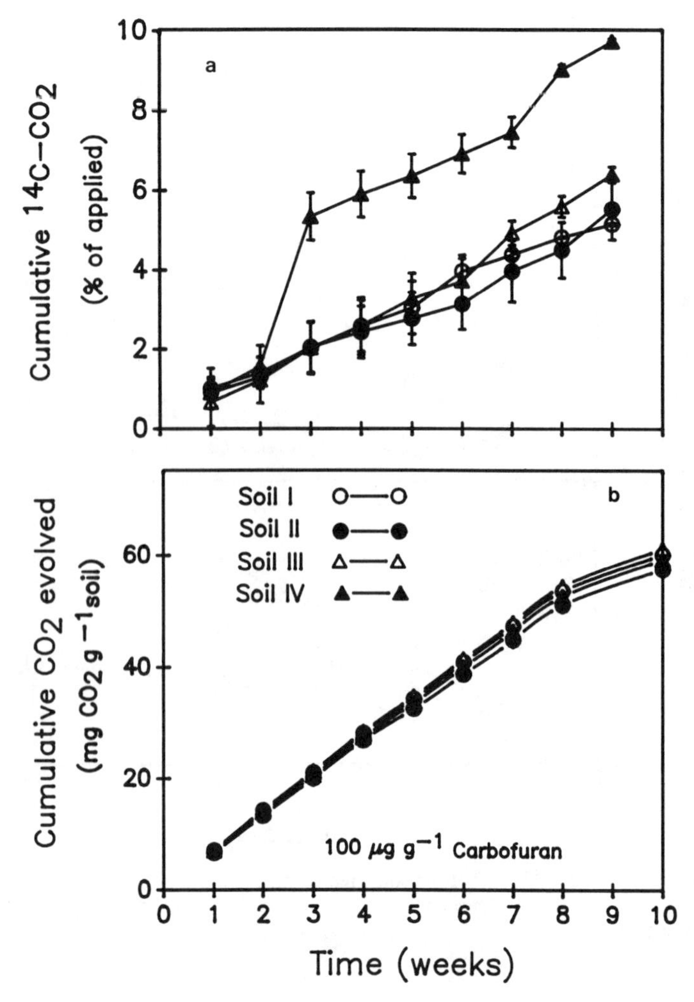

Figure 4. Evolution of (a) $^{14}C\text{-}CO_2$ or (b) CO_2 from soils treated with 100 $\mu g\ g^{-1}$ carbofuran and incubated 10 weeks. (Adapted from Ref. 24.)

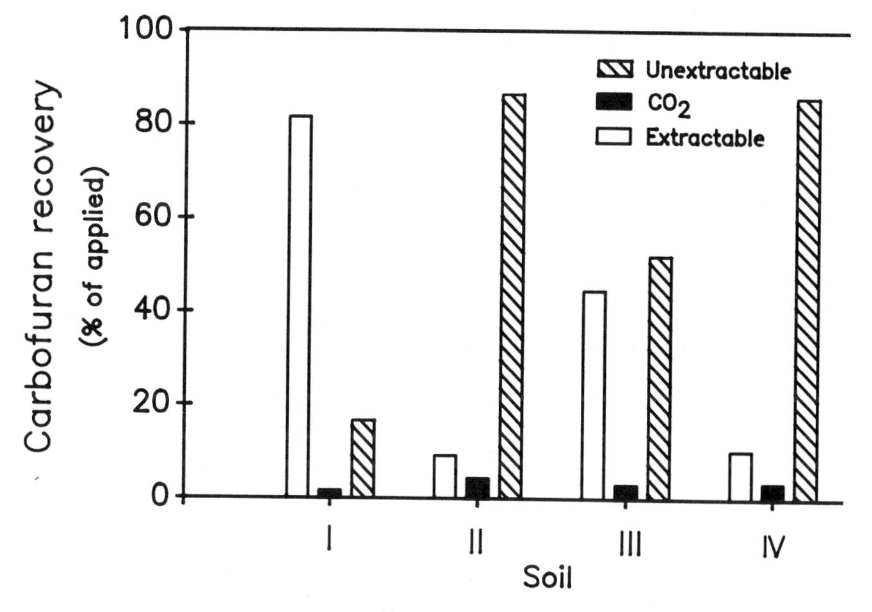

Figure 5. Percentage distribution of ^{14}C-radioactivity in three fractions from four soils after treatment with ^{14}C-ring carbofuran. (Reprinted with permission from Ref. 5. Copyright 1990 Pergamon Press.)

account loss of CO_2, the side chain would support about 5.8×10^5 microbes g^{-1} soil (assumes 1×10^8 bacteria mg^{-1} biomass). Direct counts indicate that the average population size is 1.9×10^9 cells gr^{-1}. A 5.8×10^5 gr^{-1} soil change is less than 0.05% increase in total numbers. Simple enumerations have failed to elaborate the subtle changes that are occurring. A better approach is to estimate the influence of carbofuran on the microbial biomass in the system.

Biomass Estimations

The microbial biomass is defined as the weight or standing crop of living cells in a given soil. Between 2-3% of the organic C in a soil can be associated with the biomass (12). Within this unit or crop, there are members that are more or less affected by a pesticide (Table I). Why a soil becomes enhanced for pesticide degradation has never been defined. It is clear, however, that the transition to an enhanced system is coupled to changes in the population. This results in a decreased lag period for pesticide degradation and the ability of the population to maintain the trait.

We have applied the substrate response method (13) to the study of the biomass in enhanced soils (5). A 100 μg g^{-1} soil application of carbofuran reduced the size of the active biomass in the enhanced soils as compared to either the untreated control or the treated non-enhanced soils (Fig. 6a,b). A return of the biomass to a size similar to the untreated control may be coupled to uptake of the side chain and adsorption of the carbofuran phenol by the soil (Fig. 6c). No difference in the response of the biomass within the non-enhanced soil was found. This indicates that carbofuran itself is not toxic, but one of the primary metabolite (carbofuran phenol or methylamine) may have a suppressive nature. Using a microtox test developed for Photobacterium phosphoreum, (14) the toxicity of carbofuran, carbofuran phenol and methylamine has been evaluated. They reported that concentrations in excess of 20 μg ml^{-1} carbofuran were needed to suppress 50% of the light production by P. phosphoreum. Moreover, they reported methylamine to be twice as toxic as carbofuran phenol but less toxic than carbofuran. Kale and Raghu (15) reported that carbofuran phenol was inhibitory to soil microbial numbers and soil microbial respiration. More importantly the effect was transient and most likely diminishes as carbofuran phenol is adsorbed to soils. The carbofuran-phenol is not degraded, but its adsorption protects the biomass and may prevent further degradation. This repression contrasts with the findings of Sparling et al. (16), where a series of phenols were applied to soil and biomass size increased over a 10 day period. However, these soils had not been previously exposed to phenol. Following addition of hexachlorocyclohexane or prometrine to soil, Ana Yeva et al. (17) reported an increase in biomass size. They had utilized the substrate addition method to quantify the changes in the population. Their work was conducted with formulated product and the effect of carrier materials was not differentiated. Anderson et al. (18) applied the respiratory response method to study the impact of three fungicides on microbial biomass. They found that even at very low application rates (5 μg g^{-1}) the materials radically suppressed and altered the population. The suppression was coupled to a shift in the dominant population away from fungi towards bacteria. Duah-

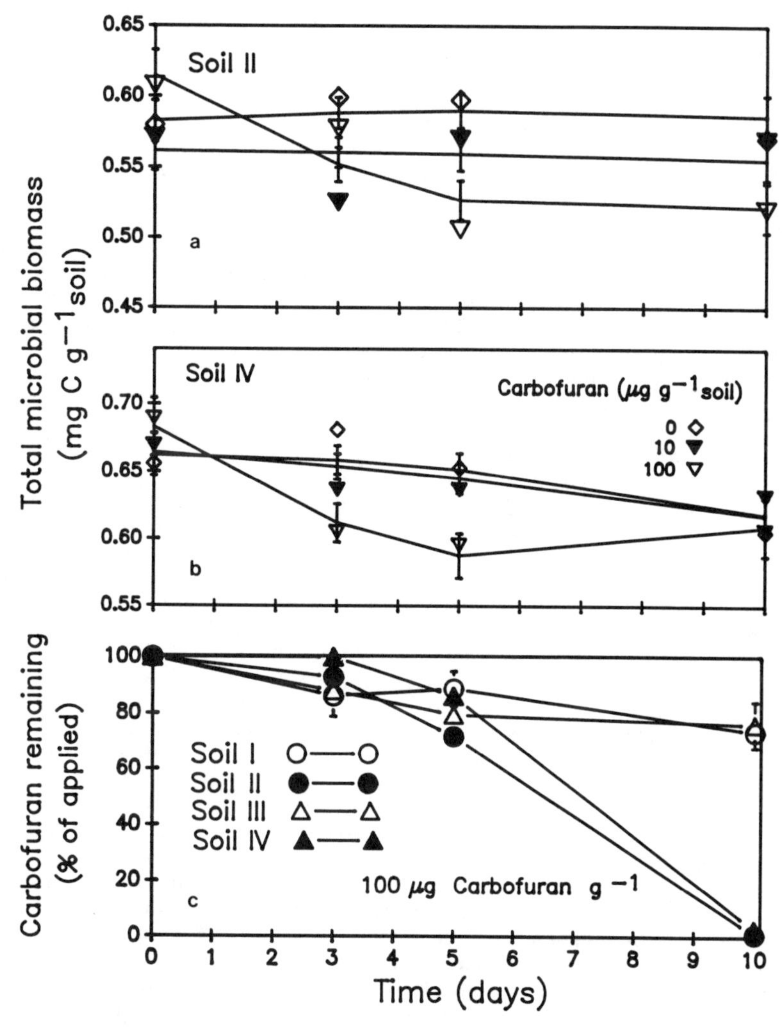

Figure 6. Changes in total biomass size for Soils II(a) and IV(b) following addition of 0, 10, or 100 μg carbofuran g^{-1} soil. Bar is on LSI ($P<0.05$) for regression. (c) Recovery of carbofuran from the biomass study soils, 100 μg g^{-1} starting application. Bar is one standard error. (Reprinted with permission from Ref. 5. Copyright 1990 Pergamon Press.)

Yentumi and Johnson (6) reported that frequency of pesticide applications shifted the population, and repeated applications of paraquat lowered the size of the microbial biomass as determined by plate counts. Moorman (19) approached the estimation of pesticide degrader biomass size using an MPN procedure. He found that following high rate (60 $\mu g\ g^{-1}$) applications of EPTC in soils enhanced for EPTC degradation, the enhanced biomass was increased as compared to the control soil.

Pesticide-Degrading Biomass Estimation

The substrate addition method coupled to a pesticide application defines how the biomass as a whole is responding. It does not define the fraction of the population that is degrading the pesticide. Efforts to understand the size of the specialized biomass have relied on the use of selective agar or MPN estimations (19). However, our findings as well as others (9) have shown the plating approach to be of limited value. Most probable number procedures give an estimate of the population that can degrade the pesticide and do not follow the general biomass. A more direct approach is to quantify the biomass size in situ.

Soulas et al. (20) proposed a modification of the work of Kassim et al. (21-22) as a way to monitor a specialized soil population. In essence, the biomass fumigation procedure of Jenkinson and Powlson, (23) is altered to follow incorporation of radiolabel into the specialized portion of the population. Following an incubation with the pesticide, the soils are fumigated and the fraction of the biomass containing label is determined. Soulas et al. (20) working with 2,4-D reported that 13.4% of the applied materials were incorporated into soil biomass. From an initial application rate of 3.7 mg 2,4-D kg^{-1}, the pesticide-specific biomass increased in size to 0.372 mg kg^{-1} soil or 3.7×10^4 microbe g^{-1} soil. Kassim et al. (21) showed that after 12 weeks, a readily usable material like acetate would be taken up into as much as 70.4 % of the biomass. However, after this length of incubation the likelihood that other forms of carbon derived from acetate are actually cycling in the biomass is great.

We used a specific labelling approach to estimate the size of the degrader population in our enhanced soil that could use carbofuran as a carbon source (24). Our system differs somewhat from that described by Soulas et al. (20) in that only carbofuran side-chain carbon appears to be available to the biomass. This may reflect the fact that carbofuran is serving as both a nitrogen and carbon source (11). Less than 3% of the ring carbon, as compared to 74 % of the side chain is evolved as CO_2 within the 10 day time period. Because carbofuran ring breakage is so limited, only incorporation of carbonyl-labeled C from carbofuran was studied. To insure maximal activity in the population, soils I and II were pretreated with 25 μg carbofuran g^{-1} soil and adjusted to -0.33 bar water content 10 days before application of the ^{14}C materials. A total of 100 ug carbofuran g^{-1} soil was applied and the soils incubated at 25°C (24). At days 1, 5, 10 and 15, three soil samples were removed, fumigated with $CHCl_3$, incubated for 24 h, and the fumigant removed with evacuation. At sampling, the NaOH traps in all other samples were renewed. Fumigated samples were transferred to 2-l glass jars and the evolved CO_2 and $^{14}C-CO_2$ trapped in NaOH. Traps were exchanged weekly for one month. Biomass size was calculated using the technique of Soulas et al. (1984) where $^{14}C-CO_2$ or CO_2 evolved in the first 14 days is corrected by subtracting the $^{14}C-CO_2$ or CO_2evolved from day 21 to 27.

Incorporation of the ^{14}C-C into the biomass averaged 0.87% of applied ^{14}C for non-enhanced soil I. Incorporation of ^{14}C-C averaged 3.76 % in the enhanced soil II. Overall the biomass size was about 0.77 mg Cg^{-1} soil. Incorporation of ^{14}C-C was 6.0 % of applied ^{14}C-carbofuran at day 1 diminishing to 2.0 % by day 15. In the enhanced soils, total biomass size dropped from day 1 to 5 and recovered by day 10 (Fig.7). This corresponds to the changes in biomass size observed with the substrate addition method (5). A 100 µg g^{-1} addition of carbofuran would supply 10.28 µg C g^{-1} from the side chain. At day 1, approximately 17% of the applied materials had evolved as CO_2, leaving 8.53 µg g^{-1} for use in the biomass. Of the remainder, 6% was incorporated into biomass. This corresponds to about 1.02×10^5 cells g^{-1} soil. By day 5, 65% of the pesticide had been lost as CO_2 and biomass had declined to contain only 5% of the applied materials. By day 15, the biomass containing ^{14}C had declined to less than 2% of the applied pesticide. This corresponds to about 5.1×10^3 cells g^{-1} soil. The initial population size was most likely elevated in response to the 25 µg g^{-1} pre-application of carbofuran (Fig. 7). Of more significance was the fact that the population was induced for the enzymes needed for the degradation process (25). The lack of a lag phase is indicative of this induction.

Utilization and uptake of the pesticide was rapid within the first 5 days. At day five utilization and uptake decline as the substrate became limited. The population dropped in response to the loss of the carbon supply. Soulas (26) had developed a theoretical model that predicted a short lag phase and a rapid increase and subsequent decline in biomass size in systems adapted for growth on a particular pesticide. Our data conforms to the curves predicted by the model. However, the model indicated complete conversion of the applied pesticide. This was not the case in the enhanced soils. Buildup of the metabolite may have decreased the activity of the population. This was not considered in Soulas (26) model. The estimated 5.14×10^3 cells gr^{-1} soil found at day 15 implies that the degradation ability is not widespread and is harbored in a distinct subsection of the population.

Conclusions

The substrate addition method for biomass estimation relies on the fact that the biomass response to the introduction of the pesticide is reflected in the response to the glucose. Fumigation relies on the conversion of the specifically labeled biomass to $^{14}C-CO_2$. From both estimations it is clear that the size of microbial biomass that can use the carbon in the carbofuran side-chain is small. Karns et al. (11) has shown that carbofuran may serve as nitrogen source for some organisms. Our estimates would overlook this N utilization. It is likely that organisms that utilize methylamine, the primary degradation product, for nitrogen would overlook this source of carbon. Data on bacterial use of methylamine is, however, limited.

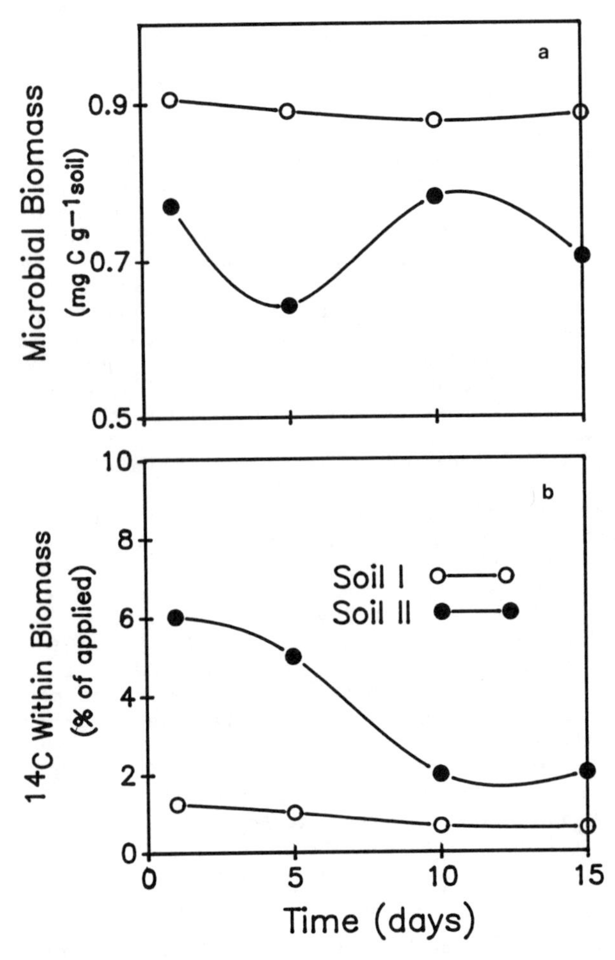

Figure 7. (a) Biomass size as indicated by fumigation; (b) incorporation of ^{14}C into biomass. (Adapted from Ref. 24.)

Results from either of the two systems are at best estimates of the process that is occurring. A more direct estimation would be to follow the changes in the genetic materials regulating the bacterial response to pesticide application.

Recently it has become possible to extract from soil restrictable DNA (27-29). If a gene probe for a specific function was derived from an isolate taken from the enhanced soil (30), then changes in the frequency of that gene sequence in a population of bacteria could be established. This information would allow conclusions about microbial growth and function in soil. The method would not rely on subculturing from the treated soil but provide direct insight into changes in the population.

Acknowledgment

This work was supported by a Grant from the North Central Region Pesticide Impact Assessment Program. Paper No. 12,353 of the Purdue University Agricultural Experiment Station Series.

Literature Cited

1. Felsott, A.; Maddox, J. V.; Bruce, W. Bull. Environ. Contam. Toxicol. 1981, 26, 781-788.
2. Suett, D. L. Crop Prot. 1986, 5, 165-169.
3. Harris, C. R.; Chapman, R. A.; Harris, C.; Tu, C. M. J. Environ. Sci. Health Part B 1984, 19, 1-11.
4. Harris, C. R.; Chapman, R. A.; Morris, R. F.; Stevenson, A. B. J. Environ. Sci. Health Part B 1988, In Press.
5. Turco R. F.; Konopka, A. Soil Biol. Biochem. 1990a, In Press.
6. Duah-Yentumi, S.; Johnson, D. B. Soil Biol. Biochem. 1986, 18, 629-637.
7. Ingham, E. R.; Coleman, D. C. Microb. Ecol. 1984, 10, 345-358.
8. Mathur, S. P.; Hamilton, H. A.; Greenhalgh, R.; MacMillan, K. A.; Khan, S. U. Can. J. Soil Sci. 1976, 56, 89-96.
9. Dzantor, E. K.; Felsot, A. S. J. Environ. Sci. Hlth B 1989, (In press).
10. Ramsay, A. J. Soil Biol Biochem 1984, 16, 475-481.
11. Karns, J. S.; Mulbey, W. W.; Nelson, J. O.; Kearney, P. C. Pest. Biochem. Physiol. 1986, 25 211-217.
12. Jackson, D. S.; Ladd, J. N. Soil Biochemistry 1981, 5, 415-473.
13. Anderson, J. P. E.; Domsch, K. H. Cand. J. Microb. 1975, 21, 314-322.
14. Somasundaram, L.; Coats, J. R., Racke, K. D.; Stahr, H. M. Bull. Environ. Contam. Toxicol. 1989, In press.
15. Kale, S. P.; Raghu, K. Chemosphere 1989, 18, 2345-2351.
16. Sparling, G. P.; Ord, B. G.; Vaughn, D. Soil Biol. Biochem. 1981, 13, 455-460.
17. Ana'Yeva, N. D.; Strekozov, B. P.; Tyuryukanova, G. K. Agrokhimiya 1986, 5, 84-90
18. Anderson, J. P. E.; Armstrong, R. A.; Smith, S. N. Soil Biol. Biochem. 1981, 13, 149-153.
19. Moorman, T. B. Weed Science 1988, 36, 96-101.

20. Soulas, G.; Chaussod, R.; Verguet, A. Soil Biol. Biochem. 1984, 16, 497-501.
21. Kassim, G.; Martin, J. P.; K. Haider Soil Sci. Soc. Am. J. 1981, 45, 1106-1118.
22. Kassim, G.; Stott, D. E.; Martin, J. P.; Haider, K. Soil Sci. Soc. Am. J. 1982, 46, 305-309.
23. Jenkinson, D. S.; Powlson, D. S. Soil Biol. Biochem. 1976, 167-177.
24. Turco R. F.; Konopka, A. 1990b, In Preparation.
25. Kaufman, D. D.; Katan, Y.; Edwards, D. F.; Jordan, E. G. Agricultural Chemicals of the Future; BARC Symposium 8; James L. Hilton, Ed.; Rowman and Allanheld: Totowa.
26. Soulas, G. Soil Biol. Biochem. 1982, 14, 107-115.
27. Holben, W. E.; Jansson, J. K.; Chelm, B. K.; Tiedje, J. M., Appl. Environ. Microbiol. 1988, 54, 703-711.
28. Sayler, G. S.; Sheilds, M. S.; Tedford, E. T.; Breen, A.; Hooper, S. W.; Sirotkin, K. M.; Davis, J. W. Appl. Environ. Microbiol. 1985, 49, 1295-1303.
29. Steffan, R. J.; Goksoyr, J.; Bej, A. K.; Atlas, R. M. Appl. Environ. Microbiol. 1988, 54, 2908-2915.
30. Pace, N. R.; Stahl, D. A.; Lane, D. J.; Olsen, G. J. Adv. Microb. Ecol. 1986, 9, 1-55.
31. Fournier, J. C.; Codaccioni, P.; Soulas, G. Repiguet C. Chemosphere 1981, 977-984.
32. Kunc, F.; Rybarova, J. Soil Biol. Biochem. 1983, 15, 141-144.
33. Loos, M. A.; Schlosser, I. F.; Mapham, W. R. Soil Biol. Biochem. 1979, 11, 377-385.
34. Williams, I. H.; Depin, H. S.; Brown, M. J. Bull. Environ. Cont. Toxic. 1976, 15 244-249.

RECEIVED February 8, 1990

Chapter 13

Adaptation of Microorganisms in Subsurface Environments

Significance to Pesticide Degradation

Thomas B. Moorman

Southern Weed Science Laboratory, Agricultural Research Service, U.S. Department of Agriculture, Stoneville, MS 38776

The degradation of pesticides in subsurface soils and aquifer environments proceeds at relatively slow rates. The lack of adaptation to pesticides in the subsurface environment may be due to both physiological and environmental contraints. Adaptation of microorganisms to pesticides in surface soils is a process that includes examples of population growth and enzyme induction as contributing factors. Introduced microbial agents have some potential for detoxification of pesticides in groundwater, but the low level of nutrients in aquifers contaminated with trace levels of pesticides limits microbial activity and may reduce the effectiveness of introduced microorganisms.

Groundwater aquifers are an important national resource providing a significant portion of our population with clean drinking water as well as water for agricultural and industrial purposes. The U.S. Environmental Protection Agency has recently reported the presence of 46 different pesticides in the groundwater of 26 states as being attributable to agricultural activities (1). The majority of these chemicals were herbicides. In addition, numerous other xenobiotics of industrial origin have been detected as groundwater contaminants. Accordingly, there is considerable research interest in the transport and biodegradation of pesticides and other xenobiotics in subsurface environments. The objective of this discussion is to review the factors affecting the adaptation of microorganisms to pesticides with the intent of evaluating the prospects for enhancing the activity of pesticide-degrading microorganisms in the subsurface environment. Certain research indicates that bioremediation approaches to the cleanup of industrial toxicants is possible (2), but there may be additional obstacles to the use of these approaches for pesticides.

Pesticides enter the subsurface environment through three possible routes: (a) as nonpoint-source pollutants moving through the soil and vadose zone; (b) as point-source pollutants from spills or disposal sites; (c) as contaminants in surface water entering ground water at aquifer recharge sites. The relative contribution of entry through recharge areas is not well known. Groundwater contaminated by nonpoint-source typically contains pesticides at levels below 100 ppb (1). Pesticide concentrations tend to be greater after point-source contamination. As pesticides leach through the subsurface soil and vadose zone materials they are degraded by soil microoganisms, but the rates of degradation tend to decline with increasing depth. In aquifer materials degradation rates may be very slow for pesticides.

Adaptation describes the process whereby microorganisms gain an increased ability to degrade a compound as a result of prior exposure to the compound. The adaptation process itself may result from genetic, physiological, or ecological changes in the microbial community. Microbial adaptation to pesticides is most obvious in soils which have received repeated applications of pesticides, as indicated by the expanding number of pesticides that are subject to the enhanced biodegradation phenomenon. For biodegradation to occur in the natural environment certain criteria must be met: (a) the existing microbial population must contain the genes coding for the appropriate enzymes and (b) those genes must be expressed. There is a quantitative relationship between populations of degraders, the concentration of pesticide in the environment, and the kinetics of degradation (3). Conceivably, changes in any one, or combinations of these processes, could be responsible for the adaptation phenomenon.

Unfortunately, the term adaptation has been used to describe results that are functionally similar, but are probably dissimilar in their causes and mechanisms. Presumably, the adaptation phenomenon does take place in subsurface environments under some conditions. However, experimental data concerning adaptation to pesticides in subsoils and aquifer environments are generally scanty. Evidence for adaptation includes the shortening of lag periods before degradation or mineralization begins, changes in degradation kinetics, or comparative evidence that previous exposure to the pesticide results in greater rates of degradation (Figure 1). The same factors (degrader populations, genetic capabilities, substrate concentrations) that limit adaptation in the surface environment are potentially limiting in the subsurface environment.

Mechanisms of Adaptation

Gene Transfer and Expression. This topic is a subject of considerable interest. A detailed review and analysis of the genetics of pesticide degradation was prepared by Karns (4) as part of this volume. The genetic and physiological regulation of the pesticide degradation may be important components of the adaptation process. If the genes necessary for pesticide degradation are inducible, then a lag period exists before degradation begins. Isolated microorganisms studied *in vitro* have lag-times ranging from a few hours to several days (5, 6, 7, 8). Presumably the induction process is longer in soil systems, based on the lower relative activity of microorganisms in soil systems. The common use of first-order kinetics to describe pesticide degradation in soils often reflects the dissipation of pesticide by physical and chemical processes occurring simultaneously to the biodegradation process. Also, inadequate sampling during the initial period following pesticide addition to the soil may obscure short lag-periods.

Factors affecting the induction of genes coding for degradative enzymes can also affect the pattern of degradation. For instance, the degradation of 2,4,5-T by *Pseudomonas cepacia* was inhibited when cells were cultured in the presence of succinate, glucose, or lactate, indicating catabolite repression (6). Pesticides may differ substantially in their ability to induce degradative enzymes. The aryl acylamidase from *Bacillus sphaericus*, which hydrolyzes a wide variety of phenylamide pesticides, was also induced by a number of different compounds. However, some of these compounds elicited only a tiny fraction of the enzyme

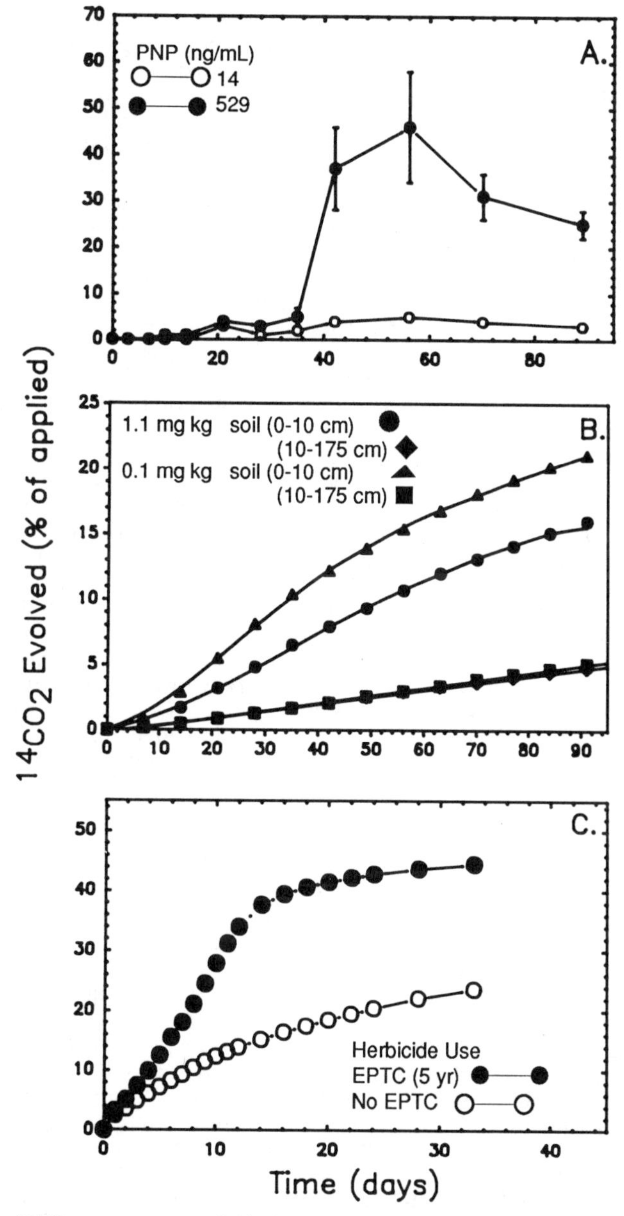

Figure 1. Different patterns of biodegradation resulting from microbial adaptation. (A) Adaptation to *p*-nitrophenol in Lulu aquifer samples at 529 ng/mL, but not at 14 ng/mL (43). (B) Adaptation to the triazinone-ring of metribuzin, evidenced by an increasing mineralization rate over time in the surface soil, but not in the subsurface soils (35). (C) Adaptation to EPTC after long-term EPTC use (12). (Reproduced with permission from Ref. 12, 35, 43. Copyrights 1988, 1989, 1987: Weed Science Society of America, American Society of Agronomy, and American Society for Microbiology, respectively)

activity in comparison to linuron (7). This microorganism was unable to utilize these pesticides for growth, and chloroanilines accumulated in the culture media. Another bacterium (*Pseudomonas* sp.), mineralized greater amounts of ^{14}C-3,4-dichloroaniline when induced by 4-chloroaniline than when induced by 3,4-dichloroaniline (8). This was also demonstrated in a soil system. Adaptation has also been observed in anaerobic systems. Linkfield et al. (9) reported acclimation periods of 20 to more than 170 days for a variety of halogenated benzoates in anaerobic lake sediments. The periods required for adaptation (acclimation) were characteristically associated with chemical structure, and not with environmentally influenced variables such as population or nutrient status. It was concluded that the acclimation periods preceeding degradation of these compounds was due to induction of necessary enzymes. These results illustrate the fact that enyme induction and regulation remains a potential cause of adaptive responses in microorganisms.

Little is known about how long microorganisms remain induced, particularly in natural ecosystems. If enzyme induction is the sole mechanism explaining adaptation then it is logical to expect microbial populations to revert to the unadapted state once the substrate (or other inducer) is removed from the environment. A chloropropham-degrading *Pseudomonas alcaligines* added to soil lost the ability to degrade chloropropham after 7 days (10). In other studies, soils with microbial populations adapted to the herbicide EPTC remained adapted after a year without exposure to EPTC (11, 12). Degradation rates were reduced, but not to the levels found in unadapted soils.

There has been considerable speculation on the possible role of plasmid-mediated transfer of pesticide-degrading genes as a mechanism of adaptation. This may be particularly important in situations where microorganisms are subjected to intermittent exposure to the substrate, or in cases where long exposure times are necessary for adaptation. This may involve evolution of new enzymatic forms. In the case of 2,4-D, the plasmid-coded enzymes which degrade the chlorinated ring differ from the isozymes coded on the chromosome which degrade unchlorinated benzoic acids (13). It has not yet been demonstrated that the accelerated rates of 2,4-D degradation in soils repeatedly treated with 2,4-D are in fact due to spread of this plasmid (vs. growth of strains with the plasmid), but this remains an interesting possibility.

The importance of plasmid-mediated transfer of degradation genes depends to some degree on the maximal frequency of transfer in the natural environment and the selective pressure for the degradation genes. There is some indirect evidence to suggest that plasmids carry degradation genes in groundwater bacteria. Plasmids were found in a greater frequency in groundwater bacteria from a contaminated site than in bacteria from pristine aquifers (14). However, it was not determined if the plasmids carried degradative genes. Jain et al. (15) determined that the TOL plasmid (toluene/xylene catabolism) was stable in the absence of selection pressure (toluene contamination) in groundwater microcosms. Interestingly, there was no detectable background hybridization to TOL in the control microcosms, indicating that the indigenous groundwater bacteria did not carry this plasmid. In contrast, the TOL^+ population in uncontaminated freshwater sediments was greater than 10^3/g sediment (16).

Dynamics of Degrader Populations. The potential for growth of microorganisms in response to pesticides or other xenobiotics would appear to be a powerful selective pressure favoring strains that carry genes necessary for metabolism. Population growth has often been put forward as an explanation for adaptive responses. The growth of microorganisms in culture and in soil or water systems can be predicted using some variant of the Monod equation (17), where B is microbial biomass, S is substrate, u_m and K_s are the maximal growth rates and substrate affinity parameters, respectively, Y is the yield coefficient and m is the maintenance coefficient.

$$dB/dt = \frac{u_m[B][S]}{(Y(K_s+[S]))} - m[B]$$

There are potential problems with the extrapolation of this relationship into natural environments. Our knowledge of the range of u_m and K_s values for pesticides is certainly not complete enough to have broad predictive value. These equations generally describe growth when a pesticide (i.e. carbon source) is the limiting factor, but in nature microorganisms probably consume multiple substrates simultaneously. Also, in nutrient-limited environments such as soils, and to an even a greater extent unpolluted groundwaters, significant amounts of the available carbon may be utilized as maintenance energy. Experimental determinations of maintenance energy requirements in soils are so high as to exceed annual C inputs into surface soils, suggesting that much of the live biomass is dormant at any particular time (18). Maintenance coefficients for dormant biomass are far lower (19). The theoretical amounts of C necessary to support active microbial biomass are shown in Figure 2 in comparison to the C supplied by pesticide degradation. These sorts of calculations suggest that the supply of C as pesticide can at least partially fulfill requirements for maintenance and growth of pesticide-degrading bacteria, and may support growth of populations that are initially small.

How well do existing data support these premises? Populations of microorganisms capable of degrading pesticides have been measured in soils using a variety of techniques. Measurement of bromoxynil, 2,4-D, dalapon, MCPA and isofenphos degraders indicates that in most instances pesticide degraders comprise only a small fraction of the total microbial population (20, 21, 22, 23, 24, 25). Although population growth has been cited as one factor in the adaptation process in waters, sediments, and soils, growth may not be possible, or continue only at slow rates in response to pesticide applications. Presumably this is due to the quantity and quality of substrate (pesticide) applied. Racke and Coates (22) examined populations of isofenphos-degraders using a MPN technique that specifically detected microorganisms capable of mineralizing isofenphos. Populations following 2 yr of isofenphos use were above 6×10^3 degraders/g soil while populations were below detectable limits in similar soils without the isofenphos exposure (22). Moorman (12), using similar techniques, found no difference in populations of EPTC-degraders in soils with differing histories of EPTC use and degradation rates (Figure 1). Populations in all soils were above 5×10^4/g soil. EPTC-degraders increased in response to 60 mg/kg soil of EPTC, but not 6 mg/kg in the soil with previous exposure to EPTC, suggesting that low rates of EPTC would not sustain population increases. In other studies with carbamothioate herbicides, populations of actinomycetes and bacteria increased in some instances after repeated exposure to EPTC, vernolate, butylate, pebulate, and cycloate (26). After 31 years of exposure to 2,4-D, populations of degraders were roughly twice that of the untreated soil (20). In contrast, two soils treated with the relatively high concentration of 10 mg 2,4-D/g soil resulted in population increases of approximately 1000-fold (21). Schmidt and Geir (27) found that 0.1 mg/kg soil of dinitrophenol did not support increased DNP-degrader populations, but concentrations of 1 mg/kg or more did support growth above the low initial level of 5.6×10^3 degraders/g soil. Thus, the general pattern that emerges is that significant increases in degrader populations can occur in response to pesticides, particularly if initial populations are low. Pesticides applied at rates commonly used in agriculture are unlikely to support large populations of degraders for sustained periods.

Measurements of the incorporation of ^{14}C-labeled pesticides tend to support the concept of limited microbial growth from pesticide degradation. In order for a pesticide to support support substantial growth, concomitant levels of mineralization and incorporation into biomass should be observed. Studies with ^{14}C-2,4-D, labeled in either the acetic acid or ring structures show that more ^{14}C from the acetic acid is incorporated into biomass than from ring-labeled, depending upon soil type and the time after addition of the ^{14}C to the soil (Table 1). Even lower amounts of carbofuran were incorporated into biomass than were observed for 2,4-D. There was not a strong relationship between mineralization and incorporation into biomass.

Table 1. Incorporation of ^{14}C from ^{14}C-pesticides into microbial biomass in different surface soils

Substrate (Pesticide)	Soil[1]	Distribution of ^{14}C (% of applied)		
		Respired	Biomass	Time (d)
2,4-D (^{14}C-acetic)	Steinbeck l	80	15	180
	Sorrento l	87	14	180
	Fallbrook sl	32	3	180
	Greenfield sl	82	3	180
	Cambisoil	85	11	45
2,4-D (^{14}C-ring)	Steinbeck l	62	5	180
	Sorrento l	79	6	180
	Fallbrook sl	22	2	180
	Greenfield sl	67	2	180
	Cambisoil	47	12	45
Carbofuran	Norfolk ls	10	0.2	10

[1]Data compiled from references 28, 29 and 30.

Several interrelated factors complicate the relationship between growth and degradation: multispecies interactions, cometabolic degradation, and substrate utilization. If multispecies consortia are required for degradation, then the amount of pesticide incorporated into microbial biomass of a particular species may be even lower. Multispecies interactions have been investigated; Lappin et al. (31) isolated a consortium of 5 bacteria that metabolized the herbicide mecoprop in culture after a lag period of 20 hr. The growth of the consortium was in contrast to the failure of individual species to grow on mecoprop. The extent of these kinds of interactions in soils is essentially unknown.

Cometabolism, the degradation or transformation of pesticides by reactions which do not support microbial growth, has also been documented. The chloroacetanilide herbicides appear to be degraded by this means. Indigenous microorganisms in lake water transformed ^{14}C labeled alachlor and propachlor into several organic products, but no $^{14}CO_2$ was produced (32). Only small amounts of $^{14}CO_2$ were evolved from alachlor-treated soils, even though some soils had been pretreated with alachlor (33). In the same study up to 63% of the applied propachlor was mineralized in soil, indicating that this herbicide was metabolized directly. Studies with isolated microorganisms have generally resulted in production of metabolites from alachlor, but mineralization and growth does not occur (34). The enzymes involved in cometabolism may transform a number of related pesticides, such as the enzyme from *Pseudomonas alcaligenes* recently described by Marty et al. (5). The inducible, intracellular, amidase hydrolyzed the carbamate herbicides CIPC, BIPC, IPC and Swep with the subsequent accumulation of metabolites. In soil these metabolites are subsequently degraded with production of $^{14}CO_2$, presumably by other microorganisms. Clearly, the adaptation process in microorganisms that exhibit cometabolism, either singly or as part of a multispecies consortium, is a more complex process than in microorganisms that realize some metabolic advantage.

Adaptation in Subsurface Environments

In general, pesticide degradation in subsurface environments, even in shallow subsoils, proceeds at rates that are considerably lower than in surface soils. This is illustrated for the herbicide metribuzin in Figure 3. Although all studies showed a trend of increased persistence with depth, the variability between soil types is large. In many instances, the pesticide degradation can be described with kinetic models that do not account for a lag period (e.g. first-order kinetics). This may indicate that an adaptation period is not required for degradation or that short lag periods were not detected because of the sampling strategies that were used (35, 36). Furthermore, abiotic processes may also be involved in the degradation process, as in the case of metribuzin. In subsurface soils and groundwater the herbicides alachlor, 2,4-D, metribuzin (see Figure 1) and atrazine were not mineralized to any significant extent (33, 35, 37, 38). Approximately 60% of the alachlor added to a groundwater sample was converted to metabolites over 47 days (33). The absence of mineralization should be taken as evidence that these pesticides are not supporting significant growth of the populations, because the degradation is apparently by cometabolic means. While there is little kinetic evidence to show that microorganisms are adapting to these pesticides, adaptation cannot be discounted with a high degree of certainty for other pesticides.

Reduced microbial activity and lower metabolic diversity are likely reasons for the reduced biodegradation of herbicides in subsoils and groundwater aquifers. Although the total microbial population in most aquifers range between 10^4 and 10^7 cells/g aquifer material (39), the proportion of these populations that can degrade xenobiotics appears to be low. In a pristine aquifer from Lulu, Oklahoma, degrader populations measured by ^{14}C-MPN were less than 5 cells/g for chlorobenzene, napthalene, and toluene (40). Recent studies indicate that surface bacteria utilize greater numbers of carbon substrates than do deep groundwater bacteria (41). Presumably, more complex xenobiotics, such as herbicides, will be considerably more persistent in subsurface environments than in the surface soil due to the lack of degraders.

Research on the degradation of industrial pollutants indicates that growth of degraders and nutrient limitations both modulate the adaptive response in the subsurface. Many studies with contaminated aquifer materials, where long-

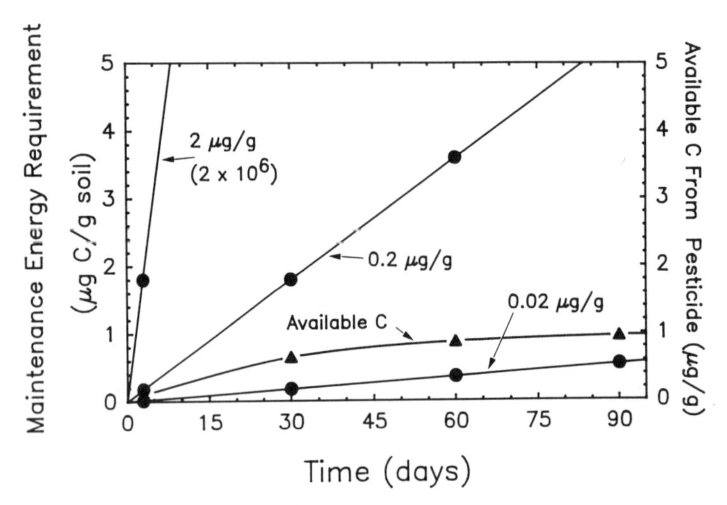

Figure 2. Theoretical estimates of C required to meet maintenance energy demands (solid circles) by different sized microbial populations calculated using a maintenance coefficient of 0.0126 μg C/μg biomass-C/hr derived from Anderson and Domsch (18). Cumulative C requirements are compared to the cumulative C supplied during first-order decomposition of a pesticide having a 20-day half-life applied at 1 μg C/g soil. Limited growth would be possible when carbon supply exceeded the maintenance requirement.

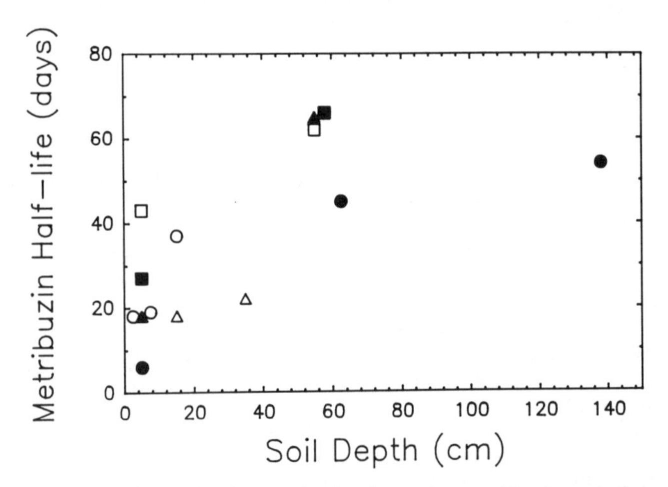

Figure 3. The effects of soil sample depth on the aerobic degradation of metribuzin in 6 different soils under comparable conditions in the laboratory. Different symbols indicate different soil types. When the degradation kinetics were not first-order, the time required for 50% degradation is shown instead. (Data compiled from Ref. 35, 36, 59, 60.)

term exposure can be assumed, have shown considerable metabolic activity with respect to common chemical pollutants (2). Wilson et al. (42) provided indirect evidence for the adaptation of aquifer microorganisms to several aromatic creosote wastes present at concentrations ranging from 100 to 1000 ng/mL. Rates of degradation in spiked samples from the contaminated aquifer were more than an order of magnitude greater than in samples from a nearby uncontaminated aquifer. Oxygen concentrations were also decreased in the contaminated aquifer, indicating increased microbial activity. Aelion et al. (43) attributed the adaptation of aquifer microorganisms to *p*-nitrophenol to an increase in the degrader population from undetectable levels to numbers exceeding 10^3/g aquifer sediment. Adaptation was expressed as a rapid increase in the rate of *p*-nitrophenol mineralization, which generally occurred after 35 days incubation (Figure 1). No adaptation was observed at the lowest *p*-nitrophenol concentration of 14 ng/g. Presumably this concentration did not support sufficient growth of degraders. This suggests that adaptation to pesticides in groundwater may not occur, since pesticide concentrations are generally below 25 ng/g soil. Increased concentrations resulted in increased adaptive responses with *p*-nitrophenol and methyl parathion in river waters and sediments (44). In other studies long-term exposure to high concentrations of pollutants (creosote wastes) appeared to result in increased growth of aquifer bacteria, based on ATP levels (45).

Inherent biodegradabilty also appears to affect the adaptation process. Substrates which are easily metabolized may not exhibit an adaptive response because initial rates of degradation are sufficient. For instance, 2-chlorophenol was readily metabolized in both unexposed and preexposed groundwater microcosms (46). More recalcitrant compounds require an adaptive period. Thus, 2,4,5-trichlorophenol added to microcosms similar to those used for the 2-chlorophenol experiments was degraded rapidly after a 1-day lag in the preexposed microcosm. A 6-day lag was required in the unacclimated microcosm. Other substrates show no adaptive response, such as chlorobenzene or 1,2,4-trichlorobenzene (43).

The low nutrient levels of deeper subsurface strata and aquifers may limit biodegradation. Nutrient additions may affect degrader populations in several ways. Initially, degradation may be reduced via catabolite repression or an analogous process. Alternatively, degrader populations may be increased by the nutrients and active degradation may begin once nutrients are exhausted. Additions of inorganic nutrients (N, P, K salts) to groundwater microcosms increased metabolism of *p*-nitrophenol, but addition of glucose or amino acids delayed mineralization (47). The effect of organic nutrients on *p*-nitrophenol degradation appears to be attributable to catabolite suppression since adaptation was required, even in unammended aquifer samples. When amino acids were applied with mineral salts and vitamins the time required for adaptation to *p*-nitrophenol was shortened relative to the control. This effect was explainable by microbial growth preceeding degradation (47). Organic nutrients also reduced the mineralization of toluene, but the mechanism in this case was less clear (47). In other studies, additions of organic nutrients increased the metabolism of alachlor in deep vadose zone and aquifer samples, but it was not determined if the increase was an adaptative response (48).

Adaptation and Bioremediation

Because approximately half of the detections of pesticides in groundwater are attributable to point-source contamination, the possiblity of removing these contaminants with biological agents should be considered. The concept of bioremediation has received considerable attention in recent years as a means of cleaning up aquifers contaminated with large amounts of gasoline or industrial solvents (2, 49). Two processes have been examined more than others: the *in situ* remediation method using indigenous microorganisms or the withdrawal and treatment method using pumps and specialized wastewater treatment systems. The second of these alternatives does not appear to be especially suited to use with pesticides due to the considerable expense involved.

Previous *in situ* remediation efforts have utilized the addition of inorganic nutrients and oxygen to stimulate biodegradation (2). In most cases, the contaminated aquifers contained relatively large concentrations of hydrocarbon pollutants which acted as a carbon sources for the microorganisms. Application of this technology to pesticides may require extensive modification for several reasons. Concentrations of pesticides in groundwater from point sources such as contaminated wells are low; therefore, the extent and duration of adaptation and potential for microbial growth are probably reduced. Some pesticides or their metabolites may not be metabolized by the subsurface microbial community or metabolized only at very slow rates. The low nutrient levels in the subsurface environment may also limit those organisms that degrade pesticides via cometabolism. The introduction of specialized microbial strains may overcome this problem.

Attempts to augment indigenous populations in soil or surface water to enhance pesticide degradation illustrate several interesting points. Early experiments by McClure showed that a mixed population of IPC-degrading microorganisms inoculated into soil could protect plants against the phenyl carbamate herbicide IPC (50). If the IPC addition was delayed, the protective effect disappeared after 30 days, suggesting that either the microorganisms could not survive, or that they had lost their adaptation. Chaterjee et al. (51) inoculated soil with a *Pseudomonas cepacia* strain capable of degrading 2,4,5-T. Populations initially declined, but after addition of 1 mg 2,4,5-T/g soil, populations increased concomitantly with 2,4,5-T degradation. Even extremely recalcitrant substrates may be degraded by inoculated microorganisms. The adaptation period before mineralization of the polychlorinated biphenyl Arochlor 1242 was decreased by addition of an *Acinetobacter* strain (Figure 4). Addition of the substrate analog biphenyl was necessary to obtain any activity (52). The length of the adaptation period was related to the level of *Acinetobacter*. Parathion degradation in soil was enhanced by the inoculation of an adapted two-member mixed culture (53). This system is unique in that one microorganism hydrolyses the parathion while the second degrades the resulting p-nitrophenol (54). This consortium degrades parathion without a detectable acclimation period in soil. These examples illustrate the potential utility of using introduced microorganisms to clean up soils or aquifers polluted with pesticides. To an unknown extent the success of such efforts may depend on how well suited the introduced strains are to existence in the subsurface environment. Other environmental factors such as spread of the microorganisms within the aquifer matrix may also be important. Limited evidence shows surprisingly extensive bacterial movement in aquifer systems (55, 56).

Conclusions

Efforts should be made to distinguish the different mechanisms that contribute to adaptation. Short time periods required for adaptation are probably a combination of the time required for enzyme induction and growth of degrader

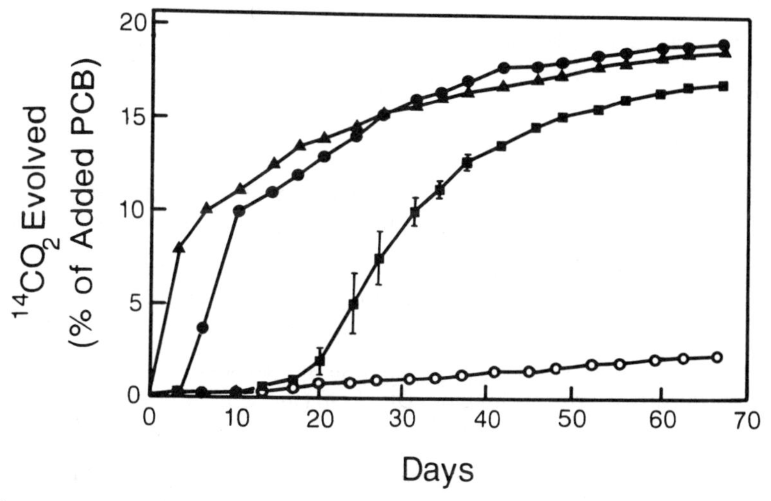

Figure 4. The effect of analog substrate (biphenyl) addition and introduction of *Acinetobacter* P6 on the degradation of PCB Arochlor 1242 (100 mg/kg) in soil. Addition of biphenyl with P6 at 10^9 (triangles) or 10^5 (closed circles) cells/mL inoculum reduced the adaptation period compared to noninoculated soil (squares). *Acinetobacter* P6 added to soil without biphenyl had no effect on degradation (open circles). (Reproduced with permission from Ref. 52. Copyright 1985 American Society of Agronomy.)

populations. The time required for growth of pesticide-degrading microorganisms has been put forward as an explanation for the adaptation response in aquatic systems, where a variety of substrates have been examined (57, 58). This may also be true in soils and subsoils when initial populations of degraders are small. Other evidence suggests that enzyme induction is at least partially responsible for the adaptation responses to some pesticides. In situations where long-term exposure is required to elicit an adaptive response the mechanisms involved may include genetic transfer and selection of adapted strains prior to growth of degrading strains or consortia.

Pesticide-degrading bacteria in the subsurface environments are likely to face unusual contraints in terms of nutrient availability and a more limited range of metabolic capabilities than surface bacteria and fungi. Current experimental evidence shows that subsurface microorganisms adapt to a broad variety of substrates, but the limited evidence pertaining to pesticides suggests that microorganisms will adapt very slowly or not at all to these compounds. It may be possible that the barriers to adaptation in the subsurface environment can be overcome by introduction of specific degraders or the addition of nutrients.

Literature Cited

1. Williams, W.M.; Holden, P.W.; Parsons, D.W.; Lorber, M.N. Pesticides in Groundwater Data Base: 1988 Interim Report; U.S. Environ. Protect. Agency, Washington DC, 1988.
2. Lee, M.D.; Thomas, J.M.; Borden, R.C.; Bedient, P.C.; Wilson, J.T.; Ward, C.H. CRC Critical Rev. Environ. Control. 1988, 18, 29-89.
3. Simkins, S.; M. Alexander. Appl. Environ. Microbiol. 1984, 47, 1299-1306.
4. Karns, J.S. In Enhanced Biodegradation of Pesticides; American Chemical Society: Washington, DC, 1990 (this volume).
5. Marty, J.L.; Kahif, T.; Vega, D.; Bastide, J. Soil Biol. Biochem. 1986, 18, 649-653.
6. Karns, J.S.; Duttagupta, S.; Chakrabarty, A.M. Appl. Environ. Microbiol. 1983, 46, 1182-1186.
7. Englehardt, G.; Wallnofer, P.R., Plapp, R. Appl. Microbiol. 1973, 26, 709-718.
8. Zeyer, J.; P.C. Kearney. Pestic. Biochem. Physiol. 1982, 17, 224-231.
9. Linkfield, T.G.; Suflita, J.M.; Tiedje, J.M. Appl. Environ. Microbiol. 1989, 55, 2773-2778.
10. Milhomme, H.; Vega, D.; Marty, J.L.; Bastide, J. Soil Biol. Biochem. 1989, 21, 307-311.
11. Lee, A; Rahman, A.; Holland, P.T. New Zealand J. Agric. Res. 1984, 27, 201-206.
12. Moorman, T.B. Weed Sci. 1988, 36, 96-101.
13. Ghosal, D; You, I.-S.; Chatterjee, D.K.; Chakrabarty, A.M. Science 1985, 228, 135-142.
14. Ogunseitan, O.A.; Tedford, E.T.; Pacia, D; Sirotkin, K.M.; Sayler, G.S. J. Ind. Microbiol. 1987, 1, 311-317.
15. Jain, R.K.; Sayler, G.S.; Wilson, J.T.; Houston, L.; Pacia, D. Appl. Environ. Microbiol. 1987, 53, 996-1002.
16. Sayler, G.S.; Shields M.S.; Tedford, E.T.; Breen, A.; Hooper, S.W.; Sirotkin, K.M.; Davis, J.W. Appl. Environ. Microbiol. 1985, 49, 1295-1303.

17. Pirt, S.J. Principles of Microbe and Cell Cultivation; Blackwell Scientific Publications: London, 1975.
18. Anderson, T.-H.; Domsch, K.H. Soil Biol. Biochem. 1985. 17, 197-203.
19. Anderson, T.-H.; Domsch, K.H. Biol. Fert. Soils 1985, 1, 81-89.
20. Cullimore, D.R. Weed Sci. 1981, 29, 440-443.
21. Ou, L.-T. Soil Sci. 1984, 137, 100-107.
22. Racke, K.D.; Coats, J.R. J. Agric. Food Chem. 1987, 35, 94-99.
23. Cullimore, D.R., Kohout, M. Can. J. Microbiol. 1974, 20, 1449-1452.
24. Burge, W.D. Appl. Microbiol. 1969, 17, 545-550.
25. Loos, M.A.; Schlosser, I.F.; Mapham, W.R. Soil Biol. Biochem. 1979, 11, 377-385.
26. Meuller, J.G.; Skipper, H.D.; Lawrence, E.G.; Kline, E.L. Weed Sci. 1989, 37, 424-427.
27. Schmidt, S.K.; Gier, M.J. Microbial Ecol. 1989, 18, 285-296.
28. Soulas, G.; Chaussod; Verguet, R. Soil Biol. Biochem. 1984, 16, 497-501.
29. Stott, D.E.; Martin, J.P.; Focht, D.D.; Haider, K. Soil Sci. Soc. Amer. J. 1983, 47, 66-70.
30. Hendry, K.M.; Richardson, C.J. Environ. Toxicol. Chem. 1988, 7, 763-774.
31. Lappin, H.M; Greaves, M.P.; Slater, J.H. Appl. Environ. Microbiol. 1985, 49, 429-433.
32. Novick, N.J.; Alexander, M. Appl. Environ. Microbiol. 1985, 49, 737-743.
33. Novick, N.J.; Mukherjee, R.; Alexander, M. Appl. Environ. Microbiol. 1986, 34, 721-725.
34. Tiedje, J.M.; Hagedorn, M.L. J. Agric. Food Chem. 1975, 23, 77-81.
35. Moorman T.B.; Harper, S.S. J. Environ. Qual. 1989, 18, 302-306.
36. Kempson-Jones, G.F.; Hance, R.J. Pestic. Sci. 1979, 10:449-454.
37. Ward, T.E. Environ. Toxicol. Chem. 1985, 4, 727-737.
38. Wehtje, G.R.; Spalding, R.F.; Burnside, O.C.; Lowry, S.R.; Leavitt, J.R.C. Weed Sci. 1983, 31, 610-618.
39. Ghiorse, W.C.; Wilson, J.T. Adv. Appl. Microbiol. 1988, 33, 107-172.
40. Swindoll, C.M.; Aelion, C.M.; Dobbins, D.C.; Jaing, O.; Long, S.C.; Pfaender, F.K. Environ. Toxicol. Chem. 1988, 7, 291-299.
41. Balkwill, D.L.; Fredrickson, J.K.; Thomas, J.M. Appl. Environ. Microbiol. 1989, 55, 1058-1065.
42. Wilson, J.T.; McNabb, J.F.; Cochran, J.W.; Wang, T.H.; Tomson, M.B.; Bedient, P.B. Environ. Toxicol. Chem. 1985, 4, 721-726.
43. Aelion, C.M., Swindoll, C.M., and Pfaender, F.K. Appl. Environ. Microbiol. 1987, 53, 2212-2217.
44. Spain, J.C.; Pritchard, P.H.; Borquin, A.W. Appl. Env. Microbiol. 1980, 40, 726-734.
45. Wilson, J.T.; Miller, G.D.; Ghiorse, W.C.; Leach, F.R. J. Contam. Hydrol. 1986, 1, 163-170.
46. Suflita, J.M.; Miller, G.D. Environ. Toxicol. Chem. 1985, 4, 751-758.
47. Swindoll, C.M.; Aelion, C.M.; Pfaender, F.K. Appl. Environ. Microbiol. 1988, 54, 212-217.
48. Pothuluri, J.V.; Moorman, T.B.; Obenhuber, D.C. Abst. 89th Ann. Mtg. Amer. Soc. Microbiol. 1989, p 297.
49. Lee, M.D.; Ward, C.H. Environ. Toxicol. Chem. 1985, 4, 743-750.
50. McClure, G.W. J. Environ. Qual. 1972, 1, 177-180.
51. Chatterjee, D.K.; Kilbane, J.J.; Chakrabarty, A.M. Appl. Environ. Microbiol. 1982, 44, 514-516.
52. Brunner, W.; Sutherland, F.H.; Focht, D.D. J. Environ. Qual. 1985, 14, 324-328.

53. Daughton, C.G.; Hsieh, D.P.H. Bull. Environ. Contam. Toxicol. 1977, 18, 48-56.
54. Daughton, C.G.; Hsieh. D.P.H. Appl. Environ. Microbiol. 1977, 34, 175-184.
55. Gerba, C.P. In Groundwater Pollution Microbiology; Bitton, G.; Gerba C.P. ed.; John Wiley and Sons: New York, 1984; .p. 225-233.
56. Jenneman, G.E.; M.J. McInerney; Knapp, R.M. Appl. Environ. Microbiol. 1985, 50, 383-391.
57. Chen, S.; Alexander, M. J. Environ. Qual. 1989, 18, 153-156.
58. Wiggins, B.A.; Jones, S.H.; Alexander, M. Appl. Environ. Microbiol. 1987, 53, 791-796.
59. Ladlie, J.S.; Meggitt, W.F.; Penner, D. Weed Sci. 1976, 24, 508-511.
60. Bouchard, D.C.; Lavy, T.L.; Marx, D.B. Weed Sci. 1982, 30, 629-632.

RECEIVED January 22, 1990

Chapter 14

Microbial Adaptation in Aquatic Ecosystems

J. C. Spain

U.S. Air Force Engineering and Services Center, Tyndall Air Force Base, Panama City, FL 32403–6001

Aquatic microbial communities can adapt to degrade a novel substrate if the chemical can serve as a source of nutrients. We have studied biodegradation of p-nitrophenol (PNP) in laboratory test systems and in the field. Adaptation in freshwater microbial communities required from 3 to 10 days and biodegradation rates increased as much as a thousandfold at PNP concentrations as low as 10 mg/liter. Populations of PNP degraders increased dramatically during adaptation and no evidence of genetic change associated with adaptation was detected. Bacteria able to grow at the expense of PNP were detected in microbial communities prior to treatment with PNP. They did not begin to grow and degrade PNP until several days after initial exposure. Results indicate that accurate prediction of biodegradation rates requires a better understanding of the factors that control adaptation in the environment. These include: enzyme induction, genetic changes, increases in microbial populations, predation, and nutrient limitation.

Factors that control the fate and transport of organic compounds in aquatic systems include: photolysis, hydrolysis, sorption, hydrodynamics, and biodegradation. Biodegradation is the least well understood of these processes, primarily because it is difficult to predict the effects of adaptation, or acclimation, of microbial populations in response to specific compounds. Adaptation is characterized by an increase in the rate of biodegradation of a compound as a result of exposure of the microbial population to the compound (1). The ability to predict such increases in biodegradation rates is critical to understanding of the behavior of chemicals in the field. For example, accuracy in prediction of the onset of rapid biodegradation would allow accurate assessment of the impact of accidental release in a river that flows into a sensitive estuary. In groundwater, accurate information about adaptation would allow prediction of whether and when a plume of contamination might affect a source of drinking water. A partial list of compounds known to elicit adaptation in aquatic systems is provided in the following table.

Table 1. Chemicals that cause adaptation in aquatic systems

Compound	Reference
p-Cresol	Lewis et al. (2)
Nitrilotriacetic acid	Pfaender et al. (3)
2,4-Dichlorophenoxyacetic acid	Audus (4)
Phenol	Murakami and Alexander (5)
4,6-Dinitro-o-cresol	Gundersen and Jensen (6)
p-Nitrophenol	Spain et al. (7)
Methylparathion	Spain et al. (7)

The common characteristic of compounds that elicit adaptation is the ability to serve as a growth substrate for bacteria. Thus it seems to be the response of a specific population of bacteria to a growth substrate that is responsible for the adaptation. To my knowledge, there are no reports of adaptation to compounds that do not serve as growth substrates. Gratuitous inducers are known but they are not degraded more rapidly as a result of adaptation.

Patterns of Adaptation

The course of adaptation involves several distinct phases: a) Before exposure to the chemical, the population of bacteria able to degrade the chemical is small and inactive. There may be no organisms in the microbial community that can synthesize the enzymes necessary for growth on the compound in question. b) Upon addition of the chemical, there is a variable lag period during which specific bacteria become active. This period involves induction of inducible enzymes and the onset of growth when specific degraders are present. The factors that control the length of this lag period are not clear and will be discussed in detail below. c) Growth of the bacteria at the expense of the added chemical results in the rapid depletion of the chemical and an enrichment of the population that is able to derive carbon and energy from the chemical. The resultant population can respond rapidly without a lag period to subsequent additions of the same chemical. If the concentration of the added compound is low, there may not be an increase in the total number of heterotrophic bacteria present, whereas the number of specific degraders may increase several orders of magnitude. Very low concentrations might cause increases in degradation rates without detectable increases in population size.

A variety of factors influence the length of the lag period noted above. Very long and variable lag periods (Figure 1A) have been taken to indicate that no organisms able to degrade the chemical were present in the initial inoculum. The lag would be the time required for the degradative ability to evolve through mutation or genetic exchange. Wyndham (8) described a classic example of a genetic change that allowed an Acinetobacter to evolve the ability to degrade high concentrations of aniline in river water. Wiggins and Alexander (9) suggested a similar explanation for the variable lag period involved in adaptation of sewage and lake water communities to biodegrade nitrosalicylates. Lag periods of up to a year have been noted in enrichment cultures set up for isolation of bacteria able to grow on chlorinated aromatic compounds (10, 11, 12, and 13). Kellogg et al. (10) suggested that genetic exchange during ten months in a chemostat was responsible for the evolution of a pseudomonad able to degrade 2,4,5-trichlorophenoxyacetic acid. Krockel and Focht (14) have recently demonstrated more rapid evolution of the ability to degrade chlorobenzenes through genetic exchange in a novel chemostat system.

Shorter and more reproducible lag periods (Figure 1B) seem to reflect the time required for growth of a small population of bacteria. Wiggins et al. (15) noted that even if the population starts to use the chemical for growth immediately, there may be a significant apparent lag before enough specific degraders accumulate to cause a detectable loss of the chemical. A variety of as yet poorly understood factors can affect the growth of the initial small populations and thus extend or shorten the lag period. Wiggins and Alexander (16) provided evidence that predation by protozoa extended the acclimation period before biodegradation of p-nitrophenol (PNP), 2,4-dichlorophenoxyacetic acid (2,4-D), and 2,4-dichlorophenol. In another series of experiments (15), the same authors found that the addition of inorganic nutrients shortened the lag period for biodegradation of PNP in lake water and sewage. Similarly, Lewis et al (2) demonstrated that p-cresol was degraded much more quickly in pond water supplemented with nitrogen and phosphorous. Low rates of p-cresol transformation were detectable immediately in their experiments but added nutrients dramatically increased the rate. Zaidi et al (17) clearly demonstrated that addition of inorganic nutrients stimulated growth of the specific degrader population when degradation rates were enhanced. These results suggest that any factor that affects the growth of the specific population involved in degradation can affect the length of the apparent lag period.

In some instances there is an extended lag period before a population of specific degraders begins to grow and metabolize the chemical of interest (Figure 1C). The appropriate bacteria are present but do not respond to the presence of the added chemical. This situation has been observed several times with PNP in our

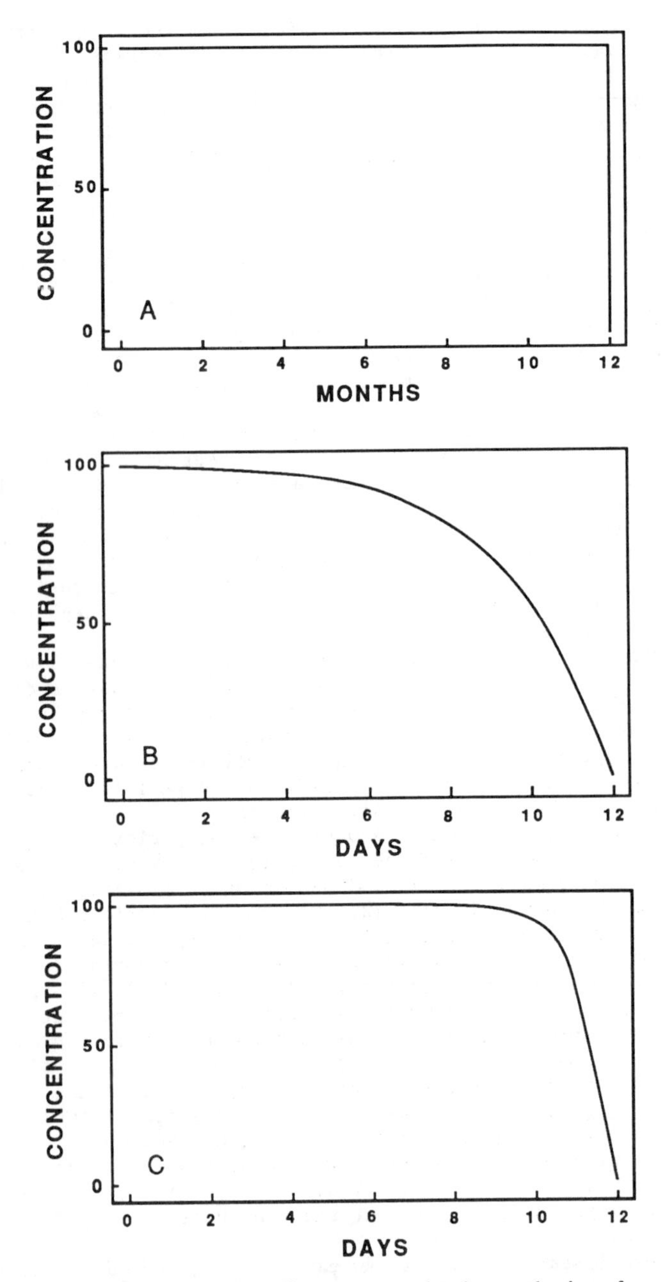

Figure 1. Patterns of substrate disappearance related to mechanism of adaptation. Genetic changes such as mutation, plasmid exchange, or recombination (A). Growth of a small population of bacteria able to degrade the test chemical immediately (B). Delayed induction of enzymes or activation of specific organisms (C).

laboratory (18). A variety of explanations have been offered for the failure of PNP degraders to respond immediately. Induction of the appropriate enzymes is generally necessary before metabolism and growth at the expense of a new substrate can begin. Induction of catabolic enzymes has been studied extensively in the laboratory with large populations of bacteria and with high concentrations of substrate. Induction of enzymes under such laboratory conditions typically occurs within minutes to hours. However, little is known about the induction of enzymes under natural conditions where small populations of starved cells might be exposed to low concentrations of a substrate under less than ideal conditions.

Hoover et al. (19) have described several anomalies in mineralization of organic compounds present at low concentrations. Boethling and Alexander (20) reported the existence of threshold concentrations below which mineralization could not be detected. Spain and Van Veld (1) reported a threshold of 10 ug per liter for adaptation of river water communities to PNP and the extent of induction was proportional to PNP concentrations above the threshold. Wiggins and Alexander (9) noted that the length of the acclimation period for PNP mineralization in sewage was directly proportional to the concentration of PNP. In contrast, Schmidt et al (21) and Button and Robertson (22) have suggested that induction of enzymes can lead to adaptation at substrate concentrations that are too low to support growth of the bacteria. It has also been suggested that the presence of inhibitors might delay induction and the onset of biodegradation (9). An inhibitor might be toxic to the specific population under consideration, or it might serve as an alternate, preferred substrate.

Enhanced Degradation of PNP

PNP is the classic example of compounds that elicit adaptation of aquatic microbial communities after a lag period of several days. The typical degradation curve is similar to the one shown in Figure 1C. We became interested in PNP because it is a hydrolysis product of methylparathion in aquatic systems. When we studied the biodegradation of PNP in laboratory test systems containing river water and sediment (7), no degradation was detectable for 3-5 days, after which the PNP was mineralized rapidly (Figure 2). When PNP was added again, it was degraded rapidly and without a lag. Microbial populations in control samples receiving PNP for the first time underwent a lag period similar to that in samples exposed at the beginning of the experiment. Inclusion of such controls in laboratory tests is essential if the increased degradation rate is to be rigorously attributed to the presence of PNP. It is clear from this example that adaptation can effect a dramatic increase in biodegradation rates; the rate in the adapted system is over a thousandfold higher than that in the control.

Subsequent experiments (Figure 3) indicated that microbial communities in the laboratory remained adapted to PNP for up to six weeks after exposure to PNP. All freshwater communities tested showed the ability to acclimate to PNP, whereas none of the marine communities could (1). This suggests that PNP released into a river

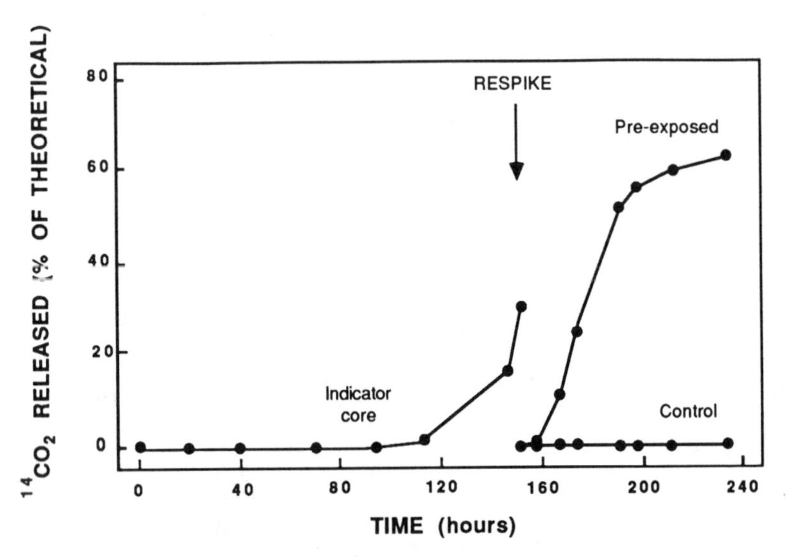

Figure 2. PNP biodegradation in sediment and water from the Escambia River. Radiolabeled PNP was added to the indicator core at a concentration of 150 ppm. Unlabeled PNP was added to the "pre-exposed" core at the same time. After 143 h radiolabeled PNP was added to the pre-exposed and control cores. (Reprinted with permission from ref. 7. Copyright 1980 American Society for Microbiology.)

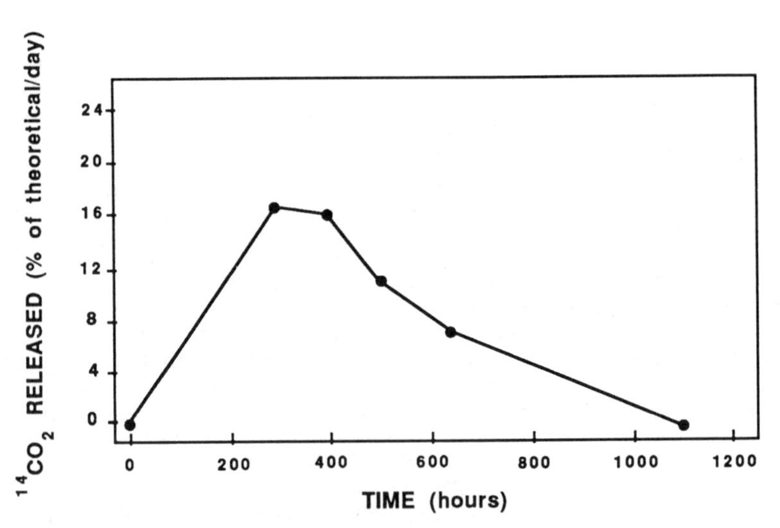

Figure 3. Duration of adaptation in samples from the Escambia River. Sediment and water samples were allowed to biodegrade 150 ppm PNP, then tested at intervals for the ability to mineralize radioactive PNP. (Reprinted with permission from ref. 1. Copyright 1983 American Society for Microbiology.)

would be degraded after several days. However, if the release occurred at a point near the mouth of the river, the PNP would enter the estuary where it would persist. Thus, the accurate prediction of the length of the acclimation period is essential to the accurate assessment of the fate and effects of compounds that can be biodegraded only after an acclimation period.

We subsequently conducted a series of experiments to determine whether laboratory test systems could be used to predict acclimation in the field (18). PNP was added simultaneously to a freshwater pond and to several types of laboratory test systems which contained sediment and water from the pond. The laboratory test systems included: shake flasks with pond water, shake flasks with water and sediment, eco-cores, small (1.5-liter) microcosms, and large (390-liter) microcosms.

After 6 days with no detectable biodegradation, the PNP disappeared rapidly from the pond. A second addition of PNP was degraded rapidly and without a lag, whereas in a control pond no degradation was detected for 6 days (Figure 4). Each of the laboratory test systems that contained sediment underwent a similar adaptation period that paralleled that in the pond. Therefore, any of the laboratory systems could have been used to predict the behavior of PNP in the pond. The microbial community in the shake flask that contained pond water without sediment adapted much more slowly, which suggested that sediment plays an important part in the adaptation process. The mechanism by which sediment exerts its effect is largely unknown; it could provide biomass, nutrients, or surfaces important to the adaptation process.

Enumeration of bacteria in the pond (Figure 5) revealed that although there were 100 specific PNP degraders per milliliter, they did not begin to increase in number for six days. They could mineralize PNP in the MPN assay and could form colonies on PNP agar, but they did not grow and degrade PNP in the pond. This provides a classic example of the type of lag period described above (Figure 1C) where the appropriate bacteria were present but did not respond to the added chemical for an extended period. The results suggest that the appropriate enzymes were not induced or that growth was somehow inhibited.

A variety of possible explanations can be offered to explain the extended lag period (see above). Diauxie, toxic organic compounds, predation by protozoa, lack of inorganic nutrients or inappropriate concentration of PNP could inhibit induction of enzymes or growth. The major shortcoming of each of these arguments is that it is difficult to explain how any of these conditions would be relieved reproducibly after 6 days.

We isolated from the pond two strains of bacteria able to grow on PNP. One of the strains, a *Pseudomonas*, undergoes an extended lag period under certain conditions in the laboratory. We are currently studying this strain to obtain information about its metabolic pathway for PNP, and the factors that regulate induction of enzymes and growth. Preliminary results indicate that the time required for induction of catabolic enzymes is inversely related to both the concentration of cells and the PNP concentration. The mechanism of such effects is unknown.

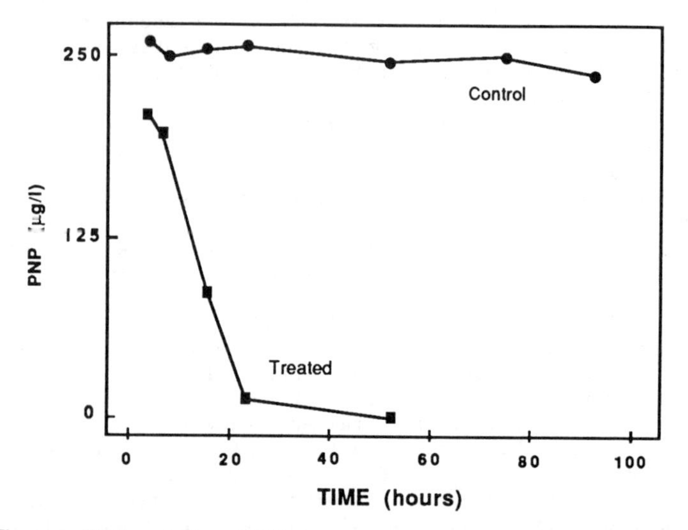

Figure 4. Biodegradation of PNP in previously treated and control ponds. The treated pond was exposed to PNP for 10 days, then both treated and control ponds received 250 ppm PNP. (Reprinted with permission from ref. 18. Copyright 1984 American Society for Microbiology.)

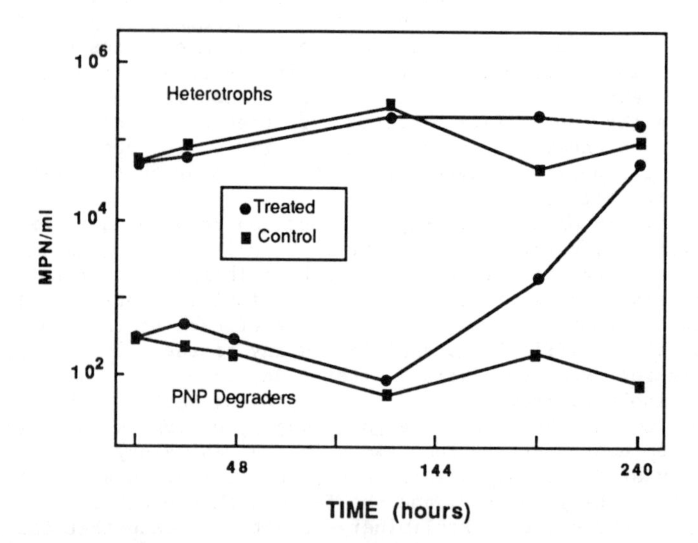

Figure 5. Changes in microbial biomass during adaptation to PNP. (Reprinted with permission from ref. 18. Copyright 1984 American Society for Microbiology.)

Conclusion

Enhanced degradation of a chemical added to an aquatic system is the result of microbial adaptation to degrade the chemical. A variety of changes in the microbial community including mutation, induction of enzymes, and growth of an initially small population can lead to the enhancement. In some instances, all of these events may play a role in the adaptation process. Thus, mutations or rearrangement of genes might allow induction of enzymes which would lead to growth of a small population. Any factor, such as nutrient limitations, that affects the metabolism and growth of the organisms involved, could affect the rate of the adaptation. Each of these processes is relatively well understood in pure cultures of bacteria under laboratory conditions. However, very little is known about how they interact to produce dramatic increases in biodegradation in the field. Future research should focus on the microbial ecology of mixed microbial communities challenged with low concentrations of organic pollutants.

Literature Cited

1. Spain, J.C.; Van Veld, P.A. Appl. Environ. Microbiol. 1983, 45,428-435.
2. Lewis, D.L.; Kollig, H.P.; Hodson, R.E. Appl. Environ. Microbiol. 1986, 51,598-603.
3. Pfaender, F.K.; Shimp, R.J.; Larson, R.L. Environ. Toxicol. Chem. 1985, 4,587-593.
4. Audus, L.J. J. Sci. Food Agric. 1952, 3,268-274.
5. Murakami, Y.; Alexander, M. Biotechnol. Bioeng. 1989, 33,832-838.
6. Gundersen, K.; Jensen, H.L. Acta. Agric. Scand. 1956, 6,100-114.
7. Spain, J.C.; Pritchard, P.H.; Bourquin, A.W. Appl. Environ. Microbiol. 1980, 40,726-734.
8. Wyndham, C. Appl. Environ. Microbiol. 1986, 51,781-789.
9. Wiggins, B.A.; Alexander, M. Appl. Environ. Microbiol. 1988, 54,2803-2807.
10. Kellog, S.T.; Chatterjee, D.T.; Chakrabarty, A.M. Science 1981, 214,1133-1135.
11. Spain, J.C.; Nishino, S.F. Appl. Environ. Microbiol. 1987, 53,1010-1019.
12. Haigler, B.E.; Nishino, S.F.; Spain, J.C. Appl. Environ. Microbiol. 1988, 54,294-301.
13. Reineke, W.; Knackmuss, H.-J. Appl. Environ. Microbiol. 1984, 53,2470-2475.
14. Krockel, L.; Focht, D.D. Appl. Environ. Microbiol. 1988, 53,2470-2475.
15. Wiggins, B.A.; Jones, S.H.; Alexander, M. Appl. Environ. Microbiol. 1987, 53,791-796.
16. Wiggins, B.A.; Alexander, M. Can. J. Microbiol. 1988, 34,661-666.
17. Zaidi, B.R.; Murakami, Y.; Alexander, M. Environ. Sci. Technol. 1988, 22,1419-1425.

18. Spain, J.C.; Van Veld, P.A.; Monti, C.A.; Pritchard, P.H.; Cripe, C.R. Appl. Environ. Microbiol. 1984, 48,944-950.
19. Hoover, D.G.; Borgonovi, G.E.; Jones, S.H.; Alexander, M. Appl. Environ. Microbiol. 1986, 51-226-232.
20. Boethling, R.S.; Alexander, M. Appl. Environ. Microbiol. 1979, 37,1211-1216.
21. Schmidt, S.K.; Alexander, M.; Schuler, M.L.; J. Theor. Biol. 1985, 114,1-8.
22. Button, D.K.; Robertson, B.R. Mar. Ecol. Prog. Ser. 1985, 26,187-193.

RECEIVED January 22, 1990

IMPLICATIONS AND SOLUTIONS

Chapter 15

Evaluation of Some Methods for Coping with Enhanced Biodegradation of Soil Insecticides

A. S. Felsot[1] and J. J. Tollefson[2]

[1]Center for Economic Entomology, Illinois Natural History Survey, 607 East Peabody Drive, Champaign, IL 61820
[2]Department of Entomology, Iowa State University, Ames, IA 50011

Several proposed strategies for either preventing or circumventing enhanced biodegradation of corn rootworm soil insecticides have been studied in Iowa and Illinois. Operational strategies included cultivation-time treatments, suppression of adults, and insecticide rotations; a technological strategy was the use of controlled-release formulations. In 9 of 13 field experiments, carbofuran (2,3-dihydro-2,2-dimethyl-7-benzofuranyl methylcarbamate) applied at cultivation prevented root damage better than when applied at planting. Root damage following adult control with carbaryl (1-naphthyl methylcarbamate) the preceding season was less than or similar to root damage following the use of soil insecticides, which suggested that corn rootworms could be controlled without use of a soil insecticide. Cultivation-time treatments and adulticiding depended on accurate timing of insecticide applications. Under laboratory conditions, rotations of carbofuran with isofenphos (1-methylethyl 2-[[ethoxy[(1-methylethyl)amino] phosphinothioyl]-oxy] benzoate) slowed the rate of metabolism of carbofuran but did not prevent the development of enhanced biodegradation. In the field, the results of rotations of carbofuran with fonofos (O-ethyl S-phenyl ethylphosphonodithioate) or terbufos (S-[*tert*-butylthio)methyl]O,O-diethyl phosphorodithioate) were inconclusive. Multi-year treatments of fonofos resulted in enhanced biodegradation, but its rate of degradation was slowed by rotations with carbofuran or terbufos. When carbofuran was applied as a controlled-release formulation, its degradation was

0097-6156/90/0426-0192$06.50/0

significantly slowed compared to the commercial sand granule; however, efficacy in the field was not improved.

In the United States, the greatest proportion of insecticides are used on corn by virtue of the vast acreages devoted to production (1). Most of the insecticides are applied at planting time to control root and seedling feeding damage by soil insects, especially the western and northern corn rootworm (*Diabrotica virgifera* and *D. barberi*, respectively) and the black cutworm (*Agrotis ipsilon*). Because soil insecticides are applied long before the insect pest appears, chemical persistence and bioavailability is critical to successful reduction of potentially damaging population levels. Although a myriad of environmental and biological factors interact to affect the ultimate bioactivity of a soil insecticide (2,3), the effect of enhanced biodegradation on efficacy of crop protection has received much attention in the 1980's (4-6).

With few reliable corn rootworm (CRW) insecticides on the market, the pressure to use the same product year after year has increased over the last decade. The enhanced biodegradation of the methyl carbamate insecticides, such as carbofuran and trimethacarb (7-10), and the organophosphinothioate insecticide, isofenphos (11-13) is well known and has been linked to crop protection failures (6,14). Enhanced biodegradation of the organophosphorus soil insecticides, fonofos, terbufos, and chlorpyrifos has been suggested in several reports (15-17), but little data has been generated to definitively determine the susceptibility of these compounds to microbial adaptations. With the assumption that chemical control of corn rootworm feeding damage will remain the predominant management strategy in continuous corn fields over the next decade, it is important to develop best management practices for the long-term use of the registered soil insecticides.

Part of the long-term management strategy would be avoiding the development of enhanced biodegradation, because this phenomenon can significantly affect insecticide efficacy. Another possibility is employing methods that circumvent enhanced biodegradation after it has developed. The techniques available and proposed for coping with the adverse effects of enhanced biodegradation, which can be classified as either operational or technological, have been discussed elsewhere (6). Briefly, operational strategies for coping with enhanced biodegradation rely on management techniques based on biological principles whereas technolgical strategies require alterations in formulation chemistry or structural chemistry of the insecticide. The operational alternatives include conservation of pesticides (e.g., use of integrated pest management [IPM] strategies such as scouting followed by control, if necessary, to reduce economic infestations), crop rotation (e.g., corn/soybeans), proper calibration of application equipment, altered timing of applications (use of cultivation-time treatments when larvae are actively feeding), and chemical rotations (use of a chemical every other year or at longer intervals). The technological alternatives include extenders and inhibitors (to delay biodegradation), new formulation technology (e.g., controlled release), and directed chemistry (synthesis of insecticides possessing higher toxicities in soil).

Few of the proposed management strategies have been experimentally investigated. The objective of this paper is to present results of previously unpublished experiments that have evaluated several proposed strategies for coping with enhanced biodegradation of soil insecticides used to control CRW, namely, cultivation-time treatments as opposed to planting-time treatments, suppression of adults to prevent egg-laying and larval damage the following spring, chemical rotations, and controlled-release formulations.

Experimental Design

Cultivation-Time Treatments. Compared to applying an insecticide a month or more before CRW eggs hatch, cultivation-time application would expose actively feeding larvae to a much higher dose of insecticide. In Iowa, applications of insecticides at cultivation time have been routinely compared with applications at planting-time since 1975, resulting in a total of 19 experiments across the state.

The experimental design in all experiments was a randomized complete block with four replications. Treatment units were single rows, 11 to 15 m long depending on the location. The likelihood of having a corn rootworm infestation was increased by either selecting sites that had a documented infestation of CRW or by planting a "trap crop". A trap crop consisted of corn planted later than normal during the previous season to provide a food source that attracted and accumulated ovipositing female beetles while surrounding fields were senescing.

Cultivation treatments of carbofuran (Furadan 15G, 1.12 kg AI/ha) were made four to six weeks after planting, when CRW larvae were actively feeding on corn roots. Plastic tubes, positioned directly in front of the cultivator sweeps, directed the insecticide to both sides of the corn row where soil was thrown over the granules to effect incorporation. For planting-time applications, Furadan 15G (1.12 kg AI/ha) was placed in an 18-cm band in front of the firming wheels and incorporated with a drag chain.

Insecticide efficacy was evaluated by digging five roots from each treated row and untreated row in mid to late July after the majority of larval feeding had occurred. The soil was removed using an air-pressurized water spray, and the feeding damage was rated using the Iowa 1 to 6 rating scale where 1 = no to minor and 6 = severe damage (18).

To determine if applying an insecticide at cultivation circumvented enhanced biodegradation, the root damage ratings of the five roots from an experimental unit (i.e., each treated and untreated row) were averaged. The most commonly accepted economic threshold (i.e., when losses due to damage equal the cost of control, in this case insecticide cost) for root ratings is between 2.5 and 3.0 (19). In addition root damage in the untreated area must exceed 3.0 before differences between insecticide treatments will become apparent. For these two reasons, six experiments that had root ratings of 3.0 or less in the untreated areas were deleted from further consideration. Treatment means in the remaining 13 experiments were analyzed using analysis of variance techniques (20). Duncan's multiple range test (21) was used to separate the treatment means.

Adult Suppression to Prevent Larval Damage. Another operational strategy for coping with enhanced biodegradation would be controlling adults rather than larvae so that the insecticides do not come into contact with the soil. If female beetle numbers could be sufficiently reduced during the egg-laying period (i.e., during August), a damaging population of CRW larvae should not develop the following season. To test this hypothesis, six CRW-infested cornfields in Iowa during 1977 were chosen. An ultra-low volume application of carbaryl (Sevin 4-Oil) at the rate of approximately 2.5 L/ha and containing 1.12 kg AI was aerially applied to half of each field and the remainder left untreated. The adulticide sprays were applied to a field when beetle counts reached one beetle per plant and 10% of the females were gravid.

During 1978 the design was modified to evaluate timing of adulticides to determine if a longer beetle-free period would improve subsequent root protection and to compare adult suppression with the preventative use of soil insecticides applied at planting the subsequent year. Two timings and multiple applications of carbaryl were achieved by dividing four Iowa cornfields in half and treating one side with a broadcast aerial application. Treatments were applied when beetle numbers reached one beetle per plant and 10% of the females were gravid. When beetle numbers recovered to 0.5 beetles per plant, the field was divided again perpendicular to the first division and one-half was treated. The protocol for insecticide application divided the fields into quadrants, where one quadrant received an early spray, one quadrant received a late spray, the third quadrant received both an early and late spray, and the final quadrant received no treatment. The following season (1979), the farmers applied their soil insecticide of choice to the fields at planting. They were asked to leave two strips across the field, one in each half, untreated with the soil insecticide. Each strip bisected two quadrants providing soil insecticide treated and untreated areas in all four adult suppression treatments. The experiment was repeated during 1979 in three Iowa cornfields. Efficacy of the adulticide and larvicide treatments was evaluated by digging 10 roots at each of five sites in all treatments. The damage was rated as previously described.

Laboratory Evaluation of Chemical Rotations. It was hypothesized that use of a soil insecticide every other year might slow the development of enhanced biodegradation. To test this hypothesis under laboratory conditions, the effect of rotations of carbofuran with isofenphos was studied using methods described by Dzantor and Felsot (10). Over a period of 181 days 5 kg soil (Flanagan silt loam, Table I) in individual buckets was pretreated in three 2-month cycles with granular carbofuran (Furadan 15G), isofenphos (Amaze 20G), an alternation between caroburan and isofenphos, or no insecticide (Table II). Following the third pretreatment cycle, soil was removed and treated with technical carbofuran to monitor chemical persistence. Another batch of soil was placed in biometer flasks (22) and treated with ^{14}C-carbonyl-labeled carbofuran (FMC Corp., radiochemical purity=93%) to monitor production of $^{14}CO_2$.

Field Evaluation of Chemical Rotations. During 1985-1988, experiments were conducted at the Univ. of Ill. Northwest Illinois Agricultural Research and Development Center near Monmouth, Illinois. A single 2-acre field was used over the 4 years of research. The soil was a Muscatine silt loam and the slope was ≤2% (soil 69, Table I). The cropping history prior to 1985 was: soybeans, 1980-82; wheat, 1983; corn, 1984. No soil insecticide had been used for at least three years prior to 1985. Chlorpyrifos (Lorsban 4E) was used for adult corn rootworm control during 1984.

The entire field was divided into four replicate blocks. Each block was randomly assigned 22 insecticide and check treatments (Table III). Each treatment consisted of four 20-m long rows of corn. Blocks and treatment rows were separated from one another by untreated corn rows, which served as buffer zones. The plot was chisel-plowed during the fall and field-cultivated in the spring. All tillage operations were

Table I. Characteristics of Soils Used in Studies of Chemical Rotation and Controlled-Release Formulations

Code	Soil Type or Assoication	Insecticides Used	Texture	pH	% Org. Carbon
11	Flanagan	none, UI borrow pit	silt loam	7.0	3.5
29	Andres-Reddick	carbofuran	clay loam	6.5	1.8
44	Westville-Pecatonica	phorate	silt loam	6.1	1.1
46	Plano-Proctor	carbofuran, terbufos	clay loam	5.7	1.5
52	Drummer Catlin	mixed, UI exp'tl. plot	silty clay loam	4.6	1.9
69	Muscatine	none prior to 1985	silt loam	5.7	2.5

Table II. Experimental Design for Testing the Effects of Insecticide Rotation on Persistence of Carbofuran

Days of Incubation	Insecticide Pretreatment Regime in Bucket No. 1	2	3	4	5
53	none	carb.	carb.	isofen.	isofen.
75	none	carb.	isofen.	carb.	isofen.
53	none	carb.	carb.	isofen.	isofen.
Code	NI	CCC	CIC	ICI	III

conducted along the north-south axis to avoid contamination of adjacent treatments with soil. Alachlor (Lasso 4EC) herbicide was applied preemergence at a rate of 2.8 kg AI/ha during all years, and atrazine was also used during 1988.

The insecticides were applied during planting in an 18-cm band over the furrow and incorporated with spring tines. The insecticides, which were applied at a rate of 1.12 kg AI/ha, included terbufos (Counter 15G), fonofos (Dyfonate 20G), and carbofuran (Furadan 15G). The sequence of multi-year and rotational treatments is shown in Table III. By 1988, the following histories had been established: (1) two, three, and four years of continuous use of a product; (2) first year use of a product on previously untreated plots; (3) all combinations of two

Table III. Insecticide treatments and coding [a]

	Insecticide History			
	Year			
Insecticide	1985	1986	1987	1988
Counter	C	CC	CCC	CCCC
		UC	UCC	UCCC
		FC	UUC	UUCC
		DC	CFC	UUUC
			CDC	FCFC
			CUC	DCDC
Dyfonate	D	DD	DDD	DDDD
		UD	UDD	UDDD
		FD	UUD	UUDD
		CD	DCD	UUUD
			DFD	CDCD
			DUD	FDFD
Furadan	F	FF	FFF	FFFF
		UF	UFF	UFFF
		CF	UUF	UUFF
		DF	FCF	UUUF
			FDF	CFCF
			FUF	DFDF
Untreated	U	UU	UUU	UUUU
		CU		CUCU
		DU		DUDU
		FU		FUFU

[a] Letters in each code represent insecticide treatments in each succeeding year.

chemical rotations over a four year period. During each year of the study, untreated plots served as controls for root-feeding damage.

Root damage ratings (18) were measured during July, and yields were determined by machine-harvesting in October. During 1987 and 1988 soil samples were collected for biological and chemical assay. Immediately following application and approximately 1.5 months later, four soil cores (5-cm diam. x 10-cm long) were randomly collected from between the second and third row of each treatment replicate (n=88 samples/day). The cores from an individual treatment were bulked together and sieved through a 3-mm screen. Two 25-g aliquots from each plot were weighed into 125-mL plastic cups and frozen until bioassayed by previously described methods (3,23). The remainder of the soil was frozen until extracted for quantitative analysis.

For statistical analyses, data were grouped by insecticide treatment during 1987 and 1988 and analyzed by the SAS General Linear Means Prodecure (20). If the analyis of variance was significant ($p \leq 0.1$) than the mean root ratings or insecticide concentrations were separated by Duncan's multiple range test at $p \leq 0.1$ (20).

Laboratory Testing of Controlled-Release Formulations. Two types of carbofuran formulations were tested, a gypsum-based roll-compacted granule manufactured by Union Carbide (Res. Triangle Pk., N. Carolina) and a proprietary controlled-release granule manufactured by Sierra Chem. Co. (Milpitas, CA). The persistence of carbofuran when applied as a gypsum granule (15% AI) was compared to the persistence of the commercial sand-based granule (FMC Corp, Princeton, NJ). Soil 29 (Table I), which was reported to be conditioned for enhanced biodegradation (8) was weighed (50 g oven-dry weight) into individual flasks and spiked with 0.01 g of granules to give a nominal concentration of 30 ppm. Soil was adjusted to 40% of MHC and incubated at 25°C. for up to 63 days. Triplicate flasks were removed periodically for analysis of carbofuran.

In a second laboratory experiment, the chemical persistence of carbofuran applied as Furadan 10G (the commercial sand-coated granule) and 2 Sierra controlled-release carbofuran formulations (samples no. 32-47-2, 9.1% AI and 32-48-2, 9.7% AI) were monitored for two months in three soil types, no. 29, 46, and 44 (Table I). Three-hundred forty-one grams (300 g oven-dry weight) of each soil were weighed into glass bottles and fortified with 1.5 g of one of the three formulations. The actual rates of application were 455 ppm (no. 32-47-2), 485 ppm (no. 32-48-2), and 500 ppm (Furadan 10G). The soils were thoroughly mixed and adjusted to 25% of MHC. The bottles were sealed with Parafilm and incubated at 25°C. At biweekly intervals, 20-g replicates of each soil were removed for quantitative analysis of carbofuran.

Field Testing of Controlled-Release Formulations. The efficacy of two additional Sierra controlled-release formulations (no. 32-45-2, 8.9% AI; no. 41-4-2, 9.3% AI) was tested in the field during the summer of 1982 in Kankakee Co.(soil 29) and Champaign Co.(soil 52) IL. The Sierra formulations were applied at the rate of 1.12 and 2.24 kg AI/ha in an 18-cm band over the row and lightly incorporated with a drag chain. The plots consisted of one 15.2-m long row of corn replicated 4 times in a random block design. The controlled-release formulations were compared to the commercial Furadan 15G formulation and an untreated check. In mid-July, 5 corn roots were randomly sampled from each replicate and rated for rootworm feeding damage (18). Soil samples were collected for carbofuran analysis immediately after application and 1 month later.

Analytical Methods. For all experiments, carbofuran was extracted from soil by slurrying with water and mixing with ethyl acetate (24). The extraction was normally repeated twice, but soils from the controlled release experiments were extracted three times. For the field studies of the controlled-release formulations, the extracts were passed through a 10-g Florisil column and eluted with 50 mL of 30% acetone in hexane to separate atrazine. Carbofuran concentration was determined by gas chromatography employing several different columns. For laboratory and field studies of controlled-release formulations, carbofuran was analyzed on a 60-cm or 45.7-cm, respectively, glass column (0.2 mm i.d.) of 5% OV-225 and detected by a Packard nitrogen-phosphorus detector. Column temperature was isothermal at 170° or 185°C, and the detector and injector were held at 250°C. For the rotation studies, the GLC column was 90 cm long x 0.2 cm i.d. and packed with 5% Apiezon + 0.125% DEGS as described by Felsot et al. (24). Isofenphos, fonofos, and terbufos were extracted and analyzed as described by Felsot et al. (25-26).

Results and Discussion

Cultivation-Time Studies. In nine of the 13 experiments analyzed, there was less root damage when carbofuran was applied during cultivation than when applied during planting (data not shown), although all root damage ratings were less than the economic threshold of 3.0. In three of the 13 experiments, carbofuran applied at planting time did not provide acceptable control, i.e., root damage ratings were greater than 3.0 (Table IV). Because the two Ames locations had been in commercial corn production and had received a planting-time banded application of carbofuran two of the preceding four years, it was assumed that the fields were conditioned for enhanced biodegradation. The Newell location was a test-plot site and only a few rows had received a carbofuran application in alternate years, but it seemed that the soil was also conditioned for enhanced biodegradation. In general, poor efficacy of carbofuran due to enhanced biodegradation could be circumvented by applying carbofuran at a time when the larvae were actively feeding.

Adult Suppression to Prevent Larval Damage. The 1977 adulticide treatments reduced root damage during the subsequent year in all six fields (Table V). In only one field (Mead Farm) was the 1978 root damage economically significant. The previous year's adulticide treatment, however, did provide good suppression of larval damage in this field.

Table IV. Average Root-Damage Ratings in Iowa Cornfields Treated with Carbofuran Applied During Planting and Cultivation[a]

		Time of Application		
Year	Location	Cultivation	Planting	Untreated
1976	Newell	3.00 a	3.70 a	4.52 b
1983	Ames	2.25 a	3.05 b	4.00 c
1985	Ames	2.65 a	3.30 a	3.75 a

a Means within a row followed by the same letter are not significantly different according to Duncan's Multiple Range Test ($p \leq 0.05$).

Table V. Average Root-Damage Ratings in Iowa Cornfields Following Adulticiding the Preceding Year to Suppress Adult Corn Rootworm Populations

		Root Damage Rating	
Location	Application Date	Adulticide	No Adulticide
Ranshau Farm	July 24	1.5	1.8
Grevengoed Farm	July 24	1.7	2.0
Goode Farm	July 29	2.2	2.4
Mead Farm	July 30	2.3	4.4
Hahn Farm	August 1	2.2	2.3
Perkins Farm	August 10	2.4	2.7

In the 1978 and 1979 experiments (Table VI), the root damage rating corresponding to no adulticide/no soil insecticide represented the root damage when no insecticides were used. The rating corresponding to no adulticide/soil insecticide represented the root protection provided by the soil insecticide alone at planting and served as a reference point for assessing the efficacy of adulticide treatments.

During 1978 two of the four fields sprayed (the Paysen and Harksen farms) had economic root damage the year following the adulticide applications (i.e., root ratings $\geq$ 3.0 in the no adulticide/no soil insecticide treatment). All but one of the adulticide treatments provided sufficient suppression of egg laying to avoid economic root damage the following year. In the Paysen field, the July 28 application of an adulticide reduced the subsequent damage, but not to the acceptable level of 3.0. While the subsequent root damage in the rest of the adulticide treatments was slightly greater than a soil insecticide alone, it was below the economic level. In fact, control was sufficiently good that the addition of a planting-time soil insecticide the following year only improved root protection slightly.

Table VI. Comparison of Adulticide Applied the Preceding Year with Soil Insecticides Applied During Planting for Protection from Corn Rootworm Larval Damage in Iowa

Location and Adulticide Treatment	Root Damage Rating	
	No Soil Insecticide	Soil Insecticide
1978		
Seesor Farm		
July 14 treatment	2.5	2.0
July 28 treatment	2.6	2.0
early + late treatment	2.0	1.5
no adulticide	2.7	1.9
Paysen Farm		
July 16 treament	2.4	2.4
July 28 treatment	3.8	1.9
early + late treatment	2.4	1.8
no adulticide	4.3	2.1
Barr Farm		
July 14 treament	2.6	--
August 2 treatment	2.4	--
early + late treatment	1.7	--
no adulticide	--	1.8
Harksen Farm		
July 19 treament	2.2	1.9
August 10 treatment	2.0	1.7
early + late treatment	2.0	1.7
no adulticide	3.4	1.7
1979		
Oster Farm		
July 28 treatment	2.2	1.9
no adulticide	2.1	2.6
Harksen Farm		
July 28 treatment	2.2	1.8
no adulticide	2.7	2.3
Shaffer Farm		
July 30 treament	1.9	1.8
August 2 treatment	--	--
early + late treatment	1.6	1.4
no adulticide	2.2	1.7

The subsequent year's root damage in all three fields that received an adulticide in 1979 was below economic levels. Under the light infestations, the prevention of larval damage by the adulticide sprays equalled the planting-time application of soil insecticides. Although carbofuran degradation was not characterized following treatment, the data suggested that larval damage could be successfully controlled without a soil insecticide if the adult population was suppressed the preceding year. The success of such a stratey would be highly dependent on scouting techniques to determine when beetle numbers had reached an economic threshold.

Laboratory Testing of Chemical Rotations. Carbofuran degraded most slowly in pretreatments of NI and III (Figure 1). The estimated 50% disappearance time (DT50%) based on the evolution of radiolabelled CO_2 was 22-24 days. In soil pretreated three times with carbofuran, DT50% dropped to 5 days (Figure 1), and after 42 days only 7% of the initially recovered carbofuran remained. When carbofuran was alternated with isofenphos, the DT50% was 8 days for the CIC rotation and 12 days for the ICI rotation (21-23% recovery of carbofuran based on chemical assays). Carbofuran biodegradation was still enhanced in pretreatments with insecticide rotation when compared to pretreatments without carbofuran, but the persistence was lengthened relative to the pretreatments with carbofuran alone.

Field Testing of Chemical Rotations. During 1987 and 1988, chemical assay of soils approximately 1.5 months after application was used as a benchmark for determining if multi-year use of an insecticide resulted in enhanced biodegradation. This time period should have coincided with active feeding by corn rootworm larvae, and therefore, effective insect control was expected to coincide with insecticide residues that were at least as high as published concentration-response estimates (3,23). Laboratory bioassays, root damage ratings, and yields were used to assess the effect of multi-year treatments and rotations on insecticide efficacy.

Recovery of terbufos and toxic metabolites during 1987 and 1988 did not differ significantly among any insecticide treatment history (Table VII, VIII). Percentage mortality did not differ among treatments, but the mortality in all treatments was noticeably lower during 1987 than during 1988, even though the soils were about 40% drier in 1988 (a drought year) than in 1987. This discrepancy can be attributed to the much lower recoveries during 1987 of parent terbufos (data not shown), which is more toxic than its metabolites (27-28). Root damage ratings were below the economic threshold of 3.0 in all treatments, and yields did not differ significantly from one another nor from the untreated check.

During 1987, the highest concentrations of fonofos were recovered from the first year (UUD) and rotational treatments (DFD, DCD, DUD), and the lowest concentrations were recovered from the three-year treatment (DDD) ($p \leq 0.055$, Table VII). These data suggested the development of enhanced biodegradation. The trend in residue recovery was more definitive of enhanced biodegradation during 1988 when 42 days after application nearly 3 to 5 times as much Dyfonate was recovered from first-year (UUUD) and rotational (CDCD, FDFD) treatments as from multi-year (DDDD, UDDD, UUDD) treatments ($p \leq 0.001$, Table VIII).

The pattern in percentage mortality during 1987 did not reflect the occurrence of enhanced biodegradation ($p \leq 0.478$) because fonofos concentrations were still at least equal to or greater than the published LC_{95} (1.88 ± 0.43) for third instar southern CRW (23). In 1988 mortality was lower in multi-year treatments than in first-year or rotational treatments ($p \leq 0.077$). The numerical trend in root damage ratings during 1988 reflected the evidence for enhanced biodegradation that was observed in the chemical and biological assays. Yields did not differ significantly among fonofos treatments nor from the untreated check.

In all treatments during 1987, nearly 90% of the applied carbofuran degraded within 45 days. The three-year carbofuran treatment (FFF) had the lowest residues, the lowest mortality, and the highest root damage ratings (Table VII). These data provide only weak evidence for enhanced biodegradation, however, because insecticide recovery in the two-year Furadan treatments (UFF, FUF) did not differ significantly from the first-year treatment (UUF) nor from the rotational treatments (FCF, FDF, FUF). During 1988, the lack of significant differences in residue recovery among treatments (Table VIII) may have reflected the influence of the very dry soils during the drought. The bioassays in 1988 showed a significant difference between four-year and first-year treatments although the insecticide residues did not differ. Interestingly, root damage ratings also reflected the trend of the bioassays (Table VIII). Neither percentage mortality nor root damage ratings for the rotational treatments (CFCF, DFDF) differed significantly from that for the multi-year treatments. As previously observed for terbufos and fonfofos, yields did not differ significantly among treatments.

Of the three insecticides studied, enhanced biodegradation has been definitively associated with repeated applications of carbofuran (6). In our studies, the chemical assay data from 1987 only weakly suggested the development of enhanced biodegradation, and the data from 1988 indicated that biodegradation rate was not affected by multi-year carbofuran treatments. Since a severe drought occurred in 1988, a lack of soil moisture may have slowed the rate of carbofuran degradation. Chapman et al. (29) and Harris et al. (16) noted that the

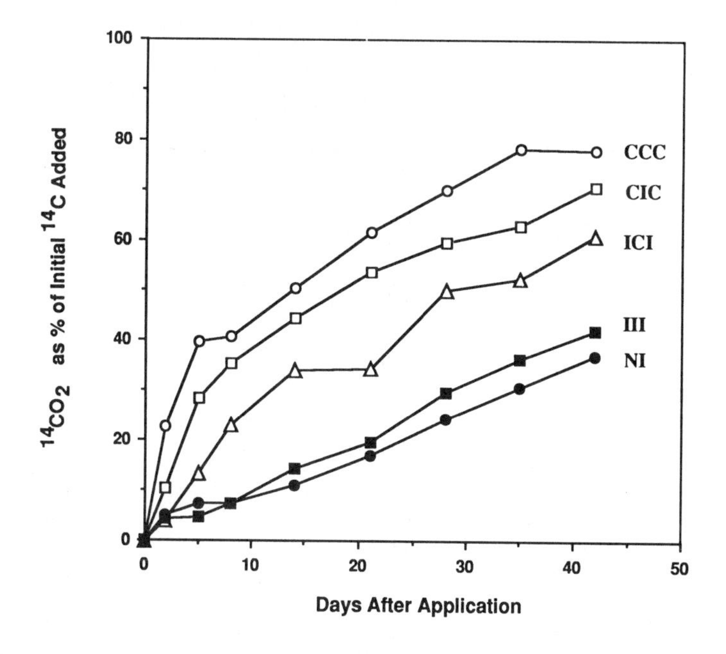

Figure 1. Recovery of $^{14}CO_2$ from ^{14}C-carbonyl-labeled carbofuran in soil 11 pretreated with carbofuran, isofenphos, or rotations of carbofuran and isofenphos. See text for treatment codes.

degradation of carbofuran in soil exhibiting enhanced biodegradation was very slow when moisture content was low.

In contrast to the chemical assay data, root damage ratings and bioassays of soils from carbofuran-treated plots strongly suggested enhanced biodegradation. The comparatively high mortalities during 1987 may be attributed to the recovery of residues at levels still close to the LC_{95} for southern CRW in soils of similar soil organic carbon content (i.e., 0.38-0.48 ppm, [3]) coupled with adequate soil moisture. The discrepancy between the chemical and biological assays during 1988 may be attributed to very low soil moisture that could inhibit the absorption of the insecticides by the larvae (2) in

Table VII. Effect of first-year use, multi-year use and rotations on efficacy and persistence of soil insecticides during 1987 at Monmouth, IL

Treatment[a]	Yield (bu)	Root Rating	45 days after application: Percent Mortality	45 days after application: Concentration (ppm)
U	131.8	4.21	18.1	0.000
CCC	137.1	2.65	76.3	1.552
UCC	148.0	2.70	86.3	1.581
UUC	134.0	2.20	70.0	1.767
CFC	136.4	2.60	87.5	1.766
CDC	134.0	2.35	81.3	2.194
CUC	139.0	2.25	70.0	1.948
$p \leq$	0.665	0.298	0.925	0.813
DDD	131.4	3.05a	100.0	1.478c
UDD	149.5	2.70abc	96.3	1.831bc
UUD	138.0	2.28bdc	100.0	2.504a
DCD	137.6	2.80ab	100.0	2.924a
DFD	144.6	2.10d	100.0	2.955a
DUD	136.7	2.15dc	88.8	2.525ab
$p \leq$	0.257	0.037	0.478	0.055
FFF	139.6	2.45a	77.5	0.244 b
UFF	137.1	2.15ab	93.8	0.383 ab
UUF	147.4	1.90b	100.0	0.388 ab
FCF	130.1	1.90b	98.8	0.308 b
FDF	141.4	2.35a	98.8	0.304 b
FUF	128.5	2.25ab	97.5	0.520 a
$p \leq$	0.359	0.098	0.142	0.043

[a] Means followed by the same letter within similar insecticide groups are not significantly different according to Duncan's Multiple Range Test ($p \leq 0.1$).

addition to inhibiting biodegradation and causing increased variability in sampling.

Regardless of whether or not enhanced biodegradation of carbofuran had really developed, rotation of carbofuran with terbufos or fonofos did not significantly improve root damage ratings. On the other hand, fonofos persistence was significantly shortened by multi-year treatments during 1987 and 1988, but rotation with either terbufos or carbofuran seemed to augment recovery of insecticide residues and root protection. During 1988, the recovery of fonofos from rotational treatments, however, was still significantly lower than from the first-year treatment.

Read (30) has observed that rotations of fensulfothion and carbofuran in rutabaga fields failed to prevent the enhancement of biodegradation of either compound. Based on efficacy tests, he has suggested alternating the use of carbofuran or fensulfothion with the use of chlorfenvinphos or terbufos (31). Another solution was a 3-4 year crop rotation with use of an insecticide only once in that interval.

Laboratory Testing of Controlled-Release Formulations. The objective of using controlled-release formulations was to slow the rate of biodegradation without affecting toxicity by more slowly releasing the insecticide into the soil water matrix. Carbofuran persisted longer when applied to soil 29 as a roll-compacted gypsum granule than when applied as the commercial sand granule (Figure 2). Within 1 month, 35% of the initial carbofuran concentration remained after treatment with the gypsum granule, but only 0.4% of the carbofuran was recovered from soil treated with the sand granule. The roll-compacted granule was made by fluxing technical carbofuran with calcium sulphate and pressing the mixture into sheets which are then dried and cut into prills. For the sand granule, carbofuran was simply coated onto the surface of the particles. The difference in persistence between the two formulations may have been due to the slightly delayed diffusion of the carbofuran from inside the gypsum granule, assuming adsorbed carbofuran was not available for biodegradation. The "inhibitory" effect of adsorption processes in soil upon biodegradation rate has been suggested by research on other pesticides (e.g., 32-36).

More persuasive evidence for the ability of delayed release rates to prolong persistence was observed in studies of two Sierra controlled-release formulations in three soil types. In soils 29 and 46, which had been pretreated in the field with carbofuran and were known for insect control failures, DT50% of carbofuran applied as a sand granule was 14 and 37 days, respectively, but increased to 52 days in soil 44 (Figure 3). In all three soils, less than 15% of the initially added carbofuran was recovered within 2 months. During the same interval, however, carbofuran concentration did not change significantly in soils treated with Sierra controlled-release formulations 32-47-2 and 32-48-2. Apparently, carbofuran was held tightly within a hydrophobic matrix that releases very slowly only small amounts of carbofuran. The released carbofuran would presumably have been degraded by adapted microorganisms, but carbofuran left in the matrix would have been biologially unavailable.

The high variability in recovery of carbofuran from the soils treated with the granular formulations reflected sampling error associated with subsampling a heterogeneous distribution of granules despite having mixed the soil before sampling. This variabilitity over the 2 month

Table VIII. Effect of first-year use, multi-year use and rotations on efficacy and persistence of soil insecticides during 1988 at Monmouth, IL

Treatment[a]	Yield (bu/acre)	Root Rating	42 days after application: Percent Mortality	42 days after application: Concentration (ppm)
U	10.0	4.26	21.0	0.000
CCCC	13.0	2.70	100.0	1.794
UCCC	15.0	2.70	100.0	3.693
UUCC	9.0	2.55	97.5	1.936
UUUC	5.9	2.70	95.0	1.681
FCFC	9.2	2.70	100.0	3.288
DCDC	14.9	2.50	97.0	1.755
$p \leq$		0.945	0.398	0.514
DDDD	9.8	3.60 a	80.0 b	0.797 a
UDDD	5.8	3.33 ab	91.3 ab	0.254 a
UUDD	11.4	3.30 ab	90.0 ab	0.664 a
UUUD	5.4	2.80 ab	99.0 a	3.578 b
CDCD	12.3	2.65 c	100.0 a	2.004 c
FDFD	11.6	2.85 bc	100.0 a	2.238 c
$p \leq$	0.469	0.015	0.077	0.001
FFFF	12.4	3.90 a	37.5 b	2.028
UFFF	7.7	4.20 a	60.0 ab	3.120
UUFF	12.6	4.15 a	36.3 b	1.320
UUUF	11.9	2.70 b	97.5 a	1.849
CFCF	12.2	3.60 a	72.5 ab	2.588
DFDF	13.1	3.85 a	56.3 ab	4.535
$p \leq$	0.890	0.002	0.011	0.408

[a] Means followed by the same letter within similar insecticide groups are not significantly different according to Duncan's Multiple Range Test ($p \leq 0.1$).

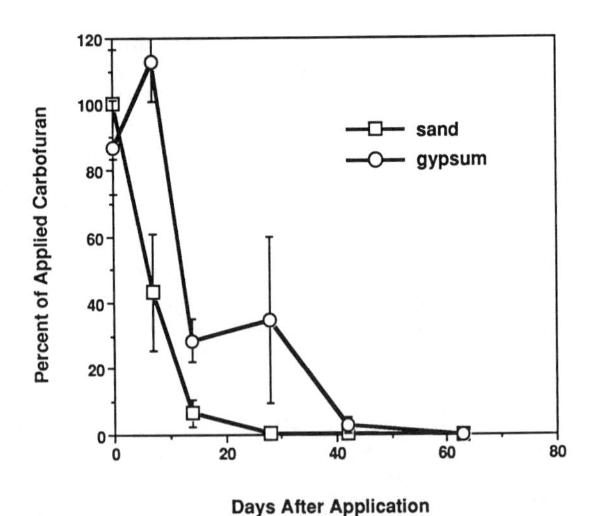

Figure 2. Loss of carbofuran from soil 29 after treatment with a coated sand formulation of carbofuran (Furadan 15G) or a gypsum-based granule.

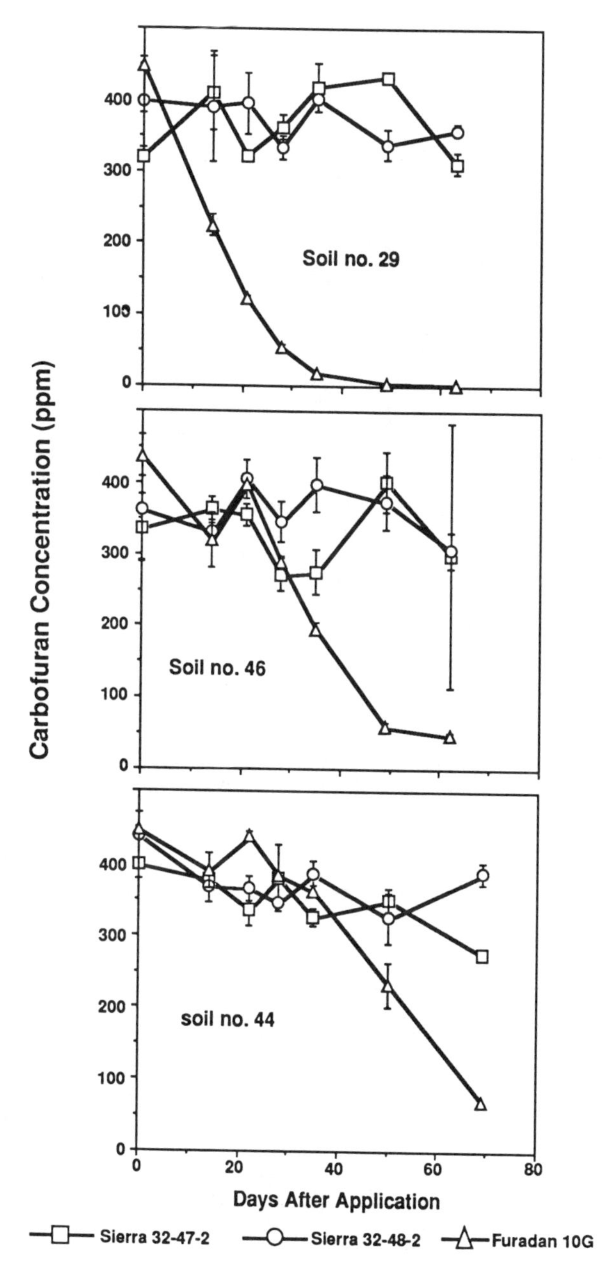

Figure 3. Persistence of carbofuran in soil 29, 46, and 44 after treatment with Sierra controlled-release formulations or Furadan 10G.

incubation period also suggested that the carbofuran was not being released into the soil solution.

Field Testing of Controlled-Release Formulations. Sampling one month after application was deemed a useful benchmark for indirectly assessing formulation efficacy because corn rootworms would have hatched and been actively feeding as first and second instar larvae. In the carbofuran-conditioned field (soil 29), about 4 times as much carbofuran was recovered from the subplots treated with Sierra formulations at 1 kg AI/ha as from the subplot treated with Furadan 15G (Table IX). In the nonconditioned soil (soil 52), carbofuran concentration in the Furadan 15G treatment did not differ significantly from the concentration in the subplots treated with Sierra formulation 32-45-2. Recovery of carbofuran was 2-3 times higher from the subplot treated with 41-4-2, but this difference may only reflect sampling variability and not necessarily a significantly different persistence pattern.

As indicated by the root damage ratings in the untreated checks, rootworm populations in field 52 were too low to assess properly the effects of controlled release on efficacy. In field 29, however, bioactivity of Sierra formulation 32-45-2 was similar to Furadan 15G at both rates, and bioactivity of Sierra 41-4-2 at the 1X rate seemed even less efficacious than Furadan 15G. The results suggested that even though concentration of carbofuran was significantly higher in subplots treated with the controlled-release formulations, the bioavailability of carbofuran for absorption by rootworms and microorganisms was restricted compared to the Furadan

Table IX. Root damage ratings and concentration of carbofuran in two soils

Formulation	Application Rate (kg/ha)	Root Rating	Concentration @ 1 mo (ppm ± std)
		soil no. 29	
Furadan 15G	1.12	2.9	0.26 ± 0.39
Sierra 32-45-2	1.12	2.5	1.01 ± 0.39
Sierra 32-45-2	2.24	3.4	5.28 ± 2.73
Sierra 41-4-2	1.12	4.0	1.20 ± 0.31
Sierra 41-4-2	2.24	2.8	3.51 ± 1.30
Check	-	4.1	< 0.02
		soil no. 52	
Furadan 15G	1.12	2.0	1.50 ± 0.63
Sierra 32-45-2	1.12	2.2	1.60 ± 0.43
Sierra 32-45-2	2.24	2.3	1.74 ± 0.73
Sierra 41-4-2	1.12	3.1	5.52 ± 1.55
Sierra 41-4-2	2.24	2.4	4.20 ± 2.10
Check	-	3.2	< 0.02

15G formulation. In related work with aldicarb, Coppedge et al. (37) concluded that slow-release formulations may extend bioactivity over a longer time but not improve efficacy of compounds with moderate or low insecticidal activity at normal application rates. Pertinently, of the currently registered soil insecticides, carbofuran is the least toxic to CRW (23), and therefore, rate of release would be critical to its bioactivity.

CONCLUSIONS

Application of carbofuran during cultivation time rather than at planting time circumvented enhanced biodegradation by providing a higher dose of chemical when the larvae were actively feeding. The success of such a practice would depend on favorable soil conditions during a relatively short time in early June. Of course, this practice would not prevent the development of enhanced biodegradation. On the other hand, the use of adulticiding during August of the preceding growing season would effectively eliminate the development of enhanced biodegradation. First, insecticide would not be applied directly to the ground, and much of it would be intercepted by a well-developed crop canopy. Secondly, the dosage per acre could probably be reduced by using carbaryl or some other insecticide with feeding arrestants or attractants (38). The experiments presented here showed that control of root damage by adult suppression was as effective as the use of soil insecticides.

Multi-year insecticide treatments did not affect the biodegradation of terbufos but definitely enhanced the biodegradation of fonofos. The effect of multi-year carbofuran treatments on biodegradation was inconclusive. Chemical rotations reduced the impact from enhanced biodegradation of fonofos but had no effect on carbofuran efficacy. Because corn rootworm populations were only moderate during the four years of this study, yields were unaffected by any insecticide treatments. As a long term chemical management strategy, the data suggested that insecticide rotations may be advantageous for reducing but not necessarily eliminating the impacts of enhanced biodegradation, but the rotations do not have to be strictly between classes (i.e., organophosphate/carbamate).

Controlled-release formulations extended the persistence of carbofuran in soils exhibiting enhanced biodegradation. Owing to the comparatively low susceptibility of CRW larvae to carbofuran, and the apparently very slow release of carbofuran from the Sierra controlled-release granules, carbofuran efficacy at normal application rates was not significantly improved from that of the coated sand granule.

In developing practices for the long-term management of pesticide use and avoidance of enhanced biodegradation, it must be recognized that microbial adaptations for metabolism are natural processes that cannot be eliminated completely. Problems with this generally beneficial phenomenon arise only when crop protection practices are adversely affected, and at best we can only hope to cope with enhanced biodegradation. The experiments presented here show that some of the proposed strategies for coping have validity and need further exploration under a wide variety of field conditions.

Literature Cited

1. Adkisson, P. L. Bull. Entomol. Soc. Am. 1986, 32, 136-41.
2. Harris, C. R. Ann. Rev. Entomol. 1972, 17, 177-98.
3. Felsot, A. S.; Lew, A. J. Econ. Entomol. 1989, 82, 389-95.
4. Walker, A.; Suett, D. L. Aspects Appl. Biol. 1986, 12, 95-103.
5. Suett, D. L.; Walker, A. Aspects Appl. Biol. 1988, 17, 213-222.
6. Felsot, A. S. Ann. Rev. Entomol. 1989, 34, 453-76.
7. Felsot, A.; Maddox, J. V.; Bruce W. Bull. Environ. Contam. Toxicol. 1981, 26, 781-8.
8. Felsot, A. S.; Wilson, J. G.; Kuhlman, D. E.; Steffey, K. L. J. Econ. Entomol. 1982, 75, 1098-1103.
9. Harris, C. R.; Chapman, R. A.; Harris, C.; Tu, C. M. J. Environ. Sci. Health 1984, B19, 1-11.
10. Dzantor, E. K.; Felsot, A. S. J. Environ. Sci. Hlth 1989, B24, 569-597.
11. Abou-Assaf, N.; Coats, J. R.; Gray, M. E.; Tollefson, J. J. J. Environ. Sci. Health 1986, B21, 425-46.
12. Chapman, R. A.; Harris, C. R.; Moy, P.; Henning, K.. J. Environ. Sci. Health 1986, B21, 269-76.
13. Racke, K. D.; Coats J. R. J. Agric. Food Chem. 1987, 35, 94-99.
14. Harris, C. R.; Chapman, R. A.; Morris, R. F.; Stevenson, A. B. J. Environ. Sci. Health 1988, B23, 301-316.
15. Horng, L. C.; Kaufman, D. D. Abstracts 193rd Nat'l. Mtg. Am. Chem. Soc., Denver, CO. Am. Chem. Soc.: Washington, DC, 1987.
16. Harris, C. R.; Chapman, R. A.; Tolman, J. H.; Moy, P.; Henning, K., Harris, C. J. Environ. Sci. Health 1988, B 23, 1-32.
17. Racke, K. D.; Coats, J. R. J. Agric. Food Chem. 1988, 36, 1067-72.

18. Hills, T. M.; Peters, D. C. J. Econ. Entomol. 1971, 64, 764-5.
19. Foster, R. E.; Tollefson, J. J.; Nyrop, J. P.; Hein, G. L. J. Econ. Entomol. 1986, 79, 303-10.
20. SAS Institute. SAS user's guide: statistics. SAS Institute, Cary, NC, 1982.
21. Duncan, D. B. Biometrics 1955, 11, 1-42.
22. Bartha, R.; Pramer, D. Soil Sci. 1965, 100, 68-70.
23. Sutter, G. R. J. Econ. Entomol. 1982, 75, 489-91.
24. Felsot, A. S.; Steffey, K. L.; Levine, E.; Wilson, J. G. J. Econ. Entomol. 1985, 78, 45-52.
25. Felsot, A. J. Environ. Sci. Health 1984, B19, 13-27.
26. Felsot, A. S.; Bruce, W. N.; Steffey, K. L. Bull. Environ. Contam. Toxicol. 1987, 30, 369-77.
27. Chapman, R. A.; Harris, C. R. J. Econ. Entomol. 1980, 73, 536-43.
28. Felsot, A. S. Proc. 37th Illinois Custom Spray Operators Training School, 1985, Coop. Ext. Serv, Univ. Ill., pp 134-8.
29. Chapman, R. A.; Harris, C. R.; Harris, C. J. Environ. Sci. Health 1986, B21, 125-41.
30. Read, D. C. Agric. Ecosyst. Environ. 1983, 10, 37-46.
31. Read, D. C. Agric. Ecosyst. Environ. 1986, 16, 165-73.
32. Iwata, Y.; Westlake, W. E.; Gunther, F. A. Arch. Environ. Contam. Toxicol. 1973, 1, 84-96.
33. Moshier, L. J.; Penner, D. Weed Sci. 1978, 26, 686-691.
34. Burkhard, N.; Guth, J. A. Pestic. Sci., 1981, 12, 45-52.
35. Felsot, A.; Wei, L.; Wilson, J. J. Environ. Sci. Health, 1982, B17, 649-673.
36. Steinberg, S. M.; Pignatello; Sawhney, B. L. Environ. Sci. Technol. 1987, 21, 1201-1208.
37. Coppedge, J. R.; Stokes, R. A.; Ridgway, R. L.; Kinzer, R. E. Slow-release formulations of aldicarb. Modeling of soil persistence. USDA Public. No. ARS-S-103, 1976, USDA-ARS, 6 pp.
38. Metcalf, R. L.; Ferguson, J. E.; Lampman, R.; Andersen, J. F. J. Econ. Entomol. 1987, 80, 870-875.

RECEIVED February 9, 1990

Chapter 16

Systems Allowing Continued Use of Carbamothioate Herbicides Despite Enhanced Biodegradation

Robert G. Harvey

Department of Agronomy, University of Wisconsin, Madison, WI 53706

Enhanced biodegradation of EPTC was reduced by applying the herbicide not more frequently than every other year. The extender dietholate did not prevent enhanced EPTC degradation when the two chemicals were used repeatedly together, possibly due to a demonstrated enhancement of dietholate biodegradation. Enhanced biodegradation of the experimental extender SC-0058 was not detected, and SC-0058 prevented enhanced EPTC biodegradation. Results suggest that by combining herbicide and crop rotation, use of chemical extenders, row cultivation, and late-season secondary herbicide applications, long-term control of weeds such as wild proso millet can be obtained.

Since EPTC (S-ethyl dipropyl carbamothioate) was synthesized in 1954, it has been used extensively to control annual grasses in alfalfa (*Medicago sativa* L.), edible beans (*Phaseolus vulgaris* L.), potatoes (*Solanum tuberosum* L.) and other crops (1). Commercialization of the crop protectant dichlormid (2,2-dichloro-N,N-di-2-propenylacetamide) in the early 1970's allowed EPTC to be safely applied in field and sweet corn (*Zea mays* L.) to control hard-to-kill weed species such as shattercane [*Sorghum bicolor* (L.) Moench.], wild proso millet [*Panicum miliaceum* L. spp. *ruderale* (Kitagawa) Tzevelev.], and yellow nutsedge (*Cyperus esculentus* L.) (1-3).

Discovery of enhanced EPTC biodegradation in soils previously treated with the herbicide (2,4-7) led to the commercialization of a formulated mixture of EPTC plus dietholate (O,O-diethyl-O-phenol phosphorothioate) for control of hard-to-kill weeds in corn. Dietholate extended the length of time EPTC controlled weeds by inhibiting enhanced EPTC biodegradation. Dietholate and chemicals with similar activity are thus referred to as extenders. However, EPTC applied with dietholate was also discovered to be vulnerable to

0097–6156/90/0426–0214$06.00/0

enhanced biodegradation (2,5,8). Problems with enhanced soil biodegradation of EPTC formulations made controlling hard-to-kill weeds in field and sweet corn more difficult. The following research and literature summary describes how enhanced biodegradation has affected wild proso millet control in sweet corn and proposes systems to resolve the problem.

Methodology

Wild proso millet control studies were conducted in southern Wisconsin at various locations and with small-plot procedures described elsewhere (3,5,8) (Harvey, R. G.; McNevin, G. R. *Weed Technol*., in press) (Harvey, R. G. *Weed Sci*., in press). EPTC was applied in a commercial formulation containing either dichlormid or R-29148 [3-(dichloroacetyl)-2,2,5-trimethyloxazolidine] to prevent sweet corn injury. Other herbicides included in field studies were commercial formulations of alachlor [2-chloro-*N*-(2,6-diethylphenyl)-*N*-(methoxymethyl)acetamide], ametryn [*N*-ethyl-*N*'-(1-methylethyl)-6-(methylthio)-1,3,5-triazine-2,4-diamine], cyanazine {2-[[4-chloro-6-(ethylamino)-1,3,5-triazin-2-yl]amino]-2-methyl propanenitrile}, cycloate (*S*-ethyl cyclohexylethyl carbamothioate), and pendimethalin [*N*-(1-ethylpropyl)-3,4-dimethyl-2,6-dinitro benzenamine]. EPTC was also applied with the experimental extender SC-0058 [*S*-ethyl-*N*,*N*-bis(3-chloroallyl)carbamothioate]. Herbicide and extender dissipation studies were conducted in the laboratory by mixing 6 mg/kg w/w EPTC and 1 mg/kg w/w dietholate or SC-0058 with soil samples collected from field plots with known histories of prior carbamothioate herbicide use. After incubation at 25 C for time periods determined by preliminary tests with the same soil, herbicide residues were extracted from soil by steam distillation, and analyzed by gas-liquid chromatography. Reference standards were prepared similarly, but were frozen immediately after treatment. Details of these procedures are described elsewhere (5,8). Percent herbicide or extender recovered was estimated by comparing residues in samples of the same soil incubated as described above or frozen immediately after treatment. Enhanced degradation was presumed to occur when percent recovery was lower in soils with histories of prior herbicide or extender use than those without. Data were subjected to analysis of variance and means compared using Duncan's multiple range test. In the tables which follow, all means within the same column followed by the same letter are not different at the 0.05 alpha level.

Field Studies

Location Not Previously Treated With EPTC. Studies conducted in 1978 and 1979 demonstrated that EPTC at both 4.5 and 6.7 kg ai/ha application rates provided effective wild proso millet control in sweet corn grown on soils not previously treated with that herbicide (Table I). Because of the prolonged period of millet germination and the short soil persistence of EPTC, millet control 95 days after treatment (DAT) was substantially less than 45 DAT. However, these late-season millet escapes did not cause sweet corn yield reductions. Applying EPTC with cyanazine slightly improved late-season millet control compared to applying EPTC alone. When results of all 44 treatments in the study were summarized, early- and late-season millet control and sweet corn yields in the 19 treatments including EPTC were substantially greater than in the 25 treatments not including EPTC.

Table I. Early- and late-season wild proso millet control and sweet corn yields resulting from first-time herbicide treatments

Herbicide(s)	Rate (kg/ha)	Percent millet control 45 DAT	Percent millet control 95 DAT	Sweet corn yield (Mg/ha)
No herbicide	---	0 c	0 e	0.2 c
Handweeded	---	94 a	97 a	12.1 a
EPTC	4.5	94 a	57 c	10.1 a
	6.7	96 a	63 c	10.8 a
EPTC + cyanazine	6.7+2.2	97 a	76 b	11.9 a
Average results of 19 combinations which included EPTC		97 a	70 bc	11.2 a
Average results of 25 combinations without EPTC		68 b	20 d	2.5 b

Location Previously Treated With EPTC Alone. When applied in 1980 to a field treated the four previous years with EPTC, EPTC at 6.7 kg/ha alone or followed by (fb) preemergence cyanazine provided satisfactory millet control 30 DAT, but not at 57 and 110 DAT (Table II). EPTC plus dietholate provided better season-long millet control and significantly greater sweet corn yields than did EPTC applied alone or followed by cyanazine. Dietholate obviously reduced the enhanced biodegradation of EPTC in the soil. Following an application of EPTC with either a single row cultivation or a postemergence-directed application of ametryn provided season-long millet control and resulted in sweet corn yields comparable to EPTC plus dietholate, thereby overcoming the effects of enhanced EPTC degradation.

Table II. Wild proso millet control and sweet corn yields from herbicide treatments on soil previously treated with EPTC

Herbicide(s)	Rate (kg/ha)	Percent millet control 30 DAT	Percent millet control 57 DAT	Percent millet control 110 DAT	Sweet corn yield (Mg/ha)
No herbicide	---	0 b	0 d	0 e	0.2 d
Handweeded	---	96 a	95 a	95 a	18.4 a
EPTC	6.7	73 a	8 d	8 e	1.3 d
EPTC fb cyanazine	6.7+2.2	85 a	23 c	21 d	4.7 c
EPTC plus dietholate	6.7	95 a	69 b	61 c	11.6 b
EPTC fb cultivation	6.7	80 a	64 b	50 b	11.4 b
EPTC fb ametryn	6.7+2.2	84 a	92 a	79 b	12.3 b

Location Previously Treated With EPTC Plus Dietholate. When applied in 1983 to a field treated the previous year with EPTC plus dietholate, 6.7 kg/ha EPTC with and without dietholate resulted in poor millet control 40, 70 and 100 DAT, and extremely low sweet corn yields (Table III). In this soil, enhanced EPTC biodegradation occurred whether applied with or without dietholate (8). Combining cyanazine with EPTC plus dietholate only increased millet control 40 DAT compared to EPTC plus dietholate without cyanazine, and did not improve sweet corn yield. Following EPTC plus dietholate with an early postemergence application of pendimethalin plus cyanazine resulted in millet control 40 DAT comparable to the handweeded treatment, and millet control 70 and 100 DAT and sweet corn yields greater than those resulting from EPTC plus dietholate plus cyanazine treatment. The contribution from the EPTC plus dietholate component of this sequential treatment was apparently small, however, since results were similar to those from pendimethalin plus cyanazine applied alone.

Table III. Wild proso millet control and sweet corn yields from herbicide treatments on soil previously treated with EPTC + dietholate

Herbicide(s)	Rate (kg/ha)	Percent millet control 40 DAT	70 DAT	100 DAT	Sweet corn yield (Mg/ha)
No herbicide	---	0 c	0 c	0 c	0.2 c
Handweeded	---	95 a	96 a	98 a	11.4 a
EPTC	6.7	52 b	11 c	0 c	0.2 c
EPTC plus dietholate	6.7	56 b	15 c	6 c	0.2 c
EPTC plus dietholate + cyanazine	6.7+2.2	77 a	25 c	8 c	0.2 c
EPTC plus dietholate fb pendimethalin + cyanazine	6.7 fb 1.7+2.2	94 a	67 b	50 b	7.4 b
Pendimethalin + cyanazine	1.7+2.2	95 a	67 b	36 b	5.2 b

Six-year Herbicide Rotation Study. Mean wild proso millet control obtained in a study conducted from 1984 to 1989 is summarized in Table IV. Since the field had been treated with 6.7 kg/ha EPTC in 1983, enhanced EPTC biodegradation presumably was present from the beginning of the study. When applied with cyanazine annually over the six-year period, EPTC and EPTC plus dietholate provided only 46 and 64% millet control, respectively. Laboratory studies confirmed that this poor millet control was due to enhanced biodegradation of the active ingredient (Harvey, R. G. Weed Sci., in press). Every other year and every third year applications of EPTC plus cyanazine and EPTC with dietholate plus cyanazine provided 83 and 86, and 88 and 91% millet control, respectively. Rotating EPTC plus cyanazine or EPTC plus dietholate plus cyanazine treatments with applications of cycloate plus cyanazine, or alachlor plus cyanazine reduced the biodegradation rate of EPTC applied alone or with dietholate (unpublished data).

Continuous use of cycloate plus cyanazine over the six-year period provided an average of 89% millet control even though cycloate is a carbamothioate herbicide (Table IV). Previous studies have shown little potential for enhanced biodegradation of cycloate (8). Alachlor plus cyanazine or pendimethalin plus cyanazine treatments, often considered as alternatives to EPTC for millet control, did not provide better millet control than combinations of cyanazine plus EPTC or EPTC plus dietholate when the latter treatments were used no more frequently than every other year.

Table IV. Wild proso millet control with EPTC and EPTC with dietholate in a six-year herbicide rotation study

Treatment	Average percent millet control over 6 year study
EPTC + cyanazine at 6.7 + 2.2 kg/ha:	
Six repeated annual applications	46 d
Alternate year applications	83 ab
Every third year applications	86 ab
EPTC plus dietholate + cyanazine at 6.7 + 2.2 kg/ha:	
Six repeated annual applications	64 cd
Alternate year applications	88 ab
Every third year applications	91 a
Other herbicides applied annually over 6 yr:	
Cycloate + cyanazine at 6.7 + 2.2 kg/ha	89 a
Alachlor + cyanazine at 4.5 + 2.2 kg/ha	80 abc
Pendimethalin + cyanazine at 1.7 + 2.2 kg/ha	70 bc

Table V. Wild proso millet control and sweet corn yields following three annual applications of EPTC alone and with extenders

Treatment used over 3 yr period		Percent millet	Sweet corn
Herbicide(s)	Rate (lb/A)	control 40 DAT	yield (Mg/ha)
EPTC plus dietholate	6.7+1.1	73 b	7.6 b
EPTC plus SC-0058	6.7+1.1	85 a	11.0 a

Dietholate Versus SC-0058. Effectiveness of the extenders dietholate and SC-0058 (experimental extender currently being investigated by ICI Americas Inc.) to prevent reductions in wild proso millet control due to enhanced biodegradation was evaluated. The combined results of three separate tests including plots treated annually for three-year periods with the respective herbicide - extender combinations are presented in Table V. Even after three consecutive applications, 6.7 kg/ha EPTC plus 1.1 kg ai/ha SC-0058 provided 85% millet control 40 DAT. The same rates of EPTC plus dietholate provided only 73% millet control 40 DAT. Use of SC-0058 also resulted in 45% greater sweet corn yields. Thus SC-0058 is more effective than dietholate in preventing enhanced EPTC degradation. Similar results have been reported elsewhere (Harvey, R. G. Weed Sci., in press).

Laboratory Studies

Biodegradation Of EPTC And Dietholate. Laboratory dissipation studies conducted in soils obtained from the herbicide rotation study described above demonstrated enhanced biodegradation of EPTC applied with and without dietholate, and also enhanced biodegradation of dietholate in soils previously treated with EPTC plus dietholate (Table VI). A more complete report of enhanced dietholate biodegradation appears elsewhere (Harvey, R. G. Weed Sci., in press). Enhanced dietholate degradation occurs when the extender is applied alone, with EPTC, or with butylate [S-ethyl bis(2-methylpropyl)carbamothioate]. Studies also show that enhanced dietholate degradation persists two years following a previous dietholate application (unpublished data). It has not yet been determined whether enhanced biodegradation of dietholate is responsible for poor weed control occurring when EPTC plus dietholate is applied to soils previously treated with the same combination or whether soil microbes responsible for EPTC degradation can circumvent the inhibitory effects via another mechanism. Previous studies in Nebraska (9) failed to show enhanced dietholate degradation even when enhanced degradation of EPTC applied with dietholate was detected. Soils utilized in the Nebraska study originated in an area of lower annual rainfall and had shorter histories of previous dietholate use than the soils used in Wisconsin studies. Soil drying has been shown to eliminate enhanced EPTC degradation (2), and a shorter history of dietholate use may have minimized the expression of enhanced dietholate degradation.

Table VI. Laboratory dissipation of EPTC alone or EPTC and dietholate in soils with different herbicide use histories

Prior herbicide use history	Percent EPTC and dietholate recovered		
	EPTC	EPTC	Dietholate
No herbicide	55 a	46 a	42 a
EPTC	6 b	6 b	39 a
EPTC w/ dietholate	9 b	15 b	17 b

Biodegradation Of EPTC, Dietholate And SC-0058. The percentages of EPTC and extenders persisting five days after application to soils with different histories of herbicide and extender application over the previous three years are shown in Table VII. Enhanced biodegradation of EPTC occurred when applied alone to soils previously treated with EPTC alone, with dietholate, and with SC-0058, and when applied with dietholate to soil previously treated with EPTC plus SC-0058. Enhanced dietholate biodegradation occurred when applied with EPTC to soil previously treated with EPTC plus dietholate. Enhanced EPTC biodegradation was prevented by applying the herbicide with SC-0058. Enhanced biodegradation of SC-0058 did not occur.

Table VII. Dissipation of EPTC and extenders applied to soils with different histories of previous herbicide and extender use

Herbicide and extenders applied for 3 yr previously and in dissipation study	Percent herbicide or extender persisting 5 DAT		
	EPTC	Dietholate	SC-0058
Soil previously untreated:			
EPTC alone	64 a	--	--
EPTC w/ dietholate	72 a	57 a	--
EPTC w/ SC-0058	66 a	--	74 a
Soil previously treated with EPTC:			
EPTC alone	21 de	--	--
EPTC w/ dietholate	57 a	80 a	--
EPTC w/ SC-0058	48 abc	--	65 a
Soil previously treated with EPTC w/ dietholate:			
EPTC alone	12 e	--	--
EPTC w/ dietholate	7 e	8 b	--
EPTC w/ SC-0058	35 bcd	--	53 a
Soil previously treated with EPTC w/ SC-0058:			
EPTC alone	26 cde	--	--
EPTC w/ dietholate	63 a	82 a	--
EPTC w/ SC-0058	50 ab	--	76 a

Discussion

Even when not subject to enhanced biodegradation, EPTC does not always provide season-long control of wild proso millet in sweet corn, but the herbicide is one of the better chemical options for controlling the weed. Enhanced biodegradation, however, limits use of the herbicide in consecutive seasons if acceptable millet control is desired. Use of EPTC only every other year will help maintain effective millet control with that herbicide. Such alternate year use may not solve millet problems if sweet corn is grown repeatedly in the same field unless effective alternative control tools are available. In the absence of acceptable alternatives, sweet corn could be rotated with other crops. Crops such as alfalfa, wheat (*Triticum aestivum* L.), and peas (*Pisum sativum* L.) usually compete with, and shade out wild proso millet in the absence of herbicides (10). Effective and economical postemergence herbicides are available to control millet in soybeans (10). In addition to reducing enhanced EPTC biodegradation, crop rotations can also reduce millet populations by reducing the number of plants which must be controlled with the herbicide (3). Despite crop and herbicide rotations, season-long millet control won't always be provided by EPTC. Combining EPTC treatments with cyanazine, or following with early postemergence pendimethalin plus cyanazine or postemergence-directed ametryn treatments will improve season-long millet control. Following EPTC applications with one or more row cultivations will also greatly improved EPTC performance even in the presence of enhanced degradation, and is consistent with current interests in protecting groundwater and increasing sustainability of farm operations.

The data presented from sweet corn research also are applicable to weed control problems in field corn. Herbicides used in sweet corn may also be used in field corn. Sweet corn is typically planted in wider rows and in lower populations than is field corn (3), and sweet corn hybrids are typically shorter than field corn hybrids. These factors cause sweet corn to be less competitive with weeds than field corn, and weed problems usually are more severe in sweet corn than field corn. Thus, systems effective for controlling weeds in sweet corn are usually even more effective in field corn.

Data presented do not directly address enhanced biodegradation of carbamothioate herbicides other than EPTC. However, research indicates that both dietholate and SC-0058 reduce enhanced butylate biodegradation problems (Harvey, R. G. Weed Sci., in press). Repeated annual applications of cycloate did not lead to enhanced biodegradation of that herbicide. Registration of cycloate for use in field and sweet corn could improve weed control options.

Acknowledgments

The author expresses his appreciation to John Albright and Tim Anthon for their assistance with field trials; to John Kutil, Tamara Harvey, Shelia Steele, and Kevin Buhler for their help with biodegradation studies; and to ICI Americas Inc. and the Midwest Food Processors Association for their financial support of this research.

Literature Cited

1. Gray, R. A. Abstr. Weed Sci. Soc. Am., 1975, p 145.
2. Harvey, R. G.; Dekker, J. H.; Fawcett, R. S.; Roeth, F. W.; Wilson, R. G. Weed Technol. 1987, 1, 341-349.
3. McNevin, G. R. Ph.D. Thesis, University of Wisconsin, Madison, 1982.
4. Rahman, A.; James, T. K. Weed Sci. 1983, 31,783-789.
5. Harvey, R. G.; McNevin, G. R.; Albright, J. W.; Kozak, M. E. Weed Sci. 1986, 34, 773-780.
6. Gunsolus, J. L.; Fawcett, R. S. Proc. North Cent. Weed Control Conf., 1980, 35, 18.
7. Obrigawitch, T.; Wilson, R. G.; Martin, A. R.; Roeth, F. W. Weed Sci. 1982, 30, 175-181.
8. Harvey, R. G. Weed Sci. 1987, 35, 583-589.
9. Wilson, R. G.; Rodebush, J. E. Weed Sci. 1987, 35, 289-294.
10. Doersch, R. E.; Fischer, H. L.; Harvey, R. G. North Cent. Reg. Ext. Public. 1987, No. 265, 4 p.

RECEIVED January 22, 1990

Chapter 17

Cultural Practices and Chemicals That Affect the Persistence of Carbamothioate Herbicides in Soil

Dirk C. Drost, James E. Rodebush, and Joanna K. Hsu

ICI Americas, Inc., Agricultural Products, Wilmington, DE 19897

Carbamothioate herbicides have been used in crop production for more than 17 years. Microbial degradation is the major mechanism for the dissipation of carbamothioates in soil. In some instances repeated use of carbamothioates has resulted in enhanced biodegradation. Enhanced biodegradation has been measured by a combination of field efficacy, chemical analysis, and/or soil bioassay. Crop rotation, herbicide rotation, chemical extenders, and sequential herbicide applications have reduced the impact of enhanced biodegradation and provided solutions which allow for continued use of carbamothioates in crop production. Of the many compounds evaluated to modify carbamothioate persistence in soil, R251005, an extender under evaluation by ICI Americas Inc., has been most effective in restoring and maintaining the activity of carbamothioates in soils showing enhanced biodegradation.

EPTC (S-ethyl dipropylcarbamothioate) + dichlormid (2,2-dichloro-N,N-di-2-propenylacetamide) and butylate (S-ethyl bis(2-methylpropyl)carbamothioate) + dichlormid are pre-plant incorporated carbamothioate herbicides used for annual and perennial grass, sedge, and broadleaf weed control in corn throughout the world. EPTC and butylate were formulated with dichlormid as ERADICANE® and SUTAN+®, but for purposes of convenience use of common names in this paper will assume dichlormid present. Gray (1) reported on the behavior, persistence, and degradation of carbamothioate herbicides in the environment. Carbamothioate herbicides are readily degraded in soil by microorganisms. However, their persistence in soil is normally sufficient to provide commercial weed control. Control of late

0097–6156/90/0426–0222$06.00/0

germinating seeds of weeds such as Digitaria, Panicum, Setaria, Sorghum, and Eriochloa spp. often require increased herbicide persistence.

Between 1976 and 1978 Stauffer Chemical Company researchers identified fields where EPTC failed to give expected herbicidal activity. The fields had a history of repeated annual applications of EPTC. Greenhouse bioassays with EPTC demonstrated reduced persistence as the cause of reduced weed control observed in the field. Sterilization of the soil with heat or chemicals restored herbicidal activity. This indicated that, in some soils, enhanced microbial degradation might be associated with the observed reduction in herbicidal activity (D.L. Hyzak, personal communication).

In 1977 Stauffer Chemical Company began field trials in the Missouri River bottoms of Iowa and Nebraska to determine if cultural practices and alternate herbicide treatments could increase weed control in fields where EPTC performance was less than expected. Many test fields did not develop enhanced biodegradation. Factors such as application and incorporation methods, tillage practices, herbicide rotation and use of herbicide combinations had major impacts on herbicide performance and also influenced the development of enhanced biodegradation. These results are further discussed in this paper.

In 1977, Rahman, et. al (2) noted that EPTC provided less than expected weed control at the Manutuke Research Station, near Gisborne, N.Z. They concluded that weed control with EPTC may decrease when used repeatedly on the same field. Further greenhouse studies linked the lack of weed control to increased microbial activity in the soil (3).

By 1979, researchers in the North Central Region of the U.S. had concluded that reduced performance of EPTC following repeated annual use for wild proso millet [Panicum miliaceum L. spp. ruderale (Kitagawa)] and shattercane [Sorghum bicolor (L.) Moench.] control was due to enhanced biodegradation. An excellent review of the work leading to identification of the problem of enhanced biodegradation was published by Harvey, et.al (4).

Following the confirmation of enhanced biodegradation of EPTC, over three thousand compounds were screened as extenders in an attempt to restore EPTC activity. Fonofos, an organophosphorus insecticide, (O-ethyl S-phenylethylphosphonodithioate) and dietholate (O,O-diethyl O-phenyl phosphorothioate) were identified (5,6) as compounds which could restore some of the herbicidal activity of EPTC in history soils and extend the activity of EPTC in soils without a history of EPTC use. The commercial use of tank-mixtures of EPTC and fonofos was registered in 1979 under a 24(c) registration which has now expired. Concurrently, ERADICANE EXTRA® (EPTC + dichlormid + dietholate) was developed and introduced in 1982. EPTC + dietholate remains a standard component of shattercane and johnsongrass [Sorghum halepense (L.)] control programs in the U.S.

The persistence of EPTC in different soils was investigated in

1980 and 1981. The study was based on 325 soil samples from 13 states in 1980 and 165 soil samples from 16 states in 1981. All the soil samples had one to three years of field applications of EPTC. Greenhouse results showed that the persistence of EPTC was not correlated with the number of years of EPTC use, soil moisture capacity, soil organic matter, or soil texture. The persistence of EPTC in greenhouse studies could not be used to accurately predict the performance of EPTC under field conditions. Finally, the study suggested that the frequency of enhanced biodegradation was greater in soils from fields treated with 6.7 kg ai/ha EPTC and soils from areas with high fall and spring soil moisture contents (J.H. Hsu, personal communication).

Reduced weed control with butylate in butylate history fields was reported in 1983 (7,8). Laboratory, greenhouse and field trials demonstrated that repeated annual use of butylate may cause enhanced butylate biodegradation and result in reduced weed control. Reports of reduced weed control with repeated use of butylate were much less frequent than with EPTC (Stauffer Chemical Company unpublished). Since 1986, the potential for enhanced biodegradation of butylate has been documented and described more completely by researchers in the Southeastern U.S. (9,10,11,12).

EPTC was repositioned in the marketplace as an annual grass herbicide in 1985, a rotational statement was added to prevent the repeated annual use of EPTC, and shattercane control claims were removed from the label. EPTC was targeted specifically for foxtail (*Setaria* spp.) and annual broadleaf weed control in the north where cool and wet soils prevail in the spring. EPTC has been a very successful product, provided excellent weed control, and increased in market share every year since its reintroduction as an annual grass and broadleaf herbicide. The other commercial carbamothioate herbicides, EPTC + dietholate and butylate, were positioned as difficult to control weed and southern corn belt herbicides, respectively. No rotational statements have been added to the EPTC + dietholate and butylate labels.

This paper reviews the results of previously unpublished trials conducted by Stauffer Chemical Company and ICI Americas Inc. on the effects of cultural and chemical practices on the activity of carbamothioate herbicides in soils. The inter-relationship of these factors in the expression of enhanced biodegradation and the commercial significance and management of enhanced biodegradation in the cases of the carbamothioate herbicides EPTC, EPTC + dietholate, and butylate is demonstrated.

Materials and Methods

Field Plot Technique. Herbicide treatments were made to plots measuring 3.1 to 6.1 by 13.2 to 18.3 m arranged in a randomized complete block design with 3 replications. Herbicides were applied with tractor-mounted compressed air or CO_2 backpack sprayers delivering 187 to 280 L/ha at 139 to 207 kPa. Herbicides were uniformly mixed with the top 8 to 10 cm of soil with 2 pass incorporation.

Soil extraction and gas chromatography (GC). EPTC and butylate residues remaining in the soil were determined at 1, 3, 5, 7, 10, 14, or 21 DAT (days after treatment) by randomly taking 5 to 6 sub-samples from 0 to 10 cm depth, compositing the sample, and collecting 150 gm of soil. The soil samples were frozen immediately after collection and maintained at 0 C until analyzed. Carbamothioate residues were extracted directly with toluene and analyzed by GC (13). The detection limit was 0.01 parts per million. Residue analyses were conducted by A&L MID WEST AGRICULTURE LABORATORIES, Omaha, NE. Herbicide residue data are presented as parts per million or as a percent of concentration present at 1 DAT. Weed control and residue data were analyzed using analysis of variance. Means were separated using Duncan's Multiple Range Test (DMRT) or an F-protected LSD at the 5% level. For data tables, means in a column followed by the same letter are not statistically different.

Results and Discussion

Enhanced biodegradation. Four annual applications of EPTC resulted in a significant reduction in the recovery of EPTC by GC analysis following the fifth annual application. At 5 and 7 DAT, EPTC recovery was 1.40 and 0.28 ppm, respectively. Adjacent plots which had been treated with EPTC for 2 years and rotated out to alternate herbicides for 2 years had 2.99 and 1.99 ppm at 5 and 7 DAT, respectively. These results demonstrate enhanced biodegradation of EPTC with repeated annual use (Table I). Many other researchers have demonstrated that repeated annual applications of EPTC may lead to reduced control of weeds (2-3,14-19) due to enhanced biodegradation.

Three previous applications of butylate led to a significant reduction in butylate persistence following the fourth annual application. At 7 DAT, butylate recovery was 0.72 ppm. An adjacent plot which had been treated for two years, then left untreated for one year prior to subsequent buytlate application had 2.11 ppm butylate recovered at 7 DAT. This demonstrated enhanced biodegradation of butylate with repeated annual use (Table II). Repeated annual applications of butylate do not lead to reduced weed control as frequently as is the case with EPTC, although butylate persistence may be reduced by annual repeated applications (7-12).

Other factors besides enhanced biodegradation which influence carbamothioate herbicide persistence include environmental conditions, application efficiency, soil incorporation, crop and chemical rotation, herbicide combinations, and chemical extenders (4-6,12-13,17,20). Lack of commercially acceptable weed control may be attributed to misapplication, poor soil preparation and incorporation, and high weed populations, as well as enhanced biodegradation.

Apparent Elimination of Enhanced Biodegradation. Good farming practices and favorable environmental conditions may maximize the efficacy of carbamothioates and provide acceptable weed control

Table I
Shattercane Control and EPTC Persistence on a Silty Clay Loam Soil near Scranton, KS, 1984

Herbicide	History[a]	Shattercane Control 30 DAT	EPTC recovery DAT 1	3	5	7	14
		---(%)---	----------(ppm)----------				
Control		0b	--	--	--	--	--
EPTC + atrazine	4 yr cont	95a	3.58a	2.58a	1.40b	0.28b	0.03b
EPTC + atrazine	2 yr out	95a	3.93a	3.34a	2.99a	1.99a	0.18a

(a) Annual treatments of EPTC from 1980-1983 for the 4 yr continuous treatment. The 2 yr out field treated with EPTC in 1980 and 1981 followed by alachlor in 1982 and cycloate in 1983.

Table II
Foxtail spp. Control and Butylate Persistence with Butylate + Atrazine on a Butylate History Silty Clay Loam Soil on a Soil Near Avoca, NE

1984 Herbicide Treatment[a]	Rate (kg/ha)	Foxtail control 6/19/84	pre-tmt 5/13/85	7/5/85	1985 butylate recovery 7 DAT -(ppm)-
		---------(%)--------			
Control	0.0	0c	3d	0c	2.11a
Cycloate + atrazine	4.5+1.7	99a	83a	96a	2.26a
Butylate + atrazine	4.5+1.7	75b	37bc	55b	0.72b
Butylate+atrazine+metolachlor	4.5+1.7+0.8	86ab	60ab	98a	1.32ab

(a) 1984 treatments applied May 1. Butylate + atrazine (4.5 + 1.7 kg/ha) broadcast applied on 5/28/85.

in soils with enhanced biodegradation. For example, in a Kansas field trial, GC analysis showed that seven days after uniform application and thorough, immediate incorporation, there were only trace amounts of EPTC remaining. GC results indicated enhanced biodegradation (Table I). However, visual weed control was excellent 30 days after treatment. EPTC controlled shattercane which germinated immediately after application and no further flushes emerged. This phenomenon, acceptable weed control when carbamothioate residues should not provide observed weed control, is termed 'apparent elimination' of enhanced biodegradation.

Apparent elimination of enhanced biodegradation may be due to increased incorporation efficiency. The carbamothioates are volatile materials and require soil incorporation for maximum activity. Gray and Weierick (21) showed that up to 44% of applied EPTC was lost in fifteen minutes when application was made to the surface of wet soil. EPTC had to be incorporated 5 to 8 cm deep to reduce the loss of EPTC due to volatility. Efficient carbamothioate applications and thorough, immediate incorporation reduce herbicide losses, increase herbicide concentrations in the soil, and result in better weed control (22). However, according to the EPA registration label, a four-hour delay in incorporation is allowed for applications in water or fluid fertilizer. An eight-hour, same-day delay is allowed for dry bulk fertilizer impregnated with carbamothioates and a 36-hour delay is allowed in semi-arid regions of the Pacific Northwest. In each case, the delay is allowed only if applications are made to dry soil (½" deep) free from dew or moisture.

Weed populations may influence the expression of enhanced biodegradation. In a 1984 experiment (Table II), butylate + atrazine (6-chloro-N-ethyl-N'-(1-methylethyl)-1,3,5-triazine-2,4-diamine) + metolachlor (2-chloro-N-(2-ethyl-6-methylphenyl)-N-(2-methoxy-1-methylethyl)acetamide gave 11% better control than butylate + atrazine on June 19, 1984. Residual control from the 1984 treatments was evaluated on May 13, 1985, prior to the 1985 herbi-cide application. Residual foxtail control was 60% in plots which had received combinations of butylate + atrazine + metolachlor in 1984 but only 37% in plots which had received only butylate + atrazine in 1984. GC analysis of butylate residues at 7 days after the 1985 application indicated that the tank-mixture with metolachlor in 1984 did not increase butylate persistence in 1985 (Table II). Since the addition of metolachlor did not enhance the persistence of butylate, the increased efficacy of butylate + atrazine in 1985 is attributed to reduced weed pressure in the plots treated with metolachlor in 1984 (J. E. Rodebush, personal communication). This is an example of apparent elimination of enhanced biodegradation (Table II).

In the same experiment, 1984 treatments included rotation to a weedy check or cycloate (S-ethyl cyclohexylethylcarbamothioate) + atrazine followed by a broadcast application of butylate + atrazine (4.5 + 1.7 kg ai/ha) in 1985. A one year rotation away from butylate increased the recovery of butylate from 0.72 ppm to between 2.11 and 2.26 ppm 7 days after the 1985 application and appeared to eliminate enhanced biodegradation in these treatments.

However, visual weed control on July 5, 1985 was 0% in the 1984 untreated plot and 96% in the 1984 cycloate treated plot. Under heavy weed pressure (J.E. Rodebush, personal communication), in the absence of enhanced biodegradation, butylate + atrazine (4.5 + 1.7 kg ai/ha) was not able to provide acceptable foxtail control.

Field determinations of herbicide efficacy by visual evaluations should be tempered by consideration of weed pressure and previous weed control experiences. Enhanced biodegradation may be masked by low weed pressure or erroneously identified as a cause of non-performance under situations of high weed pressure.

Weed germination and emergence may also influence the expression of enhanced biodegradation. A trial was conducted in a Nebraska field which had been treated with EPTC + dietholate in 1982 and 1983. The trial was initiated in 1984 with the treatments listed in Table III. In 1985 EPTC + dietholate was again broadcast applied to the entire experimental area. The recovery of EPTC was not significantly different at 1, 3, and 7 DAT compared to treated plots. At 14 DAT EPTC recovery was significantly lower in continuous use plots compared to plots not treated with EPTC + dietholate in 1984. However, weed control with EPTC + dietholate in 1985 was excellent in all plots, regardless of previous treatment. The excellent weed control was a result of high EPTC concentrations for the first 7 DAT (Table III), immediate and uniform weed emergence, and very little later weed emergence (J.E. Rodebush, personal communication). Environmental conditions were warm and dry and thus not favorable for further weed emergence. Therefore, the apparent differences in weed control which should have been observed at 30 DAT were masked by the environment. The impact of environmental conditions on weed germination and emergence must also be considered in field trials designed to evaluate enhanced biodegradation.

These representative field trials show that both enhanced biodegradation and reduction or elimination of enhanced biodegradation can occur in the field. Elimination of enhanced biodegradation may be apparent or real. Apparent elimination of enhanced biodegradation occurs when enhanced biodegradation is present, but weed control remains acceptable. The factors influencing enhanced biodegradation and apparent elimination of enhanced biodegradation are the same as those which affect carbamothioate herbicide persistence. Apparent elimination of enhanced biodegradation generally results from reduced evaporation loss due to incorporation, reduced weed seed pressure, and uniform weed germination and emergence after herbicide application.

Elimination or Inhibition of Enhanced Biodegradation by Rotations.

Rotation from continuous EPTC to alachlor (2-chloro-N-(2,6-diethylphenyl)-N-(methoxymethyl)acetamide) in 1982 and cycloate in 1983 increased the persistence of EPTC at 5 DAT in 1984 from 1.4 ppm to 2.99 ppm. EPTC recoveries were also significantly greater at 7 and 14 DAT. Rotation from continuous EPTC for 2 years appeared to eliminate enhanced biodegradation (Table I). In another location, a 1984 treatment of cycloate +

atrazine or untreated plot in 1984 increased the recovery of butylate at 7 DAT in 1985 from 0.72 ppm (continuous butylate) to 2.11 to 2.26 ppm (Table II). Definitive proof of elimination of enhanced biodegradation was not possible without comparison to a nonhistory field. Because the field trial was conducted in a commercial field, a nonhistory plot was not available in the experimental area.

In the third experiment, rotation away from continuous EPTC + dietholate in 1984 to no herbicide, trifluralin (2,6-dinitro-N,N-dipropyl-4-(trifluoromethyl)benzenamine) + metribuzin (4-amino-6-(1,1-dimethylethyl)-3-(methylthio)-1,2,4-triazin-5(4H) one), alachlor + atrazine, or cycloate + atrazine increased the persistence of EPTC + dietholate at 14 and 21 DAT in 1985. Rotation with trifluralin + metribuzin or alachlor + atrazine was more effective than rotation with weedy check or cycloate + atrazine. However, the differences were not statistically significant. All rotated herbicide treatments had higher EPTC recoveries than the untreated plot (Table III). The results indicated that 1 year rotations resulted in partial to complete elimination of enhanced biodegradation for this herbicide.

In another experiment, rotation from continuous butylate + atrazine to cycloate for 1 year increased foxtail control from 73% to 96% and significantly increased the persistence of butylate at 10, 14, and 21 DAT compared to continuous butylate use. Rotation with acetanilide herbicides for 1 year slightly increased the weed control. Two year rotations with acetanilides were as effective as cycloate in increasing the persistence of butylate and foxtail spp. control. It appeared that one annual application of cycloate or two annual applications of acetanilides eliminated enhanced biodegradation. Butylate residue data were not available from one year rotations with acetanilides or a non-history field to confirm the elimination of enhanced biodegradation (Table IV).

Additional research has shown one or two year rotations to untreated, alachlor, atrazine, cyanazine (2-[[4-chloro-6-(ethylamino)-1,3,5-triazin-2-yl]amino]-2-methylpropanenitrile), metolachlor, or trifluralin, may result in reduction or elimination of enhanced biodegradation (4,23).

Elimination or Inhibition of Enhanced Biodegradation by Herbicide Tank Mixtures. In one field trial, tank-mixes of atrazine with EPTC significantly increased green and yellow foxtail control at 60 DAT. In the green foxtail field, which had been treated with EPTC for 4 years, the application of EPTC provided little weed control. When EPTC was applied at 4.5 and 6.7 kg ai/ha tank-mixed with atrazine at 1.7 kg ai/ha, green foxtail control increased from 15 to 52% and from 22 to 94%, respectively. EPTC had only been applied once before to the yellow foxtail field. EPTC alone provided 68 to 72% control and tank-mixtures with atrazine provided 83 to 91% control, respectively (Table V). Atrazine tank-mixes appeared to be more beneficial in the field with the longer previous history of EPTC use.

In another field trial, butylate and butylate tank-mixes were evaluated for weed control in a field with a 3 year history of

Table III
Weed Control and EPTC Persistence
from EPTC + Dietholate + Atrazine in Rotation with
Various Herbicides on a Silty Clay Loam Soil, Otoe, NE
Field history of EPTC in 1982 and EPTC + dietholate in 1983
EPTC + dietholate broadcast applied at 4.5 kg/ha on 5/8/85

1984 Herbicide	Rate	Foxtail Control 30 DAT 1985	EPTC recovery DAT				
			1	3	7	14	21
	(kg/ha)	-(%)-	-----(% of 1 DAT)------				
EPTC + dietholate + atrazine	4.5+1.7	97a	100a	56a	45a	4c	3c
No herbicide	0.0	99a	100a	72a	87a	21ac	11ac
Trifluralin + metribuzin	1.1+0.3	99a	100a	60a	44a	33a	17a
Alachlor + atrazine	3.4+1.7	100a	100a	44a	58a	35a	15a
Cycloate + atrazine	4.5+1.7	99a	100a	74a	64a	24ab	12ab

Table IV
Weed Control and Butylate
Persistence on a Sandy Loam Soil, Atkinson, NE
Field received annual treatments of butylate from 1980 to 1982
On April 24, 1984 butylate + atrazine (4.5 + 1.7 kg/ha) was applied

Field history	Foxtail Control 45 DAT	butylate recovery DAT					
		1	5	7	10	14	21
	---(%)---	----------(% of 1 DAT)---------					
Control	0d	---	--	--	--	--	--
Butylate + atrazine	73c	100	64	56	12	2	1
2 yr out to acetanilides	92a	100	62	59	57	49	11
1 yr out to cycloate	96a	100	79	66	58	49	25
1 yr out to acetanilides	80b	---	--	--	--	--	--

butylate use. The trials were conducted in 1984 and 1985. Butylate at 4.5 kg ai/ha provided unacceptable green and yellow foxtail control in both years. Atrazine alone also provided unacceptable control. The tank-mixture significantly increased foxtail control. Butylate recovery data suggest that the tank-mix including atrazine enhanced the persistence of butylate. The results were statistically significant only at 5 DAT in 1984. However, the recovery was numerically greater with the tank-mix at each evaluation date (Table VI).

Butylate + atrazine + metolachlor tank mixtures increased foxtail control compared to continuous butylate + atrazine use in a field with a 3 year history of butylate use (Table II).

Herbicides in tank-mixture with EPTC or butylate can increase weed control in fields with accelerated biodegradation and may reduce the impact of enhanced biodegradation on weed control. The increased weed control in carbamothioate + atrazine or metolachlor tank mixtures may be due to the: 1) additive/synergistic action of the tank mix; 2) inhibition of enhanced biodegradation, or; 3) activity of atrazine as a mild extender for carbamothioate herbicides.

A Commercial Perspective on Managing Enhanced Biodegradation.

The field trials discussed above demonstrate that rotating away from the herbicide showing enhanced biodegradation for one year gives partial or complete elimination of enhanced biodegradation. Rotation away for two years nearly always completely eliminates enhanced biodegradation (Tables I-IV). Tank mixtures of carbamothioates with atrazine or metolachlor reduced the expression of enhanced biodegradation (Tables V-VI). Other rotations of crops and or chemical classes can achieve the same results (4).

From the commercial perspective, the rotational statement placed on the EPTC label in 1985 was based on experimental evidence and has been verified by independent research (4). Annual crop or chemical rotation program following EPTC use has been very successful. Nonperformance complaints have been reduced and there is minimal perception that enhanced biodegradation affects annual weed control with EPTC.

Impact of Climatic and Edaphic Conditions on Enhanced Biodegradation. The results of Obrigawitch, et. al (17) and Harvey and Kozak (24) demonstrated that high soil temperatures for 2 weeks to 2 months eliminated enhanced biodegradation whereas low temperatures or short incubation periods reduced the rate of EPTC degradation compared to the check soil. Soil moisture level may be the most important factor in the development and expression of enhanced biodegradation. This work and others supports the hypothesis that the expression of enhanced EPTC biodegradation is affected by variations in temperature and moisture which take place in the field from year to year. Edaphic conditions which favor abnormally warmer or drier soils after fall harvest and prior to herbicide application in the spring such as moldboard plowing,

Table V
Foxtail Control With EPTC and EPTC + Atrazine
On Silt Loam and Silty
Clay Loam EPTC History Soils, Hamburg and Vail, IA
The fields had a four-year and a one-year history of EPTC, respectively

Herbicide	Rate	Weed control 60 DAT	
		Green Foxtail	Yellow Foxtail
	(kg/ha)	(%)	(%)
Control	0.0	0f	0e
EPTC	4.5	15ef	68c
EPTC	6.7	22e	72c
EPTC + atrazine	4.5+1.7	52cd	83b
EPTC + atrazine	6.7+1.7	94ab	91ab

Table VI
Weed Control and Butylate Persistence From Butylate and Butylate + Atrazine
on a Silty Clay Loam Soil, Avoca, NE
Field treated with Butylate in 1982 and 1983

Herbicide	Rate	Foxtail control 30 DAT	butylate recovery DAT					
			1	3	5	7	14	21
	(kg/ha)	-(%)-	(% of 1 DAT)					
1984								
Butylate	4.5	36bc	100a	56a	40b	9b	0a	0a
Butylate + atrazine	4.5+1.7	88a	100a	64a	67a	39a	2a	0a
Atrazine	1.7	12c	---	--	--	--	-	-
1985								
Butylate	4.5	60b	100a	61a	--	24a	9a	14a
Butylate + atrazine	4.5+1.7	85a	100a	122a	--	42a	13a	7a
Atrazine	1.7	73ab	---	---	--	--	--	-

spring disking, or field cultivation may reduce the rate of EPTC degradation in soils which show enhanced biodegradation. Soils which are cold or freeze during winter may thus have a greater incidence of enhanced biodegradation with repeated use than warmer soils (24-25).

A greenhouse experiment was conducted to determine the effect of soil moisture on carbamothioate activity. When soil moisture was reduced to 5% (w/w) or less prior to application, the persistence of EPTC and EPTC + dietholate was greatly increased (J. Hsu, personal communication). This observation is substantiated by Sparling and Cheshire (26) who noted that drying the soil below 15% moisture content decreased total microbial counts, especially those of yeasts. Further drying to 1.5% moisture resulted in a further decrease in bacterial and fungal numbers. Tillage patterns or geographical locations which lead to reduced soil moisture may increase the persistence of carbamothioates irrespective of enhanced biodegradation.

The Impact of Chemical Extenders on Enhanced Biodegradation. Evaluation of chemical extenders began after enhanced biodegradation was observed and confirmed (2,14-19). Extenders are chemicals which increase the persistence if herbicides in soil. Early work with extenders revealed fonofos was effective in restoring activity of EPTC (14,27-28). Subsequent to fonofos, dietholate was discovered and developed as an extender.

The efficacy of dietholate as a herbicide extender for EPTC in soils with enhanced biodegradation is well documented (5,9,17-20,27-29). The benefits of dietholate use include restoration of EPTC activity in soils which develop enhanced biodegradation and extension of weed control in soils without enhanced biodegradation.

Indirect evidence suggests dietholate may block the hydroxylation of EPTC in soils demonstrating enhanced biodegradation without affecting the oxidation to the sulfoxide (6).

Dietholate increased control of johnsongrass [*Sorghum halepense* (L.) Pers.] (29), redroot pigweed (*Amaranthus retroflexus* L.) (Stauffer Chemical Company, unpublished), common lambsquarters (*Chenopodium album* L.), crabgrass (*Digitaria* spp.) (9), shattercane (30), wild proso millet (19,28), and foxtail spp. (27) in soils demonstrating enhanced EPTC biodegradation.

On soils without prior application of carbamothioates, the addition of dietholate increased johnsongrass, redroot pigweed, yellow nutsedge (*Cyperus esculentus* L.), and crabgrass control over EPTC applied alone (Stauffer Chemical Company, unpublished).

Tests established in fields with previous applications of EPTC + dietholate have demonstrated that the repeated use of EPTC + dietholate provided only 78, 40, and 80% of wild proso millet, woolly cupgrass [*Eriochloa villosa* (Thumb.) Kunth], or green foxtail *Setaria viridis* L.), respectively] (Table VII). Two other trials with EPTC + dietholate have confirmed that second year use at 6.7 kg ai/ha provided only 56 to 70% wild proso millet control at 42 DAT (Table VIII). These results agree with observations reported in Iowa, Minnesota, and Wisconsin (7,31-33).

Table VII
EPTC, EPTC + dietholate, and EPTC + R251005
Performance on EPTC + dietholate-history Soils (a)

Treatment	Rate	Weed control 60 DAT		
		Waterloo,WI	Marksville,WI	Guelph,ND
		Wild proso millet	Woolly cupgrass	Green Foxtail
	(kg/ha)	(%)		
Control	0.0	0d	0c	0d
EPTC	6.7	42bc	53ab	73bc
EPTC + dietholate	6.7	78a	40b	80ab
EPTC + R251005	6.7+1.1	70ab	72a	93a

(a) At Waterloo, a silty clay loam soil was treated with EPTC + dietholate in 1984 and above treatments applied May 3, 1985. At Marksville a silt loam soil was treated with EPTC + dietholate in 1984 and above treatments were applied May 1, 1985. At Guelph a loam soil was treated with EPTC + dietholate in 1984 and the listed treatments applied May 1, 1985.

Table VIII
Wild Proso Millet Control in EPTC + dietholate
History Fields with Silty Clay Loam or Loam Soils, Waterloo, WI
Treatments applied May 4, 1988 and May 3, 1989
Both Fields treated one year previously with EPTC + dietholate

Herbicide		Wild proso millet control 42 DAT	
		1988	1989
	(kg/ha)	(%)	
Control	0.0	0a	0a
EPTC + dietholate	6.7	56b	70b
EPTC + R251005	6.7+1.1	91c	93c

Conflicting results with retreatments of EPTC + dietholate suggest an interaction with weed species (12,20,29,33). For difficult-to control weeds, 2 consecutive years use of EPTC + dietholate may not provide acceptable control compared to first year use. With other annual weeds, however, high weed populations, delayed germination and emergence patterns, herbicide rotations and combinations, and edaphic conditions may affect the control obtained. Therefore, both previous carbamothioate use pattern and weed species are important in determining the weed control achieved from repeated carbamothioate use.

Because the repeated annual use of EPTC + dietholate resulted in reduced control of wild proso millet, woolly cupgrass, and shattercane and because conflicting results were obtained with giant foxtail, alternate extenders were evaluated which could increase the performance of EPTC and provide consistent weed control with repeated annual use. Several extenders were provided for evaluation by Stauffer Chemical Company in 1984 and 1985. A review of the efficacy of these extenders is contained in Harvey et.al (4).

Studies comparing the effect of different rates of R251005 (formerly SC-0058) [S-ethyl-*N*,*N*-bis(3-chloroallyl) carbamothioate] in combination with 3.4 kg ai/ha EPTC indicate increased shattercane control at 21 DAT control as the rate of R251005 increased from 0.6 to 1.1 kg/ha. Increasing the rate from 1.1 to 2.2 kg ai/ha produced a numerical increase in shattercane control but the increase was not statistically significant (Table IX). There was no additional increase in shattercane control above 1.1 kg/ha in another test (Table X). Other, unpublished trials indicate that 2.2 kg ai/ha may provide a numerical increase in control but the results are generally not statistically significant.

In soils without a history of EPTC or butylate use, R251005 increased velvetleaf (*Abutilon theophrasti* Medic.), common cocklebur (*Xanthium strumarium* L.), and Pennsylvania smartweed (*Polygonum pensylvanicum* L.) control compared to EPTC alone (Table XI). In an EPTC history soil, Texas panicum (*Panicum texanum* L.) control was also increased (Table XI).

Two consecutive annual applications of EPTC + R251005 controlled 'Roxorange sorghum' (Table IX-X).

Based on these trials, R-251005 was selected as the most effective extender evaluated in the 1984 and 1985 field evaluation program. However, in late 1985 further evaluation of R-251005 was discontinued due to resource constraints. One on-going trial with R251005, at the University of Wisconsin was maintained.

Further studies at the University of Wisconsin since 1985 demonstrated that 5 consecutive annual applications of EPTC + R251005 maintained wild proso millet control at acceptable levels. EPTC and EPTC + dietholate demonstrated reduced control with repeated use in the same experiment (R. G. Harvey, personal communication).

Recently, based on the favorable results in the annual repeat application trial, R251005 has been reevaluated. In 1988 and

Table IX
Effect of Different Rates of R251005
With 3.4 kg/ha EPTC on a Silty Clay Loam Soil, Lisbon, IA.
Treatments were applied on April 30, 1985

Herbicide	Rate (kg/ha)	'Roxorange' sorghum control 0 DAT	21 DAT	28 DAT
		------------(%)-----------		
No prior carbamothioate history:				
Control	0.0	0b	0b	0a
EPTC	3.4	99a	86a	0a
Prior butylate history (1979-1984):				
Control	0.0	0b	0b	0a
EPTC + R251005	3.4+0.6	100a	20b	0a
EPTC + R251005	3.4+0.9	100a	34b	0a
EPTC + R251005	3.4+1.1	100a	70a	0a
EPTC + R251005	3.4+2.2	100a	80a	0a

Table X
'Roxorange' Sorghum and Indigenous Weed
Control with EPTC on a Silty Clay Loam Soil, Lisbon, IA.
Field treated with EPTC in 1982 and 1983
Treatments applied June 28, 1984 and June 25, 1985

Herbicides 1984	Herbicides 1985	Rate (kg/ha)	Weed Control: 'Roxorange' Sorghum 0 DAT	Weed Control: All Weeds 14 DAT	Weed Control: All Weeds 63 DAT
			---------(%)----------		
Control	Control	0.0	0d	0e	0e
EPTC	EPTC	6.7	95ab	63c	72d
EPTC + dietholate	EPTC + dietholate	6.7+.6	97ab	35d	75cd
EPTC + dietholate	EPTC + dietholate	6.7+1.1	84bc	68bc	78cd
EPTC + R251005	EPTC + R251005	6.7+1.1	93ab	97a	90ab
EPTC + R251005	EPTC + R251005	6.7+1.7	99a	99a	93a

Table XI
Performance of EPTC + R251005 in EPTC History Clay Loam Soil, Luling, TX and Non-history Loam Soil, Waterloo, IA

Herbicide	Rate (kg/ha)	Weed control 40 DAT: Non-history, Velvet-leaf (%)	Non-history, Smart-weed (%)	Non-history, Cockle-bur (%)	History, Texas panicum (%)	History, Pig-weed (%)
Control	0.0	0a	0a	0a	0a	0a
EPTC	6.7	92a	95a	50a	62a	78a
EPTC + R251005	6.7+1.1	93a	96a	80a	83a	83a

1989, R251005 was tested on a limited basis to verify the ratio of EPTC:R251005 needed, and the ability of R251005 to restore EPTC activity in soils with enhanced biodegradation. EPTC + R251005 at 6.7 + 1.1 kg ai/ha provided significantly better wild proso millet control than EPTC + dietholate on EPTC + dietholate-history soils during 1988 and 1989 (Table VIII).

In 1989, eighteen long term, annual repeat application studies were initiated by ICI Americas Inc. using EPTC + R251005 at 6.7 + 1.1 kg ai/ha to evaluate weed control following repeated annual use against shattercane, johnsongrass, wild proso millet, woolly cupgrass, and giant foxtail. Trials were established on EPTC + dietholate history and nonhistory soils. Although repeated annual applications of EPTC + R251005 have been effective on wild proso millet, efficacy will need to be demonstrated on johnsongrass and shattercane prior to a decision to commercialize R251005.

Summary

With EPTC, the 1985 addition of an annual rotation statement and repositioning as an annual grass herbicide has been effective in preventing the redevelopment of enhanced biodegradation problems.

In addition to efficient application and incorporation the following factors have significantly increased the weed control from EPTC, EPTC + dietholate, or butylate when reduced performance due to enhanced biodegration was observed: 1) reduce weed pressure or establish conditions favorable for immediate, uniform weed emergence; 2) immediate, uniform incorporation just prior to corn planting; 3) tank mixtures of carbamothioate herbicides with atrazine, cyanazine, or metolachlor; 4) rotation for one year to no herbicide, alachlor, cycloate, metolachlor, or trifluralin; or, 4) formulations of EPTC plus R251005 at 6.7 + 1.1 kg/ha.

These cultural and chemical practices have prevented, moderated, or eliminated enhanced biodegradation in soils, and preserved the effectiveness of EPTC, EPTC + dietholate, and butylate herbicides in corn cropping systems.

Literature Cited

1. Gray, R. A. Proc. Calif. Weed Conf. 1971, p 128-141.
2. Rahman, A.; Atkinson, G. C.; Douglas, J. A. Sinclair, D. P. N.Z. J. of Agric. 1979, 139, 47-49.
3. Rahman, A.; Burney B.; James T. K. Proc. 34th N.Z. Weed and and Pest Control Conf. 1981, p 176-181.
4. Harvey, R. G.; Dekker, J. H.; Fawcett, R. S.; Roeth, F. W.; Wilson, R. G. Weed Technol. 1987, 1, 341-349.
5. Capper, B. E. Proc. N.Z. Weed and Pest Control Conf. 1982, p 222-225.
6. Misullis, B.; Nohynek, G. J.; Pereiro, F. Proc. British Crop Protection Conf.-Weeds. 1982, p 205-210.
7. Rudyanski, W.J.; Fawcett, R. S. Proc. North Central Weed Control Conf. 1983, p 39.
8. Tuxhorn, G. L.; Roeth, F. W.; Martin, A. R. Proc. North Cent. Weed Control Conf. 1983, p 35-38.
9. Skipper, H. D.; Murdock, E. C.; Gooden, D. T.; Zublena, J. P.; Amakiri, M. A. Weed Sci. 1986, 34, 558-363.
10. Dowler, C. C.; Marti, L. R.; Kvien, C. S.; Skipper, H. D.; Gooden, D. T.; Zublena, J. P. Weed Technol. 1987, 1, 350-358.
11. Mueller, J. G.; Skipper, H. D.; Kline, E. L. Pestic. Biochem. and Physiol. 1988, 32, 189-196.
12. McCusker, V. W.; Skipper, H. D.; Zublena, J. P.; Gooden, D. T. Weed Sci. 1988, 36, 818-823.
13. Subba-Rao, R. V.; Cromartie, T. H.; Gray, R. A.; Weed Technol. 1987, 1, 333-340.
14. Harvey, R. G. Proc. North Central Weed Control Conf. Res. Rept. 1980, p 37.
15. Harvey, R. G. Proc. 1980 Wisc. Fert., Aglime and Pest Management Conf. 1980, p 18.
16. Harvey, R. G.; Schuman, D. B. Abstr. Weed Sci. Soc. Am. 1981, p 124.
17. Obrigawitch, T.; Wilson, R. G.; Martin, A. R.; Roeth, F. W. Weed Sci. 1982, 30 175-181.
18. Schuman, D. B.; Harvey, R. G. Proc. North Cent. Weed Control Conf. 1980, p 19-20.
19. Warnes, D. D.; Behrens, R. Abstr. Weed Sci. Soc. Am., 1981, p 22-23.
20. Harvey, R. G. Weed Sci. 1987, 35, 583-589.
21. Gray, R. A.; Weierick, A. J. Weeds 13, 141-147.
22. Moomaw, R.; Martin, A.; Roeth, F. Proc. Crop Protection Clinics, University of Nebraska, 1983, p 163-166.
23. Drost, D. C. Proc. North Cent. Weed Contr. Conf. 1984, p 46.
24. Harvey, R. G.; Kozak, M. E. Proc. North Central Weed Contr. Conf. 1984, p 78-79.
25. Reynolds, D. A.; Dexter, A. R. Proc. North Cent. Weed Control Conf. 1984, p 83-84.
26. Sparling, G. P.; Cheshire, M. V. Soil Biol. Biochem. 1979, 11, 317-319.
27. Gunsolus, J. L.; Fawcett, R. S. Proc. North Cent. Weed Control Conf. 1980, p 18.

28. McNevin, G. R.; Harvey, R. G. Proc. North Central Weed Control, Conf. 1980, p 71-72.
29. Green, J. D.; Martin, J. R.; Witt, J. R. Proc. South. Weed Sci. 1982, p 20.
30. Obrigawitch, T.; Roeth, F. W.; Martin, A. R.; Wilson, R. G. Jr. Weed Sci. 1982, 30, 417-422.
31. Harvey, R. G. Proc. North Central Weed Control Conf. 1982, p 29-30.
32. Warnes, D. D.; Behrens, R. Proc. North Cent. Weed Control Conf. 1982, p 59.
33. Harvey, R. G.; McNevin, G. R.; Albright, J. W.; Kozak, M. E. Weed Sci. 1986, 34, 773-780.

RECEIVED January 24, 1990

Chapter 18

Spectrophotometric Methodologies for Predicting and Studying Enhanced Degradation

Joseph P. Reed[1], Robert J. Kremer[2,3], and Armon J. Keaster[1]

[1]Department of Entomology, University of Missouri, Columbia, MO 65211
[2]Department of Agronomy, University of Missouri, Columbia, MO 65211
[3]Agricultural Research Service, U.S. Department of Agriculture, Columbia, MO 65211

Currently it is difficult to assess the incidence of enhanced degradation of soil-applied pesticides at a specific location. Simple, reliable diagnostic tests are needed that can be used to evaluate potential enhanced degradation scenarios. Using standard soil extraction methods, spectrophotometry can provide simple, specific and sensitive detection of pesticides and their metabolites in soil. Such methodology will provide information useful in predicting potential enhanced degradation before pesticide application and implementing more reliable recommendations for controlling the target pest.

Enhanced degradation is a growing concern in the agricultural community because some pesticides are degraded so rapidly that their persistence is not adequate for satisfactory pest control. Soils which demonstrate an increased pesticide degradation rate following repeated application are referred to as 'problem' or 'aggressive' soils (1). Likewise, soils not displaying enhanced degradation are referred to as 'non-problem' soils even though both soils may possess the same edaphic characteristics.

The extent and importance with which enhanced degradation manifests itself appear to be dependent upon pesticide use patterns. Perception of the problem at the field level through pest control failures is of paramount importance. A survey conducted in Nebraska during 1983 and 1984 concluded that the incidence of enhanced degradation of carbamothioate herbicides was more prevalent in South Central Nebraska than all other parts of the state (2). This regional difference was presumed to be due to grower perceptions of poor control of shattercane (*Panicum miliaceum* L.), a weed endemic to this area. A 1984 survey of Missouri corn producers indicated concerns about the performance of their pesticides. Over 20% of the corn producers had experienced an outright pesticide failure and almost 30% rated performance of both herbicides and insecticides used for corn pest control as poor

0097–6156/90/0426–0240$06.00/0

(3). Surveys of grower perceptions of pesticide performance are major efforts that yield dubious results. It would be a major undertaking to research several pesticides suspected to undergo enhanced degradation using standard analytical techniques (e.g., GLC) in the laboratory. Such endeavors are not always economically feasible and would be further complicated by the difficulty of collecting accurate efficacy/field data. However, the application of a rapid and cost-effective method for field detection of soil-applied pesticides and their metabolites would be very attractive for management of enhanced degradation.

This paper briefly examines the basis of spectrophotometry as a diagnostic method. It summarizes current and future applications of the method for detecting pesticides subject to enhanced degradation in soil.

Spectrophotometric Methods

Analyses by spectrophotometric methods are advantageous because of their speed, simplicity, and precision beyond that of volumetric and gravimetric methods (4). A relatively small quantity of sample often suffices for a determination thereby reducing the amounts and costs of reagents used. Also, the sample often remains unaltered allowing recovery for confirmatory analyses by other chemical techniques including GLC and HPLC.

The spectrophotometer essentially allows a comparison of the radiation absorbed in or transmitted by a solution containing an unknown quantity of some substance and a solution containing a known quantity of the same substance, the standard (4). Different types of spectrophotometric methods are available depending on the nature of the particular substance under detection. Ultraviolet and visible spectrophotometry is a versatile, quantitative technique that is based on the characteristic absorption of a substance in the sample for identification. Commercial instruments are readily available for this technique and,depending on the optical system employed, the sensitivity of such instruments is quite high. The adaptability of such instrumentation for field use holds much promise and will be discussed later. Numerous handbooks are available on the specific aspects of this technique (5,6). Fluorescence spectrophotometry, in which light absorbed by a substance causes light of a different wavelength to be emitted (fluoresce), is applicable to many biochemical analyses. The technique requires frequent calibration checks, since loss of excitation levels with this method is common. It is not a good candidate for in-field use, although the methodology is quite adequate for laboratory use (6,7). Infrared spectrophotometry, based on characteristic absorption spectra of long wavelength, is a powerful technique useful in identification of organic compounds. Relationships between chemical structure and infrared spectra are well established. The characterization and identification of a compound are advantages of this method since quantitation is difficult. Collections of spectra are readily available (5,6). This methodology has potential for application to field situations where the presence of a specific pesticide metabolite is a determining factor.

Spectrophotometric Applications

Certain pesticides can be extracted from soil and water into a solvent and be analyzed directly by spectrophotometry. When exposed to light of different wavelengths, absorption of light will vary depending on the molecular structure of the compound. Thus, a record of absorbance at each wavelength yields an absorption spectrum characteristic for each pesticide compound. Most pesticides have absorption maxima in the wavelength range of 200 to 400 nm, which requires an ultraviolet spectrophotometer for determination (8). Other spectrophotometric methods have been developed based on color formation when an appropriate reagent is available to react with a specific moiety in the pesticide molecule. The colored complex can then be detected quantitatively using a spectrophotometer in the visible light spectrum (400 to 700 nm). For example, the reaction of triazine herbicides with ethylcyanoacetate to yield a red-colored complex has been developed for spectrophotometric detection of these herbicides in soil extracts (9).

Immunoassays, analytical procedures based on the specific binding of animal-derived antibodies to a target molecule (10), have been developed as rapid methods for detection of pesticides in soil and water samples. Field samples can be analyzed on site after water or solvent extraction using a visualization method, which generally involves reaction of the antibody-pesticide complex with an enzymatically derived colored product. The most sensitive and quantitative method for monitoring the color reaction is spectrophotometry.

Unfortunately, immunoassay systems have not been developed for all pesticides used in agriculture. Therefore, alternative methods based on biological reactions that can be coupled with spectrophotometry can be developed. It has been demonstrated that many common soil enzymes which normally function within the microbial cell can also function extracellularly and may be able to mediate certain steps in the degradation of soil-applied pesticides (11). Assays for enzymes in soil or microorganisms can be devised following standard procedures (12). After meeting conditions (activators, temperature, pH) critical for the assay, enzymatic activity can be determined by measuring substrate disappearance or product appearance using the spectrophotometer. Enzymatic assays can be accomplished by adding a substrate to a soil-buffer system which, when acted upon by the enzyme, yields a colored product that is readily detected spectrophotometrically. Spectrophotometric assays can be used to detect soil enzymes directly responsible for pesticide degradation (11) or indirectly by monitoring activity of enzymes presumed to be associated with pesticide degradation (13).

Spectrophotometric Enhanced Degradation Assays

General Assay: Urease Activity in Problem Soils. Tu (14) examined urease activity in clay loam soils incubated with various pesticides. He observed increased levels of microbial urease activity with carbofuran- and ethoprop-amended soils (Table I). These results suggest that activity levels of specific indigenous soil enzymes may be influenced by certain pesticides applied to soil or by their metabolites formed during degradation. An assay for specific enzymes could be linked to field observations of agricultural pests like that of corn root ratings use to assess control efficacy of corn rootworm (*Diabrotica* spp.) by soil-applied insecticides. For instance, recent field studies, revealed that control of corn rootworm by carbofuran and ethoprop was as poor as the untreated control treatment on carbofuran history soils (Table II), indicating the presence of soil microorganisms involved in enhanced degradation of these insecticides (15). In contrast, ethoprop provided good rootworm control on ethoprop history soils. A soil enzyme assay might have been used to verify the involvement of soil microorganisms in degradation. Continued efforts in this area may result in a definite diagnostic assay based on enzymatic activities coupled with specific pesticides.

Table I. Urease Activity as Influenced by Pesticide Amendment

Pesticide	Urease Activity* (100 ug NH_4^+/g soil)
Carbofuran	21**
Ethoprop	19**
Fonofos	15
Phorate	16
Control	17

*Determined after 7 days incubation of 10 ug a.i./g soil.
**Denotes significant difference (P=.05) from control.
Adapted from (14).

Table II. Rootworm Control in Carbofuran and Ethoprop History Soils

Insecticide	Root Rating	
	Carbofuran History	Ethoprop History
Fonofos	3.40	3.50
Phorate	4.00	3.65
Ethoprop	4.55	3.00
Carbofuran	4.75	2.63
Control	5.10	5.50

Adapted from (15).

General Assay: Enzyme Activities in an Enhanced Soil. Reed et al. (13) characterized cell-free culture filtrates of actinomycete, bacterial and fungal isolates using a spectrophotometric method for assaying the activities of rhodanese, phosphatase and phosphodiesterase. Elevated enzyme levels, expressed as specific activity, of several of the microbial isolates cultured in the presence of selected pesticides indicated a potential for biodegradation of these pesticides (Table III). Recent work of Mueller et al. (16) and Gauger et al. (17) agrees with our findings that actinomycetes and bacteria possess a diversity of metabolic pathways through which many pesticide substrates can be biodegraded.

Table III. Enzyme Activities of Microorganisms From Soils Exhibiting Enhanced Pesticide Biodegradation

Isolate	Pesticide	Specific Activity (ug product/mg protein/h @ 25 C)		
		Rhodanese	Phosphatase	Phosphodiesterase
Bacteria:				
Alcaligenes	IS[*]	8.0 a[**]	1.8 a	12.6 a
Pseudomonas #3	CB, BU	9.1 a	1.8 a	5.2 b
Pseudomonas #6	CB, IS, AL	4.3 b	1.8 a	12.1 a
Pseudomonas (R)	None	4.5 b	1.8 a	11.0 a
Actinomycetes:				
Nocardia sp.	None	0 b	12.0 a	30.4 b
Streptomyces #13	TB, AL, BU	5.5 a	11.6 a	41.2 a
Streptomyces #25	All	5.5 a	11.8 a	61.0 a
Streptomyces (R)	None	0 b	12.7 a	30.1 b
Fungi:				
Aspergillus sp.	EP, CL	6.4 b	11.8 a	26.8 b
Fusarium #25	EP, BU	8.2 a	12.1 a	18.8 b
Fusarium #210	IS	0 b	12.5 a	45.5 a
Penicillium (R)	None	0 b	11.3 a	21.2 b

[*]Pesticides metabolized by microbial isolate: AL=Alachlor; BU=Butylate; EP=EPTC; IS=Isofenphos; TB=Terbufos.
[**]Means within a column for each microbial group followed by a common letter do not differ significantly (P<0.05).
Adapted from (13).

EPTC and Butylate: Fluorescein Diacetate Assay. Spectrophotometric determinations of the hydrolysis of fluorescein diacetate have been shown to be simple, rapid, and sensitive methods for determining microbial activity in soil (18). Essentially, the hydrolytic cleavage of diacetate from fluorescein is responsible for the reaction products including fluorescein, which may be detected spectrophotometrically at 490 nm. This method is somewhat non-specific in that it is indicative of overall activity of several enzymes (protease, lipase, esterase) rather than of a specific class of enzymes. Enzyme activity may be influenced by subtle pH changes in the sample since abiotic hydrolysis of fluorescein diacetate may occur. Also, an associated lag phase in soil hydrolytic activity must be accounted for in each assay.

Reed et al. (19) adapted the fluorescein diacetate assay to qualitatively measure the potential ability of soil microorganisms to degrade carbamothioate herbicides in enhanced soils. They observed both prior treatment and herbicide concentration effects on hydrolytic activity in soils with butylate and EPTC histories and soils with no carbamothioate history. The assay was able to detect significant differences ($P<0.05$) in enzymatic activity when herbicide concentration varied in soil. This sensitivity indicated that the assay may be useful in detecting enhanced soils based on the amount of parent pesticide remaining in soil at various periods after application. A potential disadvantage in adapting this method for field studies is the time of incubation (60 min) required for the samples.

Carbofuran: Metabolite Assay. Rapid detection of enhanced degradation of carbofuran in soil can be accomplished in two different manners. The first approach is to measure the actual amount of carbofuran remaining after application and compare this value with what might be expected to remain after a selected time. This concept has been used by FMC in the form of a 'test ticket' that colorimetrically determines the amount of carbofuran based on inhibition of acetylcholinesterase. The diagnostic kit is manufactured by Enzy-Tek of Lenexa, KS (20). A disadvantage of this method is lack of specificity due to the ability of the test reagents to react with other soil insecticides and their metabolites that are also acetylcholinesterase inhibitors.

A second approach involves detection of the metabolite, carbofuran phenol, which is indicative of the rate of enhanced degradation occurring in the soil (Table IV). The phenolic metabolite is a product of degradation and, when reacted with p-nitrobenzenediazonium fluoroborate, a colored complex forms, the intensity of which can be related to the amount of carbofuran remaining in soil (21). This 'quick test' can be developed into a spectrophotometric method assuring a high degree of specificity in identifying enhanced degradation of carbofuran.

Table IV. Carbofuran Degradation and Carbofuran Phenol Production in Liquid Suspensions Inoculated with Enhanced and Non-enhanced Soil

	Amount Present After 7 Days (%)	
Source of Inoculum	Carbofuran	Carbofuran Phenol
None	90	—
Non-enhanced Soil	97-100	—
Enhanced Soil	<10	75-100

Adapted from (21)

Isofenphos: Metabolite Assay. Racke and Coats (22) discovered a new metabolite of isofenphos, isopropyl salicylate, in cultures of isofenphos-degrading microorganisms isolated from enhanced soil. Since isopropyl salicylate is closely related to aspirin, we investigated methods currently used in the medical diagnostics field. The method of Sunshine (23) is used by coroners for detection of salicylate in blood. Samples containing salicylate acidified and reacted with ferric nitrate and mercuric chloride produce a colored complex, which can be measured spectrophotometrically at 540 nm. The salicylate procedure was adapted for assays of isopropyl salicylate in soils receiving repeated applications of isofenphos under controlled conditions. Preliminary results indicate that the metabolite could be detected colorimetrically in soil extracts, suggesting the occurrence of enhanced degradation of isofenphos (Table V). These results will be confirmed using HPLC analyses. It appears that this procedure might be developed as a method for identifying potential enhanced degradation of isofenphos in problem soils.

Table V. Preliminary Results of Isopropyl Salicylate Detection Four Days After Soil Amendment with Isofenphos

Soil	Isofenphos Amendment (mg/kg)	Isopropyl Salicylate[a] (mg/kg)
Enhanced	5	3
	50	23
Non-enhanced	5	0
	50	2

[a]Detection after four days of incubation.

Iprodione and Vinclozolin: Metabolite Assay. The dicarboximide fungicides, iprodione and vinclozolin are used in the treatment of soilborne plant diseases of vegetable crops. Recent research on soils receiving repeated applications of these fungicides indicated that enhanced degradation was a common phenomenon (24). The metabolite, 3,5-dichloroaniline, is a specific product of microbial degradation of these fungicides. Walker (25) has devised a colorimetric method that can be used to detect 3,5-dichloroaniline. Soils with various enhanced degradation histories were extracted with acetone to yield the aniline metabolite. After acidification and addition of N-1-naphthyl ethylenediamine dihydrochloride, a colored complex results which can be quantified spectrophotometrically. The assay could differentiate between soils in which 90% of iprodione degradation occurred in 10 to 12 days and those in which it occurred in 20 to 25 days. For soils amended with vinclozolin, detection of the metabolite occurred at 3, 7, and 10 days. All untreated control soils yielded negative reactions for fungicide degradation. Thus, this specific assay has a potential as a diagnostic tool in identifying enhanced degradation of iprodione and vinclozolin in soil.

Summary

Soil microorganisms undoubtedly play a major role in the removal and subsequent 'safening' of pesticides in the soil environment. Ideally, a pesticide should persist long enough to exert its toxic effect on the target pest and then degrade biologically. The development of pesticides that degrade in the environment has been encouraged, however, the ability of soil microorganisms to degrade pesticides at enhanced rates was overlooked (26). The necessity of pesticides in modern agriculture has increased the need for field diagnostic assays for potential enhanced degradation. Assays based on spectrophotometry are sensitive, simple and relatively inexpensive to run. Further investigations into enzymatic assays coupled to specific determinations of pesticide metabolites may result in more accurate systems of identification and prediction of enhanced degradation in soils.

Acknowledgments

This article is a contribution of the Agricultural Research Service, U.S. Department of Agriculture and the Missouri Agricultural Experiment Station, Journal Series No. 10,933.

Literature Cited

1. Kaufman, D.D.; Edwards, D.E. In Pesticide Chemistry: Human Welfare and the Environment; Miyamoto, J.; Kearney, P.C., Eds.; Pergamon Press: Oxford, England, 1982, 177-182.
2. Roeth, F.; Wilson, R.; Martin, A.; Shea, P. Weed Technol. 1989, 3, 24-29.
3. Reed, J.P.; Keaster, A.J.; Kremer, R.J.; Kerr, H.D. Corn Pesticide Performance in Missouri 1984; MAES SR-334; Missouri Agricultural Experiment Station: Columbia, MO, 1985.
4. Umbreit, W.W.; Burris, R.H.; Stauffer, J.F. Manometric & Biochemical Techniques; Burgess Publ. Co., Minneapolis, MN, 1972.
5. Snell, F.; Snell, C. Colorimetric Methods of Analysis; Van Nostrand and Rheinhold, New York, 1970.
6. Meites, L. Handbook of Analytical Chemistry; McGraw-Hill, New York, 1963.
7. Argauer, R. In Pesticide Analytical Methodology; Harvey, J.; Zweig, G., Eds.; ACS Symposium Series No. 136; American Chemical Society: Washington, DC, 1979; Chapter 3.
8. Webber, J.B. In Research Methods in Weed Science; Camper, N.D., Ed.; Southern Weed Science Society, Champaign, IL, 1986; Chapter 9.
9. Vickrey, T.M.; Karlesky, D.L.; Blackmer, G.L. JAOAC. 1980, 63, 506-510.
10. Lankow, R.K.; Grothaus, G.D.; Miller, S.A. In Biotechnology in Agricultural Chemistry; LeBaron, H.M., Ed.; ACS Symposium No. 334; American Chemical Society: Washington, DC, 1987; Chapter 19.
11. Burns, R.G.; Edwards, J.A. Pestic. Sci. 1980, 11, 506-512.
12. Tabatabai, M.A. In Methods of Soil Analysis, Part 2; Page, A.L., Ed.; American Society of Agronomy, Madison, WI, 1982; Chapter 43.
13. Reed, J.P.; Kremer, R.J.; Keaster, A.J. Bull. Environ. Contam. Toxicol. 1987, 39, 776-782.
14. Tu, C.M. J. Environ. Sci. Health. 1981, B16, 179-191.
15. Tollefson, J. Solutions. 1986, 30(1), 48-55.

16. Mueller, J.G.; Skipper, H.D.; Lawrence, E.G.; Kline, E.L. Weed Sci. 1989, 37, 424-427.
17. Gauger, W.; MacDonald, J.; Adrian,N.; Matthees, D.; Walgenbach, D. Arch. Environ. Contam. Toxicol. 1986, 15, 137-141.
18. Schnurer, J.; Rosswall, T. Appl. Environ. Microbiol. 1982, 43, 1256-1261.
19. Reed, J.P; Krueger, H.R.; Hall, F.R. Bull. Environ. Contam. Toxicol. 1989, 43, 929-934.
20. EnzyTec Technical Bulletin; Lenexa, KS, 1987.
21. Chapman, R.; Moy, P.; Henning,K. J. Environ. Sci. Health. 1985, B20, 313-319.
22. Racke, K.; Coats, J. J. Agric. Food Chem. 1987, 35, 94-99.
23. Sunshine, I. Manual of Analytical Toxicology. Chemical Rubber Company; CRC Press, Cleveland, OH, 1971; 307-308.
24. Walker, A.; Brown, P.A.; Entwistle, A.R. Pestic. Sci. 1986, 17, 183-193.
25. Walker, A. Pestic. Sci. 1987, 21, 233-240.
26. Felsot, A.S. Ann. Rev. Entomol. 1989, 34, 453-476.

RECEIVED February 7, 1990

Chapter 19

Enhancing Biodegradation for Detoxification of Herbicide Waste in Soil

A. S. Felsot and E. K. Dzantor

Center for Economic Entomology, Illinois Natural History Survey, 607 East Peabody Drive, Champaign, IL 61820

Pesticides in soil at high concentrations have been found to be unusually resistant to normal biodegradative processes. Microbial systems have been proposed as cost effective techniques suitable for cleanup of pesticide waste. In an attempt to enhance the detoxification in soil of high concentrations of the herbicide alachlor, experiments were designed to test the effects of several factors on alachlor persistence: soil dilution, concentration, formulation, nutrient amendments, and microbial inoculation. Alachlor in soil from a waste site degraded faster when diluted with uncontaminated soil by 90% than when diluted by 10 or 50%. Alachlor was metabolized into water-soluble compounds in soil at a concentration of 10 ppm but not at 1000 ppm. Amendment of soil with ground corn or soybean stubble stimulated the biodegradation of alachlor at a concentration of 100 ppm but not at 1000 ppm. When alachlor was present as a sole carbon source, several bacterial isolates partially detoxified concentrations of 10 ppm but not 100 ppm. Fungal isolates could cometabolically degrade 100 ppm alachlor in pure culture; however, inoculation of soil with an alachlor-degrading fungus alone did not enhance degradation.

The use of pesticides has received intense scrutiny for generations. The concern about residues in food and potential adverse health effects predates the advent of synthetic organic pesticides in the late 1940's (1).

0097–6156/90/0426–0249$06.00/0

Since the passage of the Resource Conservation Recovery Act (1976), much concern has focused on pesticide waste, which is inevitably generated by agricultural and urban sectors alike (2-4). In Illinois, agrochemical retail facilities have been particularly problematic. These facilities provide farmers with a variety of services including the custom application of fertilizers and pesticides. Many chemicals are handled at one loading location where spillage is common, resulting in the accumulation of high concentrations of hazardous chemicals. Rinsing of equipment and of empty containers also produces pesticide contaminated discharges that can move off-site as runoff if not handled properly. Very similar conditions of spillage and rinseout may occur on individual farms, especially when the same site is used repeatedly for loading and cleanup. The Illinois Environmental Protection Agency has documented problems of high level pesticide contamination at various agrochemical retail facilities around the state, and the Illinois Department of Public Health has frequently detected unusually high concentrations of pesticides in on-site wells and nearby community wells (5).

Biodegradable pesticides can be extremely persistent when present in the soil at unusually high concentrations (6-10), which contributes to an increased risk of surface and ground water contamination. When a business has a major contamination incident, state or federal regulatory agencies can order a cleanup, but such action is more difficult for private farms and residences. The nature of the cleanup is more problematic. Contaminated soil is excavated and removed to a "secure" landfill. The end result is perhaps a cleaner site, but the waste has not been detoxified.

Cleanup technologies have been proposed for on-site and/or off-site destruction of pesticide wastes (4). The most reliable technologies focus on detoxification of liquid wastes. Destruction of wastes in soil is more difficult and expensive, especially for small businesses and individual homeowners.

New approaches that are consistent with the ubiquity of pesticide waste and mindful of the costs are needed for cleanup in situ, especially when soil has been highly contaminated by past disposal practices (4). Decontamination by microbial systems, which is suitable for meeting those needs, is becoming a feasible technology to clean up waste (11-13). Several strategies have been used for the development of microbial decontamination systems (i.e., bioremediation systems): (1) pretreatment of contaminants with various reagents to produce degradates more amenable to microbial mineralization; (2) treatment of wastes with microbial enzymes; (3) enrichment of wastes to stimulate indigenous microorganisms (biostimulation [14]); (4) inoculation of wastes with adapted microorganisms (bioaugmentation [14]).

Recent research with pesticide wastewater has shown that prior treatment with ozone alone (15) or UV light and ozone (16,17) facilitated the microbial metabolism of the resulting degradation products. Similarly, exposure of chlorinated phenols to UV light and hydrogen peroxide allowed microbial mineralization to proceed (18). These emerging technologies, although promising for wastewater, are not easily applicable to treatment of contaminated soil.

A second strategy that might prove useful for decontaminating pesticide wastes in soil has been the addition of hydrolytic enzymes derived from microorganisms (19,20). Organophosphorus insecticides such as parathion and diazinon are most susceptible to decontamination by enzymes. Some success has been realized with immobilization of degradative enzymes on inert surfaces through which wastewater is passed and detoxified.

The third strategy, biostimulation, stimulates degradative activities of resident microflora by enriching the environment with nutrient amendments or by changing the physical characteristics of the environment. For example, degradation of organic solvents and petroleum products can be enhanced in groundwater by addition of oxygen and nutrients (21). Chlorinated pesticides such as toxaphene can be partially decontaminated by addition of nutrients to soils maintained under anaerobic conditions (22). Recently, the degradation of a variety of pesticide classes, including phenoxyacetate, dinitroaniline, and triazine herbicides and organophosphorus insecticides was enhanced in highly contaminated soil by organic, nutrient, and mineral amendments (23). Degradation of pentachlorophenol in anaerobic (flooded) soil was enhanced by the addition of anaerobic sewage sludge (24). DDT degraded significantly faster in flooded soil amended with rice straw (25).

The bioaugmentation strategy involves the selection of adapted microbial strains that metabolize a pesticide as a carbon or nutrient source. Under these circumstances, the rate of degradation is enhanced compared to the rate normally observed in soil or water. The development of an adapted microbial strain usually begins with the enrichment and isolation of pesticide-degraders from the contaminated environment. Enrichment is relatively easy with compounds that are used as carbon or nutrient sources. Compounds that are not readily utilizable, however, require novel approaches to enrich and isolate potential degraders, which may be manipulated to enhance their degradative capabilities.

Recently, contaminant-degrading strains of microorganisms have been constructed by recruiting into a single organism the genes coding for degradative enzymes

(26,27). The success of this technique has hinged upon the knowledge that many xenobiotic-degrading genes are resident on extrachromosomal pieces of DNA called plasmids (28). Pesticide degradation plasmids were first described for 2,4-D and MCPA (29-31). Plasmids are also known to code for enzymes degrading 2,4,5-T (13) and the OP insecticide diazinon (32). Several copies of a specific plasmid occur within an individual cell, and plasmids can be transferred from a donor cell to a recipient cell. Thus, the degradative potential of a microorganism can be quickly amplified in the population. Furthermore, plasmid exchange can occur between species, although it occurs more readily within a species.

By taking advantage of bacterial transformation and conjugation, researchers have constructed strains possessing an entire pathway for xenobiotic mineralization by culturing in chemostats two or more strains possessing complementary parts of the pathway. For example, microbial inocula from hazardous waste sites were placed in a chemostat with microbial strains possessing known plasmids for aromatic hydrocarbon and chlorobenzoate metabolism (27). After cultivation with 2,4,5-T as a sole carbon source for 8 to 10 months, an isolate was produced that could utilize 2,4,5-T. Success in breeding a microbial strain with the capability of metabolizing chlorobenzene has been obtained using a three-stage chemostat system with two known isolates having complementary capabilities of chlorobenzoate metabolism (33).

Our interest in microbial systems for decontamination of pesticide waste evolved from our attempts to remediate highly contaminated soil at an agrochemical retail facility in Piatt County, IL (34). Soil containing high levels of alachlor (2-chloro-2',6'-diethyl-N-[methoxymethyl] acetanilide) (24,000 ppm in the top 7.5 cm of one location) was excavated and land-applied to corn and soybean plots in an effort to stimulate natural biodegradative mechanisms. Other contaminants included metolachlor (2-chloro-N-[2-ethyl-6-methylphenyl]-N-[2-methoxy-1-methylethyl] acetamide), atrazine (2-chloro-4-[ethylamino]-6-[isopropylamino]-S-triazine), trifluralin (α,α,α-trifluro-2,6-dinitro-N-N-dipropyl-p-toluidine), and nitrogen fertilizer. In corn and soybean plots receiving contaminated soil, alachlor and metolachlor persistence was significantly greater than in plots freshly treated with herbicide sprays comprised of equivalent concentrations (10). Soil that had been left in piles on the waste site seemed to have a reduced microbial population and a depressed enzyme activity that may have accounted for the persistence of the herbicides. In laboratory experiments, simulated spills of alachlor (10,000 ppm soil) reduced microbial bioactivity, and the pesticide did not degrade after one year (10).

In an effort to enhance detoxification of high concentrations of alachlor, we designed studies to test the effects of soil dilution, concentration, formulation, and nutrient amendments on persistence. Additionally, we developed a protocol to enrich, isolate, and screen bacteria and fungi for enhanced capabilities of alachlor degradation. Our studies represent the intital stages of the biostimulation and bioaugmentation strategies for waste cleanup.

Materials and Methods

Soil used in the alachlor persistence studies and in the enrichments for alachlor-degrading organisms was derived from two sources: a waste site at an agrochemical facility in Piatt Co., IL and a soybean plot near the waste site that was divided into replicated blocks for a land application study (34). The soil type was a mixture of Ipava silt loam (fine, montmorillonitic, mesic, Aquic Argiudolls) and Sable silty clay loam (fine silty mixed, mesic, Typic Haplaquolls) with pH 5.4-5.5. The soil at the waste site had been excavated and stored in piles which were sampled as needed for laboratory studies (waste-pile soil). Untreated check plots in the soybean field served as sources of uncontaminated soil (CHECK soil). Soils were stored at 2-4°C and passed through a 3-mm screen before use.

Soil Dilution Experiment. Waste-pile soil was mixed with CHECK soil in large plastic bags to produce dilutions of 0, 10, 50, and 90%, which yielded mean alachlor concentrations of 46.8, 47.4, 29.4, and 6.52 ppm, respectively. Thirty-gram portions (oven-dry weight) of soil were dispensed into 250 mL Erlenmeyer flasks and adjusted to 30% moisture (w/w). The flasks were covered with Parafilm and incubated at 25°C. Immediately after mixing, and at periodic intervals during the next 42 days, triplicate flasks were removed for extraction of alachlor. Activity of dehydrogenase soil enzyme was assayed (35) in soils simultaneously incubated in companion flasks.

Effect of Concentration and Formulation. Moist CHECK soil (26.4% moisture, 30 g oven-dry weight) was treated with technical grade alachlor (prepared in acetone) or an emulsifiable concentrate formulation (Lasso 4EC, 45.1% a.i., prepared in water) to yield application rates of 10, 100, and 1000 ppm soil. Stock solutions of alachlor were prepared by mixing the appropropiate amount of either the technical grade or emulsifiable alachlor with 2.6 µCi of uniformly ring-labelled ^{14}C alachlor (Monsanto Co., specific activity=13.74 mg/mCi, radiochemical purity=95%).

One hundred microliters of stock solutions were applied to the soil. After sitting for approximately 6 hours under a fume hood, the flasks were swirled by hand to mix the soil and closed off with rubber stoppers from which hung plastic center wells containing 0.5 mL of 2 N KOH, which served as traps for $^{14}CO_2$. One set of treated soils was capped immediately to determine if $^{14}CO_2$ was evolved during the 6-hour aeration period. A set of untreated soils served as controls to correct for background radioactivity. A third set of soils was treated with 80 µg of uniformly ring-labelled ^{14}C-glucose to ensure that CO_2 was being trapped in the center wells.

Soils were extracted on the same day as application and after 28 days of incubation at 25°C. Every 2-3 days during the interim, flasks were opened and the center wells were removed and placed directly in liquid scintillation cocktail (Biosafe II) for determination of radioactivity. Soils were extracted twice by stirring with ethyl acetate followed by a third extraction with a 1:1:1 mixture of hexane/acetone/methanol. After the last extraction, the soil was filtered through glass microfiber filter paper. The solvents were combined and evaporated to dryness under vacuum at 35°C.

After evaporation, the extract was partitioned between water and methylene chloride (1:1). The water phase was reextracted with CH_2Cl_2. The aqueous and organic phases were evaporated to dryness and made to a 2 mL volume with acetone and methanol, respectively. Five hundred microliters of each phase were counted by liquid scintillation spectrometry. Parent alachlor was determined in the partitioned extracts by GLC.

Enrichment and Isolation of Potential Alachlor-Degrading Microorganisms. The protocol for enriching, isolating, and screening alachlor-degrading microorganisms is summarized in Figure 1. A modified soil perfusion system (36) and soils from a simulated alachlor spill containing 10,000 ppm alachlor served as primary enrichments for selecting potential degraders. Inocula from the primary sources were used to further enrich for alachlor-degraders in chemostat and batch shake flasks (Figure 1, secondary enrichment). The perfusing medium contained mineral salts medium (MSM), alachlor (100 mg/L), dextrose (500 mg/L), and 100 mg/L of yeast extract, chloroacetate, benzoic acid and p-chloroaniline. The mineral salts medium was composed of (g per L final concentration): $MgSO_4 \cdot 7H_2O$ (0.2 g); NaCl (0.1 g); $CaCl_2.2H_2O$ (0.1 g); KNO_3 (0.5 g), and K_2HPO_4 (1.0 g). The source of soil was a soybean plot to which alachlor-contaminated soil had been land-applied.

The chemostat vessel was a modified 500-mL Virtis fermentor with a teflon impeller. The medium consisted of 350 mL MSM containing alachlor (100 mg/L), glucose (100 mg/L), and yeast extract (50 mg/L). The chemostat was inoculated with 20 mL of soil perfusate, the inoculum was allowed to grow to stationary phase as a batch culture,

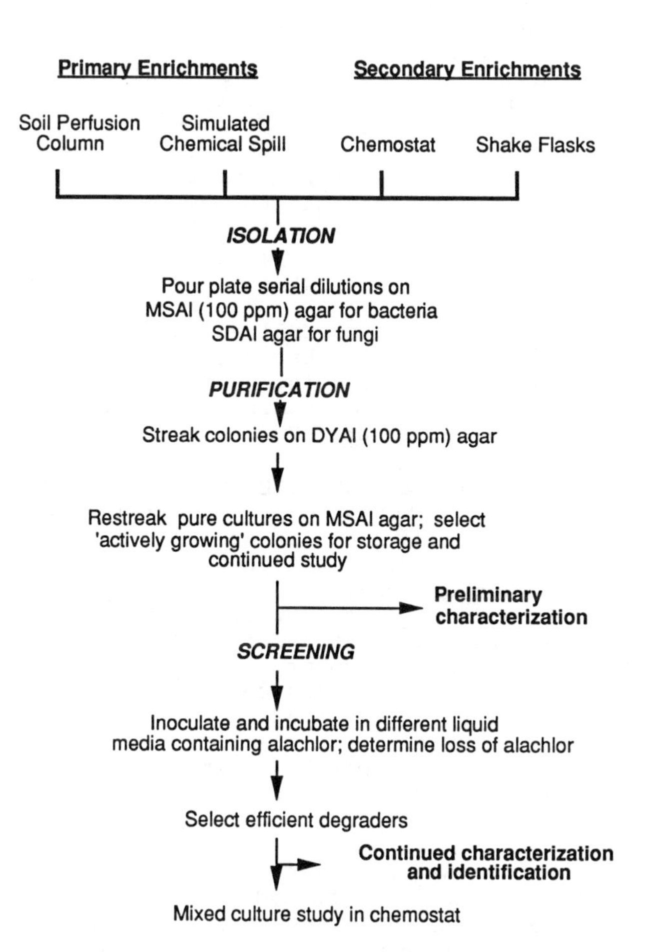

Figure 1. Protocol for enrichment, isolation, and study of microbial cultures.

and then fresh medium was pumped in at the rate of 0.1 mL/min. The chemostat was sampled periodically for microbial isolation and chemical assay of alachlor.

Batch culture systems consisted of 125-mL Erlenmeyer flasks containing 50 mL MSM and 100 mg/L alachlor. The flasks were inoculated directly with 0.5 g of soil from simulated chemical spill experiments or with 1 mL aliquots of soil suspensions (1 g soil/10 mL H_2O). The flasks were incubated on a rotary shaker at 25°C for up to 4 weeks with periodic sampling for microbial isolation.

Cultures from the enrichment systems were subsampled periodically and plated either on agar containing MSM plus alachlor (MSAl) to isolate bacteria or on Sabaroud dextrose with alachlor (SDAl) to isolate fungi (Figure 1, isolation step). The agar plates were incubated at 25°C and examined periodically for microbial growth. Bacterial colonies that formed on MSAl agar were purified by streaking on dextrose-yeast extract-alachlor (DYAl) agar (Figure 1, purification step). Purified bacterial colonies were restreaked on MSAl and isolates that grew were stored in DYAl for further characterization. Fungi were purified by hyphal tipping on SDAl. All pure isolates were maintained on their respective agar preparations and stored at 5°C.

Screening Potential Alachlor-Degrading Bacteria. Twenty-eight bacterial isolates were preliminarily screened for the ability to degrade alachlor cometabolically or as a sole carbon source. From this test, nine isolates were selected for further screening. First, the bacteria were inoculated into 20 mL of filter-sterilized MSAl containing glucose (100 mg/L) and yeast extract (500 mg/L). After incubation for 24 h on a rotary shaker, 0.2 mL of culture was inoculated into test tubes containing 5 mL of one of the following filter-sterilized media: (1) alachlor (10 or 100 ppm) in MSM (MS-Al); (2) alachlor (10 or 100 ppm) in MSM + yeast extract (50 mg/L) + dextrose (0.1 or 1.0 g/L) (YAl-D or YAl-D+); (3) alachlor (10 or 100 ppm) in soil extract + dextrose (0.0, 0.1 or 1.0 g/L) (SEAl, SEAl-D or SEAl-D+).

Soil extract was prepared by autoclaving a mixture of 1 kg CHECK soil and 1.5 L of H_2O for 30 min at 121°C. After cooling at 5°C overnight, the soil suspension was filtered through glass microfiber filters and centrifuged at 5523 x g for 5 min. The extract was buffered at pH 6.8 using K_2HPO_4 (0.5 g/L).

After inoculation, duplicate cultures were incubated in the dark without agitation at 25°C for 10 days. Four-mL aliquots of the cultures were extracted twice with ethyl acetate for analysis of alachlor. Controls consisted of uninoculated media.

Screening Potential Alachlor-Degrading Fungi. Two fungal isolates (CCF-1 and CCF-2) were tested in batch shake-flask cultures in a medium (PYAl) containing (g/L): peptone (1.0), yeast extract (3.0), dextrose (20.0), K_2HPO_4 (1.0), $CaSO_4$ (0.014) and alachlor (0.1). Inocula were prepared by growing two isolates in PYAl. After 5 days, growth was harvested by centrifugation, and the fungal pellets were aseptically macerated. One-mL aliquots of the homogenized cultures were inoculated into 50 mL PYAl or MSAl. Controls included alachlor-fortified, uninoculated flasks and inoculated flasks containing peptone-yeast extract medium without alachlor. At intervals of 3, 7, and 14 days, cultures were filtered and washed; the filtrate and washings were combined and extracted twice with ethyl acetate for alachlor analysis. Chloride released from alachlor was analyzed in part of the filtrate by using a modified ferricyanide colormetric method (37).

Effect of Nutrient Amendments and Microbial Inoculum. Batches of CHECK soil (2-mm mesh, 25% moisture) were mixed with ground (2-mm mesh) corn (CS) or soybean (SB) stubble at a rate of 20 g/kg oven-dried soil. Thirty grams of amended soils were weighed into flasks; half of the flasks containing SB were treated with a stock solution of NH_4NO_3 at a rate of 1 mg N/g (SB+N). Aliquots of formulated alachlor (Lasso 4EC) were pipetted on soil to yield concentrations of 100 or 1000 ppm. About 1-2 hours after pesticide treatment, half of the flasks assigned to each treatment were inoculated with fungal isolate CCF1 (0.5 mg fungal units/0.28 mL/g oven-dry soil), which had been blended in phosphate-buffered water (pH 7) containing alachlor (100 ppm). Unamended, uninoculated soils fortified with alachlor were controls. All flasks were closed with Parafilm and incubated at 25°C. Once a week the flasks were opened for aeration. On days 0, 14, 28, and 56, flasks were removed for extraction and analysis of alachlor, soil dehydrogenase (35), and soil esterase.(38).

Extraction and Analysis of Alachlor. Soil was slurried with 12 mL of water and extracted twice by stirring with 50 mL of ethyl acetate for 45 minutes. The solvent was decanted after each extraction and concentrated on a steam bath. The extract was rediluted with ethyl acetate and analyzed by GLC with nitrogen-phosphorus specific detection. Residues were separated on a 90-cm x 0.2 mm i.d. glass column packed with 5% Apiezon + 0.125% DEGS maintained isothermally at 190°C. Injector and detector were held at 250°C, and gas flow rates were adjusted as needed to obtain maximum sensitivity and resolution. Residues were quantitated by the method of external standards.

Results and Discussion

Effects of Soil Dilution. In previous studies at an agrochemical facility contaminated with herbicide waste, alachlor had not totally degraded after two years in soil that had been excavated and stored in piles on the ground (10). Land application has been studied as a method for stimulating biodegradation by diluting the soil on cropped land (34). Under laboratory conditions, alachlor degraded slower in waste-pile soil than it did in soil diluted by 90% (w/w) (Table I). Seventy-percent of the alachlor was detoxified in 42 days compared to ≤30% in waste-pile soil diluted up to 50%. The initial concentration of alachlor in soil diluted by 90% was 6.5 ppm, which would be similar to the concentration in a 15-cm depth of soil after a typical application of alachlor at a rate 3.36 kg AI/ha The typical half-life of alachlor in field soil ranges from 2-4 weeks (39), but high concentrations typical of waste are very persistent (40).

The activity of soil dehydrogenase at the beginning of the soil dilution experiment in 0, 10, 50, and 90% diluted waste-pile soil was 18, 0, 65, and 138%, respectively, of the activity in CHECK soil. After 21 days, soil dehydrogenase was still inhibited in the 0 and 10% diluted waste-pile soil. The inhibition of enzyme activity suggests that high concentrations of alachlor may be toxic, but microbial bioactivity can be restored if contaminated soil is diluted enough.

Table I. Percentage Recovery of Alachlor in Diluted and Undiluted Waste-Pile Soil

	% of Initial PPM Recovered at Indicated % Dilution			
Days	0	10	50	90
5	122	106	123	98
10	112	98	128	79
21	135	78	102	39
42	76	71	106	30

Effect of Concentration and Formulation. Other research has shown that high concentrations of pesticides are very persistent compared to normally applied levels (8,9). Furthermore, high concentrations can reduce microbial populations, which may explain the slower rate of biodegradation (8). Although high concentrations of alachlor seemed to reduce microbial bioactivity, we were unsure if the effect was due to alachlor itself or additives in the formulation; therefore, we conducted a study to look at the interaction of concentration and formulation (technical vs. emulsifiable concentrate) on alachlor degradation.

Six hours after application, recovery of ^{14}C in each phase as a percentage of the intially added amount was not significantly different among the various concentrations

Table II. Effect of Concentration and Formulation on Metabolism of Alachlor in Soil at 28 Days Following Application

		% of Initially Added Alachlor	
Concentration	Phase	Technical	Emulsifiable
10 ppm	^{14}C organic	25.2	23.3
	^{14}C aqueous	15.2	9.5
	$^{14}CO_2$	0.4	0.5
	^{14}C unextracted	59.2	66.7
	alachlor organic	23.5	25.5
	alachlor aqueous	0.0	0.0
100 ppm	^{14}C organic	66.9	67.2
	^{14}C aqueous	7.6	5.0
	$^{14}CO_2$	0.2	0.2
	^{14}C unextracted	25.3	27.6
	alachlor organic	69.2	76.2
	alachlor aqueous	0.1	0.0
1000 ppm	^{14}C organic	113.4	102.6
	^{14}C aqueous	0.9	1.2
	$^{14}CO_2$	0.2	0.2
	^{14}C unextracted	0.0	0.0
	alachlor organic	131.4	119.2
	alachlor aqueous	0.0	0.2

or formulations. The small percentage of ^{14}C detected in the aqueous phase (<2.5%) could have been an artifact from the partitioning process. No $^{14}CO_2$ was lost in the 6 hours allowed for aeration of the flasks containing the emulsificable concentrate formulation.

Twenty-eight days after application, cumulative $^{14}CO_2$ in the 10 ppm treatments was higher than in the 100 and 1000 ppm treatments but still averaged less than 0.5% of the added radioactivity (Table II). Recovery of radiolabel in the organic phase was significantly lower in the 10 ppm treatment than in the 100 or 1000 ppm treatment, but for any one concentration there was no significant difference between technical and emulsifiable concentrate formulations. The complete recovery of both parent alachlor and ^{14}C in the organic phase of the 1000 ppm treatment indicated that alachlor was not significantly degraded in the 1000 ppm treatment. Alachlor degraded more slowly at 100 ppm than at 10 ppm.

Partitioning of radioactivity between the organic and aqueous phase showed a nearly logarithmic increase in the ratio of ^{14}C in each phase from 10 ppm to 100 ppm regardless of formulation (Table II). In the aqueous and organic phases of the 10 ppm treatments, the absence of parent alachlor in the GLC analysis showed that the two-fold difference in ^{14}C between the two phases was in part due to production of water soluble metabolites. The 10 ppm treatment contained nearly twice as much water soluble radiolabel as the 100 ppm treatment.

Alachlor has been shown to degrade in soil by microbial cometabolism rather than by mineralization (41), and CO_2 was not produced after introduction of radiolabelled alachlor to microbial suspensions (41,42). An early investigation of alachlor metabolism by a pure culture of fungus showed significant production of 2,6-diethylaniline, which could be further metabolized (42). More recent research, however, has shown that alachlor is dechlorinated in soil followed by conjugation with glutathione (J. Malik, Monsanto Agricultural Products Co., personal communication). The conjugates are then further metabolized to yield sulfonic and oxanilic acid derivatives of alachlor (39), which would be very water soluble.

Screening of Bacteria for Degradation of Alachlor. When alachlor was present as a sole carbon source in MSM (MS-Al), three isolates, C1, SA3-2, and SA3-3, degraded 59, 45, and 27%, respectively, of a 10 ppm dose after 10 days of incubation (Table III). At a dose of 100 ppm, however, no bacterial isolate could degrade alachlor as a sole carbon source (Table IV). Isolate C1 very actively detoxified 10 ppm alachlor in the presence of additional nutrients but seemed poisoned by 100 ppm alachlor (Table III,IV). Isolates CCII, C4, and SA3-1 degraded 20-27% of 100 ppm alachlor in media containing dextrose plus soil extract or yeast extract. In other research, Novick et al. (41) were unable to isolate alachlor mineralizers after soil inocula were treated with 1 and 100 ppm alachlor. In a study with the closely related herbicide metolachlor (43), mineralizing organisms could not be isolated and it was noted that actinomycetes could not tolerate concentrations $\geq$200 ppm. These data suggest microbial toxicity at high concentrations of alachlor and metolachlor.

Screening of Fungi for Alachlor Degradation. Fungal isolates CCF-1 (tentatively identified as *Fusarium* sp.) and CCF-2 degraded more than 70% and 50%, respectively of a 100 ppm dose of alachlor after 14 days of incubation in peptone-yeast extract medium (PYAl). About 18% of the chlorine in alachlor was released as chloride in both cultures. Neither isolate degraded alachlor when present as a sole carbon source in MSM. These data show that efforts to develop microbial systems for detoxification of waste need to focus on fungi in addition to bacteria. Indeed, a white rot fungus has been recently reported to degrade several organochlorine compounds (44).

Table III. Percentage Alachlor Removed from Solution after 10 Days of Incubation with 10 ppm Alachlor in Indicated Media[a]

Isolate	MS-Al	YAl-D	YAl-D+	SEAl	SEAl-D	SEAl-D+
SA3-1	0.0	0.0	6.8	-	-	-
SA3-2	45.4	21.1	0.0	23.2	25.4	25.4
SA3-3	26.5	0.0	0.0	0.0	0.0	0.0
SA4	0.0	12.4	16.7	0.0	0.0	10.3
2A1	0.0	13.8	0.0	0.0	0.0	1.5
C1	58.5	47.4	46.5	38.2	40.7	48.0
C2	15.3	16.7	26.5	14.7	2.9	0.0
C4	13.4	12.2	56.7	12.5	23.3	2.8
CCII	12.2	6.1	30.2	12.5	57.9	12.5

[a] See text for culture media codes.

Table IV. Percentage Alachlor Removed from Solution after 10 Days of Incubation with 100 ppm Alachlor in Indicated Media[a]

Isolate	MS-Al	YAl-D	YAl-D+	SEAl	SEAl-D	SEAL-D+
SA3-1	0.0	23.0	0.0	0.0	0.0	0.0
SA3-2	0.0	0.0	0.0	0.0	0.0	5.0
SA3-3	0.0	6.5	0.0	0.0	0.0	12.3
SA4	0.0	0.9	11.5	1.8	18.2	17.7
2A1	0.0	16.2	10.8	0.0	12.2	14.3
C1	0.0	0.0	0.0	5.9	0.0	0.0
C2	0.0	16.5	0.0	9.0	0.0	9.6
C4	0.0	2.2	15.6	22.7	20.0	22.3
CCII	5.0	21.0	21.3	16.5	27.4	20.8

[a] See text for culture media codes.

Effect of Nutrient Amendments and Microbial Inoculum. Degradation of a 100 ppm dose of alachlor was enhanced when the soil was amended with ground corn and soybean stubble (Table V). After 56 days of incubation about 94 and 83% of the alachlor were degraded in CS and SB+N-amended soils respectively, compared to about 40% degradation in the unamended soils. In soils amended with SB alone, degradation was slower than in soils with CS and

SB+N but was still significantly higher than degradation in the unamended soil.

The rationale for choosing the organic amendments was based on a hypothesis that increasing the carbon to nitrogen (C:N) ratio in soil would stimulate microbial activity and cause a depletion of soil N, and thereby induce microbial attack of less readily available N sources like alachlor. The order of C:N ratio in the amendments was CS < SB+N < SB. According to the hypothesis, the order of degradation rate should have been

Table V. Degradation of 100 ppm Alachlor in Soils Amended with Corn and Soybean Stubble and Incubated with Fungal Isolate CCF-1[a]

Soil Treatment		Percentage Alachlor Recovered [a] Days of Incubation		
Amendment	Inoculation	14	28	56
none	no	97 A	63 A	59 A
corn	no	56 B	24 C	6 C
soybean	no	68 B	42 B	36 B
soybean+N	no	65 B	26 C	17 C
none	yes	84 A	63 A	57 A
corn	yes	38 C	23 C	4 C
soybean	yes	55 B	37 B	28 B
soybean+N	yes	53 B	22 C	3 C

[a] Means followed by the same letter within an inoculation group are not significantly different according to Duncan's multiple range test ($p=0.05$).

SB > SB+N > CS. In fact, the more rapid degradation of alachlor in CS- and SB+N-amended soil coincided better with indicators of microbial activity than C:N ratio. For example, soil dehydrogenase activity and soil esterase in CS- and SB+N-amended soils was usually higher than in unamended and SB-amended soils (Figure 2). Although dehydrogenase in SB-amended soils was lower than in unamended soils, soil esterase was significantly higher in SB-amended soils. Soil enzyme activity is indicative of microbial metabolic activity but is not necessarily correlated with capacity to degrade xenobiotics; the data do show, however, that stimulation of enzyme activity was coincidental with a rapid loss of alachlor.

On the basis of efficiency of alachlor removal from solution, fungal isolate CCF-1 was chosen for soil inoculation. CCF-1 alone did not enhance the degradation of alachlor, but the combination of inoculum and amendment, compared to organic amendments alone, slightly enhanced degradation. This additional enhancement was more pronounced in the SB- and SB+N-amended soils than in the CS-amended soils. To determine whether the isolate is

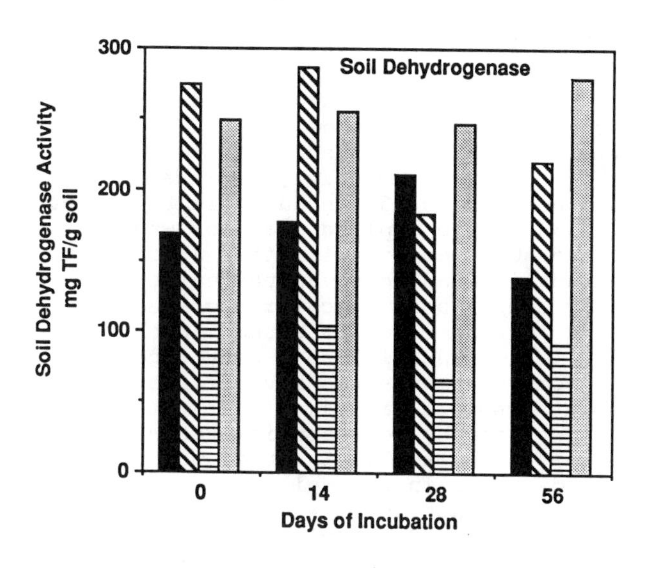

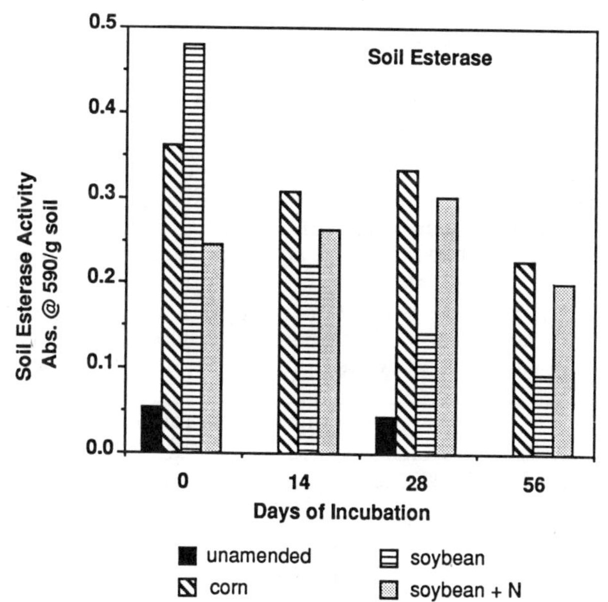

Figure 2. Soil enzyme activity in soils treated with alachlor at a rate 100 ppm, amended with corn, soybean, or soybean+N, and inoculated with fungal isolate CCF-1.

useful in the field, increased dosages of CCF-1 would have to be tested to overcome interspecific competition from other microorganisms. The research of Barles et al.(45) and Kilbane et al. (46) shows that microbial inoculation of soil can be successful if a compound is mineralizable, but whether cometabolizing organisms can compete successfully remains to be seen.

After 56 days of incubation, degradation of 1000 ppm alachlor did not exceed 35% in any soil treatment (Table VI). About 33% of the applied alachlor was degraded in the inoculated, CS-amended soil after 28 days, but no further degradation was observed thereafter. In nearly every two-way comparison, alachlor degraded significantly faster in the 100 ppm treatments than in the 1000 ppm treatments. Coincidentally, soil dehydrogenase activity in all 1000 ppm treatments was severely depressed.

The soil amendment experiments demonstrate that alachlor is degraded by cometabolism rather than mineralization in the soil. In other words, alachlor can be partially degraded in the soil (i.e., detoxified), but its metabolism does not result in procurement of energy or nutrients for growth. By supplying extra nutrients, indigenous microorganisms that have an incidental ability to detoxify a chemical can proliferate, which coincidentally causes a decline in the observed concentration of the chemical. For example, after addition of various organic amendments to soil maintained under flooded (or anaerobic conditions), other researchers have reported enhanced biodegradation of recalcitrant organochlorine pesticides such as pentachlorophenol (24), DDT (25) and toxaphene (47). Under aerobic conditions, glucose and wheat straw have enhanced the biodegradation of the asymetrical triazine herbicide metribuzin, but

Table VI. Degradation of 1000 ppm Alachlor in Soils Amended with Corn and Soybean Stubble and Incubated with Fungal Isolate CCF-1

Soil Treatment		Percentage Alachlor Recovered[a] Days of Incubation		
Amendment	Inoculation	14	28	56
none	no	96 BC	81 A	87 A
corn	no	87 C	73 A	83 A
soybean	no	116 A	79 A	84 A
soybean+N	no	105 A	99 A	88 A
none	yes	89 A	102 A	86 A
corn	yes	79 A	67 C	67 A
soybean	yes	77 A	82 BC	75 A
soybean+N	yes	96 A	89 AB	77 A

[a] Means followed by the same letter within an inoculation group are not significantly different according to Duncan's multiple range test (p=0.05).

alfalfa residue slowed biodegradation (48). This latter research suggests caution is called for when using amendments and mineral nutrients or altering environmental conditions. Recently, Winterlin et al.(23) showed that the degradation of a number of pesticides could be enhanced by certain amendments or environmental conditions, but some conditions actually inhibited degradation. The process of pesticide degradation by cometabolism underscores the need to understand the types of environmental manipulations that are necessary to optimize microbial performance.

In conclusion, enhanced biodegradation is a desirable phenomenon to be exploited for detoxification of pesticide waste. Degradation of compounds that are not mineralizable can still be enhanced if appropriate nutrients and environmental conditions are present. Thus, microbial systems for waste disposal should not be limited by cometabolic nutritional strategies but will require more intensive study of the biochemical ecology of microorganisms.

Acknowledgments

We thank M. McGiffen and L. Case for technical assistance. This research is a contribution from the Illinois Natural History Survey and Illinois Agricultural Experiment Station, College of Agriculture, Univ. of Ill. at Urbana-Champaign, and was supported in part by the Ill. Hazardous Waste Research and Information Center, project no. HWR 88-042.

Literature Cited

1. Felsot, A. S. Ill. Natural History Surv. Bull. 1985, 33, 199-218.
2. Wilkinson, R. R.; Kelso, G. L.; Hopkins, F. C. State-of-the-Art-Report Pesticide Disposal Research, U.S.EPA Report 600/2-78-183, Cincinnati, 1978.
3. Bridges, J. S; Dempsey, C. R. Proceedings: Research Workshop on the Treatment/Disposal of Pesticide Wastewater. U. S. EPA report no. 600/9-86/001, Cincinnati, 1986.
4. Seiber, J. N. In National Workshop on Pesticide Waste Disposal; JACA Corp., U. S. EPA report no. 600/9-87/001, Cincinnati, 1987; pp 11-19.
5. Long, T. Proc. Ill. Agric. Pesticides Conf. '89, Univ. Ill. Coop. Ext. Serv., Urbana-Champaign, 1989, pp 139-149.
6. Wolfe, H. R.; Staiff, D. C.; Armstrong, J. F.; Comer, S. W. Bull. Environ. Contam. Toxicol. 1973, 10, 1-9.

7. Staiff, D. D.; Comer, S. W.; Armstrong, J. F.; Wolfe, H. R. Bull. Environ. Contam. Toxicol. 1975, 13, 362-368.
8. Davidson, J. M.; Rao, P. S. C; Ou L. T.; Wheeler, W. B.; Rothwell, D. F. Adsorption, Movement, and Biological Degradation of Large Concentrations of Selected Pesticides in Soils. U. S. EPA report no. 600/2-80-124, Cincinnati, 1980.
9. Schoen, S. R.; Winterlin, W. L. J. Environ. Sci. Health 1987, B22, 347-77.
10. Felsot, A. S.; Dzantor, E. K. Proc. Ill. Agric. Pesticides Conf. '89, Univ. Ill. Coop. Ext. Serv., Urbana-Champaign, 1989, pp 150-54.
11. Koybayashi, H.; Rittmann, B. E. Environ. Sci. Technol. 1982, 16, 170A-183A.
12. Jain, R. K.; Sayler, G. S. Microbiol. Sci. 1987, 4, 59-63.
13. Chakrabarty, A. M. Biotechnol. 1987, 1, 67-74.
14. Mathewson, J. R.; Grubbs, R. B. Hazmat World 1989, 2(6), 48-51.
15. Kearney, P. C.; Muldoon, M. T.; Somich, C. J.; Ruth, J. M.; Voaden, D. J. J. Agric. Food Chem. 1988, 36, 1301-06.
16. Kearney, P. C.; Darns, J. S.; Muldoon, M. T.; Ruth, J. M. J. Agric. Food Chem. 1986, 34, 702-06.
17. Somich, C. J.; Kearney, P. C.; Muldoon, M. T.; Elsasser, S. J. Agric. Food Chem. 1988, 36, 1322-26.
18. Miller, R.; Singer, G. M.; Rosen, J. D.; Bartha, R. Environ. Sci. Technol. 1988, 22, 1215-19.
19. Munnecke, D. M. J. Agric. Food Chem. 1980, 28, 105-111.
20. Honeycutt, R.; Ballantine, L.; LeBaron, H.; Paulson, D.; Seim, V. In Treatment and Disposal of Pesticide Wastes; Krueger, R. F.; Seiber, J. N. Ed; ACS Symposium Series No. 259, Am. Chem. Soc: Wash., DC, 1984; pp 343-352.
21. Lee, M. D.; Ward, C. H. Environ. Toxicol. Chem. 1985, 4, 743-50.
22. Craigmill, A. L.; Winterlin, W. L.; Seiber, J. N. In National Workshop on Pesticide Waste Disposal; JACA Corp., U. S. EPA report no. 600/9-87/001, Cincinnati, 1987; pp 31-38.
23. Winterlin, W.; Seiber, J. N.; Craigmill, A.; Baier, T.; Woodrow, J.; Walker, G. Arch. Environ. Contam. Toxicol. 1989, 18, 734-47.
24. Mikesell, M. D.; Boyd, S. A. Environ. Sci. Technol. 1988, 22, 1411-14.
25. Mitra, J.; Raghu, K. Toxicol. Environ. Chem. 1986, 11, 171-181.
26. Reineke, W.; Knackmuss, H.-J. Nature 1979, 277, 385-6.

27. Kellogg, S. T.; Chatterjee, D. K.; Chakrabarty, A. M. Science 1981, 214, 1133-5.
28. Farrell, R.; Chakrabarty, A. M. In Plasmids of Medical, Environmental, and Commercial Importance; Timmis, K. N.; Puhler, A., Eds; Elsevier/North Holland Biomedical Press: 1979; pp 97-109.
29. Pemberton, J. M.; Fisher, P. R. Nature 1977, 268, 732-733.
30. Fisher, P. R.; Appleton, J.; Pemberton, J. M. J. Bacteriol. 1978, 134, 798-804.
31. Pemberton, J. M.; Corney, B.; Don, R. H. In Plasmids of Medical, Environmental, and Commercial Importance; Timmis, K. N.; Puhler, A. Eds; Elsevier/North Holland Biomedical Press: 1979; pp 287-299.
32. Mulbry, W.; Karns, J. S.; Kearney, P. C.; Nelson, J. O.; McDaniel, C. S.; Wild, J. R.; Appl. Environ. Microbiol. 1986, 51, 926-30.
33. Krockel, L.; Focht, D. D. Appl. Environ. Microbiol. 1987, 53, 2470-75.
34. Felsot, A.; Liebl, R.; Bicki, T. Feasibility of Land Application of Soils Contaminated with Pesticide Waste as a Remediation Practice, IL Hazardous Waste Research and Information Center Project Report No. RR 021, Savoy, 1988, 55 p..
35. Frankenberger, W. T.; Johanson, J. B. Soil Biol. Biochem. 1986, 18, 209-13.
36. Kaufmann, D. D. Weeds 1966, 14, 90-91.
37. Bergmann, J. G.; Sanik, Jr. J. Anal. Chem. 1957, 29, 241-3.
38. Schnurer, J; Rosswall, T. Appl. Environ. Microbiol. 1982, 43, 1256-61.
39. Sharp, D. B. In Herbicides: Chemistry, Degradation and Mode of Action, 3rd ed.; Kearney, P. C., Kaufman, D. D., Eds.; Marcel Dekker: New York, 1988; pp 301-33.
40. Junk, G. A.; Richard, J. J ; Dahm, P. A. In Treatment and Disposal of Pesticide Wastes; Krueger, R. F.; Seiber, J. N., Eds; ACS Symposium Series No. 259, Am. Chem. Soc: Wash., DC, 1984; pp 37-67.
41. Novick, N. J.; Mukherjee, R.; Alexander, M. J. Agric. Food Chem. 1986, 34, 721-5.
42. Tiedje, J. M.; Hagedorn, M. L. J. Agric. Food Chem. 1975, 23, 77-81.
43. Saxena, A.; Ahang, R.; Bollag, J.-M. Appl. Environ. Microbiol 1987, 53, 390-6.
44. Bumpus, J. A.; Tien, M.; Wright, D.; Aust, S. D. Science 1985, 228, 1434-6.
45. Barles, R. W.; Daughton, C. G.; Hsieh, D. P. H. Arch. Environ. Contam. Toxicol. 1979, 8, 647-60.
46. Kilbane, J. J.; Chatterjee, D. K., Chakrabarty, A. M. Appl. Environ. Microbiol 1983, 45, 1697-1700.

47. Mirsatari, S. G.; McChesney, M. M.; Craigmill, A. C.; Winterlin, W. L.; Seiber, J. N. J. Environ. Sci. Health 1987, B22, 663-690.
48. Pettygrove, D. R.; Naylor, D. V. Weed Sci. 1985, 33, 267-270.

RECEIVED February 8, 1990

Chapter 20

Implications of Enhanced Biodegradation for the Use and Study of Pesticides in the Soil Environment

Kenneth D. Racke

Environmental Chemistry Laboratory, DowElanco, 9001 Building, Midland, MI 48641–1706

Microbial adaptation for biodegradation is a common microevolutionary phenomenon that occurs wherever biodegradable pesticides or other xenobiotics contact microbial populations. The accelerated biodegradation of pesticides in agricultural systems has certainly had a substantial impact on pesticide users, researchers, and the agrochemicals industry. Major advances in our understanding of the theoretical and practical aspects of enhanced biodegradation have come about as a result of the recent flurry of research activity generated by contemporary occurrences associated with pest control failure. Future challenges and opportunities that remain relate to the development of conceptual predictive models, the design of biodegradable yet adaptation-resistant pesticides, clarification of the soil microbial ecology of biodegradation, and the application of adapted microbial strains for environmental decontamination and hazardous waste detoxification.

Enhanced pesticide biodegradation has recently generated much interest and concern. The preceeding chapters have systematically outlined the phenomenon, detailed the microbiological and chemical interactions involved, and discussed means of coping with enhanced biodegradation as it occurs in agroecosystems. The purpose of this concluding chapter is to place enhanced pesticide biodegradation in perspective by discussing the recent impact of enhanced biodegradation, major research advances that have been made, and the future research challenges that need to be addressed.

The Impact of Enhanced Biodegradation

In a frustrating way the impact of enhanced biodegradation of pesticides in agricultural systems is very hard to quantify. Part of the problem has been the difference of opinion that has existed regarding its true significance. For example, an industry spokesman

0097–6156/90/0426–0269$06.00/0

was quoted as saying: "From a theoretical standpoint, this is an interesting problem. From a practical standpoint, it's not a great problem. In what we're seeing, there's no major economic impact at all..." (1). Despite this type of hyperbole, many researchers in the private and public sectors have felt that the contemporary occurrence of enhanced biodegradation has indeed had a substantial impact (2,3,4). Consider the recent statement of one reviewer of enhanced degradation: "The case histories presented earlier clearly show that this natural phenomenon has adversely affected crop protection practices in certain agroecosystems." (4). Who has been affected by enhanced degradation?

Pesticide Users. The most direct losers in the enhanced degradation game have been growers who actually use pesticides and rely on them for economically significant protection from the ravages of insect, weed, and fungal pests. Many diverse agroecosystems have been touched by the effects of pest control failures caused by enhanced degradation, and instances have been documented for such diverse crops as corn, rice, sugar beets, onions, avocados, and potatoes (3,4,5,6,7). The geographical distribution of these observations (North America, Europe, Asia, The Middle East) implies that adapted microorganisms are no respecters of national boundries!

Actual estimates are rare of the percentage of growers in any one area experiencing enhanced degradation. A survey of 360 Iowa corn growers in 1985 regarding experiences with rootworm insecticide control failures revealed that an average of 19% of the growers experienced poor performance of rootworm insecticides during 1984, and an average of 12% during the preceeding 5 years. For failures within the last 5 years, the greatest percentage of failures by number used was 23% for isofenphos, but the variable behavior of other compounds (0-18%) indicates that many factors may be influencing perceived rootworm control efficacy (8). A similar survey of 378 Nebraska farmers for carbamothioate weed control performance gave more distinct results. In South Central Nebraska in 1983 and 1984 60% of EPTC users and 40% of butylate users indicated that they were dissatisfied with the control of weeds in corn. Analyses of soil samples taken from surveyed fields found a significant correlation between the enhanced degradation of these herbicides (as detected by laboratory assay) and poor weed control performance (9). However, measurements of actual economic impact due to weed or insect control failure are very difficult to estimate. This is because the actual crop yield reduction is not always directly proportional to observed damage or competition caused by marauding insects and weeds. But perception is often reality, and there has certainly been a loss of confidence among growers in the ability of the agrochemicals industry to supply reliable and efficacious products. This is partially because pest control failures of both long-used, reliable pesticides (e.g., carbofuran) and highly touted new pesticides (e.g., isofenphos) often have, seemingly, taken everyone by surprise (10).

Pesticide Manufacturers. Pesticide manufacturers have also been adversely affected by the occurrence of enhanced degradation of their products. One of the most direct adverse results involves the loss of products or sales due to real or perceived pest control failures.

The Midwestern U.S. rootworm insecticide market (>200 MM $/year) supplies a litany of lost business and failed products. Products removed from this market or halted in late stages of development at least partially as a result of enhanced degradation include fensulfothion (Chemagro), bufencarb (Chevron), bendiocarb (Fisons), cloethocarb (BASF), trimethacarb (Shell), and isofenphos (Mobay) (4,10). The loss of sales due to enhanced degradation is exemplified by carbofuran (FMC) use in Iowa corn, wherein carbofuran plummeted from a 20% marketshare in 1977 (11) to less than 8% by 1985 (12).

The agrochemicals industry has also funded numerous projects related to an assessment of the potential for new and existing products to undergo enhanced biodegradation and for development of remedial measures. Whereas much of this funding has gone to outside investigators at government and university laboratories, considerable in-house effort has also been expended. The magnitude of these costs is impossible to estimate given the proprietary nature of many of these projects, but presented and published reports by industrial representatives indicate that substantial work has been conducted in industrial laboratories (13,14,15,16). This is exemplified by the work of Gray and Joo (16) of Stauffer, who demonstrated the rapid degradation of a number of herbicides manufactured by other companies as well as several of their own carbamothioates. It is generally recognized that ongoing assay of candidate soil pesticides for susceptibility to enhanced degradation is a necessary developmental screening step that no company can afford to ignore. Considerable efforts have also been devoted to developing means of coping with enhanced degradation of existing products (17). For example, consider the extensive development and testing by Stauffer of an altered formulation of the herbicide EPTC containing a microbial inhibitor (18,19,20).

A final industrial cost that must be considered is the substantial educational efforts that agrochemicals companies have directed toward pesticide users and researchers. The more comely of these efforts have revolved around attempts to supply digestable information for growers and dealers, and to facilitate information exchange between independent researchers working in the area. For example, in 1985 Union Carbide distributed an informative booklet entitled "Corn Rootworm Management: Insights from the Experts" (21) that contained a very balanced presentation of the topic of enhanced biodegradation by eight university and private researchers and extension personnel. In addition, several special conferences on soil pesticide degradation problems have been sponsored by industry, including a pair in 1985 hosted by Stauffer and Dow. The seamier side of this educational effort has involved use of enhanced degradation information for marketing purposes. These types of efforts have usually involved one of two strategies: 1. Companies marketing products with known enhanced degradation problems have sought to imply that all the competitor's products are equally susceptible to microbial adaptation (e.g., a grower brochure: "No insecticide seems immune to this problem."). 2. Companies marketing products having unknown enhanced degradation potential, or withresistance to microbial adaptation, have sought to reassure growers using their products that they need have no worry of enhanced degradation (e.g., a grower brochure: "---- resists microbial action".). Unfortunately, this marketing activity has in some cases muddled the issue with growers regarding the actual status of enhanced degradation problems with specific pesticide products.

The Research/Extension Community. University, state, and federal researchers and extension personnel have also been affected by the contemporary occurrence of enhanced degradation. The state extension services, which serve as the critical link between the heady scientific university research and the practical informational needs of growers, have often found themselves in the most difficult position in regard to enhanced degradation problems. They are the ones often called upon to deal with pest control failure problems and to make specific recommendations regarding ways to avoid or manage these problems. Part of the agony of extension workers may be attributed to a lack of progress on the research level. For instance, the earliest published reports of studies examining the possibility that enhanced degradation was the cause of reported carbofuran pest control failures in Midwestern U.S. corn presented conflicting conclusions, with some negative (22,23) and some positive (24,25) findings. Extension recommendations often directly reflect the level of research understanding available. In the early 1980's these recommendations started to address the problem of enhanced degradation. For example, an April 1982 Corn Rootworm Management bulletin from Iowa State University (26) stated: "It is recognized that continued carbamate insecticide use enhanced the speed with which certain soil microorganism break down Furadan, and that this is the cause of poor control". The recommendation to avoid using carbofuran if it performed poorly followed. But reflecting the lack of proof that organophosphorus insecticides could also undergo enhanced degradation, this statement also appeared: "Currently available ISU research data do not indicate that continuous use of an organophosphate soil insecticide significantly affects the performance of any insecticide,...". After the discovery of enhanced isofenphos degradation (10) extension recommendations were modified and reflected the uncertainty regarding the ability of any soil insecticide to maintain efficacy if used continuously (27,28). It was during the mid 1980's that both insecticide and herbicide recommendations started emphasizing the rotation or disuse of pesticides with proven enhanced degradation potential (27,28,29).

Researchers at university and government laboratories have responded to enhanced degradation by focusing considerable research efforts on unresolved issues. Researchers in many diverse disciplines have become involved as funding for enhanced degradation research became available. One of the leading sponsors of enhanced degradation research in the U.S. has been the Pesticide Impact Assessment Program (PIAP) of the U.S. Department of Agriculture. After sponsoring a workshop on the problem in St. Louis in 1984, the PIAP has allocated considerable funding to enhanced degradation research. For example, the North Central Region PIAP allocated over $300,000 for these projects between 1982 and 1989 (A.C. Waldron, personal communication). State funding has also been available for this type of research, as exemplified by projects supported by the Nebraska Corn Development Board (30), the Iowa Corn Promotion Board (31), and the Ontario Pesticides Advisory Committee (32). But this support, although significant, has often proven inadequate for the task, and funding from the agrochemicals industry has been used to fill this gap. Information on the magnitude of industrial support is unavailable. However, some idea of the extent of this support can be inferred from publication acknowledgements that include mention of BASF (6,33), Dow (34), FMC (23), Mobay (35), Stauffer (36,37,38), and Union Carbide (39) as contributing to financial support.

The research efforts that have been stimulated by the phenomenon of enhanced degradation have resulted in the creation of a substantial knowledge base regarding microbial adaptation for pesticide degradation and associated pest control failures. The body of scientific literature on the subject has grown rapidly and is now substantial. Recent reviews of enhanced degradation and microbial adaptation for pesticide degradation by Kaufman (2) (59 ref.), Kearney and Kellogg (40) (108 ref.), Roeth (3) (167 ref.), Sandmann et al. (41) (161 ref.), Suett and Walker (5) (37 ref.), and Felsot (4) (117 ref.) attest to the growth in published information. In addition to published communication of research results, a number of conferences have facilitated information exchange. In addition to U.S. Department of Agriculture and industry sponsored meetings, the American Chemical Society (1984, 1989), the Weed Science Society of America (1986), and the Entomological Society of America (1984) have sponsored special symposia on enhanced degradation, and the Soil Science Society of America has been the forum for numerous papers on the topic during the 1980's.

What advances in our understanding of enhanced pesticide biodegradation have accompanied this contemporary flurry of research activity?

Advances in Research Understanding

Major advances in our scientific understanding of enhanced degradation may be related to the following definition offered for enhanced biodegradation of a pesticide (or other organic):

> "Enhanced biodegradation is the accelerated rate of pesticide degradation, by adapted soil microorganisms, observed in soil following previous application of it or a similar pesticide. Enhanced degradation may result in pest control failure and crop yield reduction."

Accelerated Pesticide Degradation Rate. The acceleration in degradation rate due to microbial adaptation has been well characterized. For pesticides that are somewhat environmentally unstable (e.g., EPTC) the increase in degradation rate associated with enhanced degradation reduces an inherently short soil half-life (13 days) to an extremely short one (4 days) (42). For other pesticides (e.g., isofenphos), enhanced degradation results in more spectacular decreases (12 weeks to <1 week) in observed soil half lives (32,43). A common characteristic of enhanced degradation that has been experimentally observed is the slight lag in degradation followed by a rapidly accelerating rate of degradation that declines when little pesticide remains (10).

There have also been major breakthroughs in the recognition of the differential susceptibility of pesticides to microbial adaptation based on chemical structure and availability of an accessible microbial catabolic pathway. Some classes of pesticides (e.g., carbamate insecticides, carbamothioate herbicides, phenoxyalkanoic herbicides) have proven especially susceptible to enhanced degradation (3,4,44). Apparently, the basic structure of these families of pesticides is conducive to microbial adaptation and catabolism. However, even within these groups there are compounds that are relatively resistant to enhanced degradation. For instance, cycloate and molinate are carbamothioate herbicides that are relatively resistant to the development of enhanced rates of degradation (3,16), as is the phenoxyalkanoic herbicide 2,4,5-T (3). But certainly all pesticides are not equally susceptible to enhanced degradation under normal agricultural use practices. Some of the physical and chemical features associated with resistance to the adaptation that leads to enhanced degradation in soil that have been recognized include lack of microbial availability due to high sorption (e.g., chlorpyrifos) (45), toxicity of intermediate pesticide metabolites (e.g., 2,4,5-T) (45), and exotic saturated ring structures (e.g., cycloate, molinate) (16).

Degradative Cross-Adaptations. Major advances have also been made regarding the specificity or lack of it associated with the enhanced degradation of specific pesticides and pesticide classes. In some cases a pesticide may be applied to soil for the first time and be rapidly degraded by populations of microorganisms that have adapted due to previous application of a similar pesticide. For instance, numerous interactions in soil for one pesticide that predisposes a soil to degrade another have been noted within the carbamate insecticide (33,46,47) and carbamothioate herbicide (42,48) groups. The basis for these observations has been confirmed by work with isolated pesticide-degrading microorganisms. Carbamate insecticide and carbamothioate herbicide-degrading microorganisms isolated from soil have proven capable of degrading a variety of similar pesticide structures within a given family (37,49). In contrast, the lack of degradative interactions observed in soil for organophosphorus insecticides has been attributed to the specificity shown by isolated pesticide-degrading microorganisms (50,51,52).

Some evidence indicates that these cross-adaptations may depend somewhat on structural similarity. For example, soils that rapidly degrade aryl carbamate insecticides (carbofuran, carbaryl) do not rapidly degrade oxime carbamates (aldicarb, oxamyl) (33,46,47). Work with carbofuran hydrolase, the enzyme used by carbofuran-degrading microorganisms to initiate carbofuran degradation, has indicated that although it had a rather high affinity for aryl carbamates, it had a much lower affinity for oxime carbamates such as aldicarb (53). Interactions in soil degradative behavior between herbicides, insecticides, and fungicides, and different classes within these groups have not been commonly observed (33,46,54). Apparently, the enzymes and metabolic pathways used by adapted soil microorganisms may be somewhat specific for the type of ester linkage that must be hydrolyzed for catabolism to be initiated or the specific secondary metabolic steps that must be pursued for extraction of energetic benefit (45).

Pesticide-Degrading Microbial Populations. Innovative microbiological investigations have paved the way for increased understanding of the mechanisms of enhanced degradation. One of the major advances has involved the recognition that rather than being caused by general increases in total soil microbial populations or 'biomass' (55), the enhanced degradation of pesticides in soil is mediated by a relatively small segment of the total soil microbial community. For example, the ratio of 2,4-D-degrading microorganisms to total microorganisms in soil has been found to be between 1:6.5 x 10^2 and 1:3 x 10^5 (56) These pesticide-degrading populations of microorganisms have often been detected and enumerated in soil by using statistically-based most-probable-number estimates. The media used for these assays is often a liquid one containing the pesticide as the sole source of carbon or some particular nutrient and inoculated with aliquots of soil. In many instances, greater numbers of these pesticide-degrading microorganisms can be detected in pesticide-treated soil in which enhanced degradation has occurred than in untreated soil (31,37,56,57). However, in some cases the actual numbers of microorganisms able to degrade a particular pesticide have been substantial both in treated and untreated soils, and the observed enhanced degradation has been attributed to differential rates of microbial metabolism by microbial populations in these soils (33,38)

Another major advance has been the isolation of soil microorganisms involved in enhanced pesticide degradation and elucidation of the catabolic nature of the adaptation involved. A partial listing of pesticide-degrading microorganisms that have been isolated from soil and the principal compound metabolized is shown below:

Bacterium globiforme	2,4-D	1950	(58)
Achromobacter	2,4-D	1957	(59)
Flavobacterium	2,4-D	1958	(60)
Arthrobacter	2,4-D	1967	(61)
Pseudomonas	2,4-D	1971	(62)
Alcaligenes	2,4-D	1977	(63)
Arthrobacter	EPTC	1987	(64)
Flavobacterium	EPTC	1989	(65)
Methylomonas	EPTC	1989	(65)
Rhodococcus	EPTC	1989	(66)
Pseudommonas	Carbofuran	1981	(25)
Achromobacter	Carbofuran	1986	(49)
Pseudomonas	Carbofuran	1986	(67)
Flavobacterium	Carbofuran	1986	(67)
Corynebacterium	Isofenphos	1985	(15)
Pseudomonas	Isofenphos	1987	(31)
Arthrobacter	Isofenphos	1988	(50)

It is obvious that there are a number of different soil microorganisms, primarily bacteria, that can adapt for efficient degradation of pesticides applied to soil. A key discovery in the characterization of these isolates has been the actual catabolic use of these pesticides by the degrading microorganisms. Although in the past undue emphasis has been placed on the phenomenon of cometabolism (68), in which no benefit is gained by the degrading microorganisms, the catabolic use of these pesticides by adapted populations of soil microorganisms has been clearly established. In most cases, adapted microbial metabolism of a pesticide involves an initial ester hydrolysis followed by complete mineralization/incorporation of one or more of the hydrolysis products and concommitant microbial extraction of energy (50,64,69). These organisms can literally grow on certain pesticides supplied as sole sources of carbon! The genetic basis for these degradative abilities has been reported to hinge on the possession of specific degradative plasmids (64,65,66).

Relationship to Pest Damage and Yield. From a practical standpoint, an extremely important realization has been that the enhanced biodegradation of a given pesticide may or may not affect pest damage and crop yield. This has been particularly important in relating laboratory results to field observations. Some pesticides, notably the phenoxyalkanoic herbicides, are foliar-acting and applied post-emergently to control weeds. Therefore, "their soil residual activity is of little practical significance for weed control" (44). This may be why early observations of enhanced 2,4-D degradation (1949+) were relegated to the status of academic curiosity and the full implications for soil-acting pesticides almost completely ignored until the present.

For pesticides that are soil-acting in nature, susceptibility to the development of enhanced rates of degradation does not always result in an increased rate of degradation in the field. This can be due to the environmental and management factors that affect the activity of soil microorganisms. Environmental factors that can significantly inhibit the development or expression of enhanced degradation include temperature, pH, and moisture (70,71,72). For example, soils with pH < 5.9 did not develop enhanced rates of carbofuran degradation upon repeated carbofuran treatment unless the pH was increased by liming (71). Enhanced carbofuran degradation was also shown to be inhibited in plots that were manured, due to the acidification of the soil by the practice (73)! Management practices such as the use of granular or slow-release formulations have also proven capable of modifying the expression of enhanced degradation in soil (70,74).

Another key advance in understanding the practical significance of enhanced degradation has been that although a pesticide may be rapidly degraded by adapted microbial populations in soil, this fact alone does not directly tranlate into economically significant yield reduction due to pest control failure. One reason for a possible lack of economic impact can be that the degree of persistence observed even under the limitations imposed by enhanced degradation may be sufficient for pest control. For instance, although isofenphos has been shown to undergo enhanced degradation in turf thatch, efficient control is experienced as long as insect grub pests are present and active at application (35). Another reason for lack of economic impact may be the oftentimes poor correlation observed between pest activity and crop yield. Soil-dwelling rootworm pests damage corn by pruning roots, and the degree of pruning observed has been used as an indication of insecticide efficacy. However, due to the interaction of several significant environmental factors with corn physiology, a rather poor correlation between root damage and crop yield is obtained ($R^2<0.1$) (75). Yield reduction resulting from enhanced degradation of the carbamothioate herbicides seems closely linked to the presence of several pernicious weed species that are non-uniformly distributed (9). If these weeds are not present, then the shorter residual activity caused by enhanced degradation may not affect control of more easily killed weeds. "Whether reduced persistence of a herbicide results in weed control failure depends on many factors including the types and numbers of weed seeds in the soil, the germination rate of the weeds compared to the herbicide's persistence or that of an active metabolite, and the coactivity of a companion herbicide applied simultaneously." (3) It should therefore never be assumed that the susceptibility of a pesticide to enhanced degradation automatically dooms it to miserable pest control failures and eventual withdrawal from the marketplace.

The Microbial Context of Enhanced Biodegradation

Contemporary researchers who are viewing enhanced degradation from agricultural perspectives too often tend to be unfamiliar with or lose sight of the significance of the phenomenon of microbial biodegradative adaptation in a microbial context. This is because, rather than feel that they are dealing with a fascinating microbial characteristic, the most prevalent emphasis is on enhanced degradation as an interfering factor having negative impact on pest control programs. It is important to recognize that the occurrence of enhanced degradation is not limited to the most typically conceived of agricultural systems or to agricultural systems at all. The premature failure of cattle dips due to microbial adaptation for degradation of the insecticide coumaphos should alert us to the extremely pervasive nature of the phenomenon (76). Microbial adaptation for pesticide degradation can happen in any environment in which biodegradable pesticides come in contact with populations of microorganisms. Microorganisms are also not limited to adaptation for pesticide degradation, and any organic chemical whether man-made or natural is fair game for catabolic attack by these ubiquitous degraders. There are numerous accounts of microbial adaptation for biodegradation of pesticides and other xenobiotics in such diverse ecosystems as groundwater (77), streams (78), wastewater treatment plants (79), oil slicks (80), and compost piles (81).

It is also important to recognize the microevolutionary nature of microbial adaptation for pesticide degradation. The soil environment itself has been viewed as a grand recycling depot, "continually receiving reduced compounds of carbon and continually oxidizing them to CO_2 and water" (82). Few energy-yielding reactions have not been exploited by microorganisms, and if energy can be extracted from the degradation of a given compound there is a good chance that microorganisms exist that can adapt to utilize it (82). We may be amazed that the addition of a pound or two of a pesticide over an acre can actually induce an adaption that leads to enhanced degradation. Yet we are guilty of viewing the situation from an anthropocentric perspective. In the microenvironment of the soil ped or sand granule the presence of a granule or droplet of pesticide may represent a tremendous selective advantage to the opportunistic microorganism that is primed to exploit it. Soil is a great repository of these types of microorganisms, and it should not surprise us when pesticides we apply are treated in a similar fashion to the incredible diversity of organic matter that is routinely added to soil and decomposed. For a more complete discussion of the micro-evolutionary aspect of microbial biodegradative adaptation and selection, see Clarke (83).

Research Challenges and Opportunities

No summary of research knowledge is ever complete without some enunciation of unanswered questions and unresolved controversies. Certainly a number of these exist in regard to the enhanced biodegradation of pesticides in soil, and some of the more significant ones will be briefly described.

Relationship of Persistence and Control. A clearer understanding of the relationship of pesticide persistence to pest control is urgently required. Tremendous knowledge of acute toxicity (bioassay) of pesticide products to pests is often generated, and considerable efforts are directed to field level testing of crop control responses to various pesticide application rates. However, an integrated understanding of the mysterious interface between pesticide application and resultant pest inhibition or mortality is lacking. The fact is that we know too little about the actual environmental concentrations of pesticides required for pest control with regard to temporal and spatial considerations. This understanding would greatly aid us in distinguishing between merely academic versus truly significant instances of enhanced degradation.

Predictive Tools. We have frequently responded to instances of enhanced degradation after products are marketed, or have recognized the susceptibility of new products in rather late stages of development. There is a great need to be able to predict microbial adaptation for degradation of candidate pesticides early in the developmental process. Current screening assays, which often rely upon years of repeated field treatments, have not reached the level of sophistication necessary for early detection diagnosis. Could some of the liquid assays developed by aquatic microbiologists for screening of biodegradability be employed (84,85), or do the known interactions of pestcides with soil surfaces render these types of assays suspect? Can the extreme of selective pressure represented by repeated pesticide treatment of laboratory soils be relied upon for relevant predictive insight? Surely the database that already exists regarding the differential susceptibility to microbial adaptation of currently used or discarded pesticides is sufficient to validate some actual laboratory assay or modelling approaches.

Pesticide Design. Rational design of pesticides that are susceptible to cometabolic or chemical degradation in soil but that do not induce microbial biodegradative adaptation needs to be put into practice. It often seems that the process of pesticide discovery is taking into account only the exploitation of novel means of bioactivity while ignoring environmental considerations until much later. At the molecular design level we need to begin to recognize structures that are associated with desired biodegradative properties. Considerable information has already been generated on structural features that are associated with biodegradability (86,87,88). Can we adapt some of these approaches for the design of soil pesticides? Although we will always have to deal with the incredible capacity of soil microorganisms to adapt, we should never again rely upon extremely recalcitrant, halogenated products.

Microbial Ecology of Degraders. It is in the realm of soil microbiology that some of the most perplexing mysteries of biodegradation remain unanswered. Although many strains of microorganisms displaying remarkable biodegradative potential in pure culture have been isolated from soil (and should be added to the American Type Culture Collection), how much do we really understand about the microbial ecology of pesticide-degrading microorganisms in soil? Are the microorganisms we have isolated and carefully characterized the major degraders active in soil? Are there populations of a number of strains of degraders active in each soil? Is there a set of common biodegradative plasmids that are promiscuously passed between microbial species and genera in the soil community? One of the greatest challenges will be moving from the realm of pure culture isolations in the laboratory to that of an understanding of population dynamics of pesticide-degrading microorganisms in the soil ecosystem. It is sad to realize that in spite of our tremendous knowledge of microorganisms of medical importance, we still know so little about the microorganisms that profusely populate the ground we tread.

Bioremediation. Exploitation of the adaptability of soil microorganisms for environmental decontamination and the destruction of hazardous wastes presents a major opportunity for positive application of the knowledge we have gained. Much work is already being done in this area (89). The pesticide-degrading microorganisms isolated, the techniques developed, and the insight gained through examination of enhanced biodegradation need to be applied to this critical area. We have unique access to a vast reservoir of microbial diversity, and as we have discovered that our carefully designed and marketed pesticides have been used as fodder for the soil ecosystem, we have hopefully begun to sense the potential that is offered.

Literature Cited

1. Fox, J. L. Science 1983, 221, 1029-1031.
2. Kaufman, D. D.; Katan, Y.; Edwards, D. F.; Jordan, E. G. In Agricultural Chemicals of the Future; Hilton, J. L., Ed.; Beltsville Symposia in Agricultural Research No. 8; U. S. Dept. of Agriculture: Totowa, MD, 1985; pp 437-451.
3. Roeth, F. W. Rev. Weed Sci. 1986, 2, 45-66.
4 Felsot, A. S. Ann. Rev. Entomol. 1989, 34, 453-476.
5. Suett, D. L.; Walker, A. Aspects Appl. Biol. 1988, 17, 213-222.
6. Walker, A.; Brown, P. A.; Entwistle, A. R. Pestic. Sci. 1986, 17, 183-193.
7. Brown, A. M.; Coffey, M. Phytopathol. 1985, 75, 135-137.
8. Coats, J. R.; Racke, K. D.; Michael, D. D. Enhanced Degradation of Organophosphorus Insecticides: Scope and Mechanisms North Central Regional Pesticide Impact Assessment Program Final Report No. 317: Wooster, OH, 1986; 41 pp.
9. Roeth, F. W.; Wilson, R. G.; Martin, A. R.; Shea, P. J. Weed Technol. 1989, 3, 24-29.
10. Racke, K. D.; Coats, J. R. This volume.

11. Jennings, V.; Stockdale, H. Herbicides and Insecticides Used in Iowa Corn and Soybean Production, 1977; ISU Bulletin Pm-845; Iowa State Cooperative Extension Service: Ames, IA, 1978; 6 pp.
12. Wintersteen, W.; Hartzler, R. Pesticides Used in Iowa Crop Production in 1985; ISU Bulletin Pm-1288; Iowa State Cooperative Extension Service: Ames, IA, 1987.
13. McCullough, J. M.; Belcher, D. W.; Little, R. J.; Devine, J. M.; Vassalotti, P. M.; Stryker, P. G.; Raemisch, D. R.; Plunge, A. G.; Kappel, C. L. Abstracts of Papers, 42nd Meeting of the North Central Branch of the ESA, Des Moines, IA; Entomological Society of America: College Park, MD, 1987: paper 7.
14. Staetz, C. A. Abstracts of Papers, 42nd Meeting of the North Central Branch of the ESA, Des Moines, IA; Entomological Society of America: College Park, MD, 1987: paper 81.
15. Murphy, J. J.; Cohick, A. D. Abstracts of Papers, 40th Meeting of the North Central Branch of the ESA, Lexington, KY; Entomological Society of America: College Park, MD, 1987: paper 23.
16. Gray, R. A.; Joo, G. K. Weed Sci. 1985, 33, 698-702.
17. Drost, D. C.; Rodebush, J. E.; Hsu, J. H. This volume.
18. Prochnow, C. L. Proc. West. Weed Sci. Soc., 1981, pp 55-56.
19. Capper, B. E. Proc. New Zealand Weed Pest Conf., 1982, pp 222-225.
20. Wilson, R. G.; Rodebush, J. E. Weed Sci. 1987, 35, 289-294.
21. Corn Rootworm Management: Insights from the Experts, Union Carbide Agricultural Products Company, 1985, 18 pp.
22. Ahmad, N.; Walgenbach, D. D.; Sutter, G. R. Bull. Environ. Contam. Toxicol. 1979, 23, 572-574.
23. Gorder, G. W.; Tollefson, J. J.; Dahm, P. A. Iowa State J. Res. 1980, 55, 25-33.
24. Greenhalgh, R.; Belanger, A. J. Agric. Food Chem. 1981, 29, 231-235.
25. Felsot, A.; Maddox, J. V.; Bruce, W. Bull. Environ. Contam. Toxicol. 1981, 26, 781-789.
26. Corn Rootworm Management; ISU Bulletin Pm-670; Iowa State Cooperative Extension Service: Ames, IA, 1982.
27. Doersch, R. E.; Doll, J. D.; Wedberg, J. L.; Grau, C. R.; Harvey, R. G.; Kenney, J. E. Pest Control in Corn 1983; University of Wisconsin Extension Service: Madison, WI: p. 36.
28. Foster, D.; Wintersteen, W. K. Insecticide Recommendations for 1986 Corn Production; ISU Bulletin IC-404; Iowa State Cooperative Extension Service: Ames, IA, 1985.
29. 1986 Herbicide Manual for Ag Chemical Dealers; ISU Bulletin WC-92; Iowa State Cooperative Extension Service: Ames, IA, 1985.
30. Obrigawitch, T.; Roeth, F. W.; Martin, A. R.; Wilson, R. G. Weed Sci. 1982, 30, 417-422.
31. Racke, K. D.; Coats, J. R. J. Agric. Food Chem. 1987, 35, 94-99.
32. Chapman, R. A.; Harris, C. R.; Moy, P.; Henning, K. J. Environ. Sci. Health 1986, B21, 269-276.
33. Racke, K. D.; Coats, J. R. J. Agric. Food Chem. 1988, 36, 1067-1072.
34. Racke, K. D.; Coats, J. R.; Titus, K. R. J. Environ. Sci. Health 1988, B23, 527-539.
35. Niemczyk, H. D.; Chapman, R. A. J. Econ. Entomol. 1987, 80, 880-882.

36. Harvey, R. G.; McNevin, G. R.; Albright, J. W.; Kozak, M. E. Weed Sci. 1986, 34, 773-780.
37. Mueller, J. G.; Skipper, H. D.; Lawrence, E. G.; Kline, E. L. Weed Sci. 1989, 37, 424-427.
38. Moorman, T. B. Weed Sci. 1988, 36, 96-101.
39. Hendry, K. M.; Richardson, C. J. Environ. Toxicol. Chem. 1988, 7, 763-774.
40. Kearney, P. C.; Kellogg, S. T. Pure Appl. Chem. 1985, 57, 389-403
41. Sandmann, E. R. I. C.; Loos, M. A.; Van Dyk, L. P. Rev. Environ. Contam. Toxicol. 1988, 101, 1-53.
42. Obrigawitch, T.; Martin, A. R.; Roeth, F. W. Weed Sci. 1983, 31, 187-192.
43. Chapman, R. A.; Harris, C. R. J. Environ. Sci. Health 1982, B17, 355-361.
44. Smith, A. E.; Lafond, G. P. This volume.
45. Somasundaram, L.; Coats, J. R. This volume.
46. Harris, C. R.; Chapman, R. A.; Harris, C.; Tu, C. M. J. Environ. Sci. Health 1984, B19, 1-11.
47. Read, D. C. J. Econ. Entomol. 1987, 80, 156-163.
48. Harvey, R. G. Weed Sci. 1987, 35, 583-589.
49. Karns, J. S.; Mulbry, W. W.; Nelson, J. O.; Kearney, P. C. Pestic. Biochem. Physiol. 1986, 25, 211-217.
50. Racke, K. D.; Coats, J. R. J. Agric. Food Chem. 1988, 36, 193-199
51. Forrest, M.; Lord, K. A.; Walker, N.; Woodville, H. C. Environ. Pollut. 1981, A24, 93-104.
52. Sethunathan, N.; Pathak, M. D. J. Agric. Food Chem. 1972, 20, 586-589.
53. Derbyshire, M. K.; Karns, J. S.; Kearnety, P. C.; Nelson, J. O. J. Agric. Food Chem. 1987, 35, 871-877.
54. Rudyanski, W. J.; Fawcett, R. S.; McAllister, R. S. Weed Sci. 1987, 35, 68-74.
55. Frehse, H.; Anderson, J. P. E. In Pesticide Chemistry: Human Welfare and the Environment; Miyamoto, J.; Kearney, P. C., Eds.; Pergamon Press: Oxford, England, 1983, 23-32.
56. Cullimore, D. R. Weed Sci. 1981, 29, 440-443.
57. Fournier, J. C.; Codaccioni, P.; Soulas, G. Chemosphere, 1981, 10, 977-984.
58. Audus, L. J. Nature 1950, 166, 356.
59. Bell, G. R. Can. J. Microbiol. 1957, 3, 821-840.
60. Steenson, T. I.; Walker, N. J. Gen. Microbiol. 1958, 18, 692-697.
61. Loos, M. A.; Roberts, R. N.; Alexander, M. Can. J. Microbiol. 1967, 13, 679-690.
62. Gaunt, J. K.; Evans, W. C. Biochem. J. 1971, 122, 529-532.
63. Fisher, P. R.; Appleton, J.; Pemberton, J. M. J. Bacteriol. 1978, 135, 798-804.
64. Tam, A. C.; Behki, R. M.; Khan, S. U. Appl. Environ. Microbiol. 1987, 53, 1088-1093.
65. Skipper, H. D. This volume.
66. Dick, W. A.; Ankumah, R. O.; McClung, G.; Abou-Assaf, N. This volume.
67. Chauhdry, G. R.; Ou, L. T.; Wheeler, W. B. Abstracts of Papers, Sixth International Congress of Pesticide Chemistry, Ottawa, Canada; IUPAC: Oxford, England, 1986; paper 6C-12.
68. Alexander, M. Science 1981, 211, 132-138.

69. Don, R. H.; Pemberton, J. M. J. Bacteriol. 1981, 145, 681-686.
70. Chapman, R. A.; Harris, C. R.; Harris, C. J. Environ. Sci. Health 1986, B21, 125-141.
71. Read, D. C. Agric. Ecosyst. Environ. 1986, 15, 51-61.
72. Obrigawitch, T.; Wilson, R. G.; Martin, A. R.; Roeth, R. W. Weed Sci. 1982, 30, 175-181.
73. Somasundaram, L.; Racke, K. D.; Coats, J. R. Bull. Environ. Contam. Toxicol. 1987, 39, 579-586.
74. Felsot, A. S.; Tollefson, J. J. This volume.
75. Foster, R. E.; Tollefson, J. J.; Nyrop, J. P.; Hein, G. L. J. Econ. Entomol. 1986, 79, 303-310.
76. Shelton, D. R.; Karns, J. S. J. Agric. Food Chem. 1988, 36, 831-834.
77. Wilson, J. T.; McNabb, J. F.; Cochran, J. W.; Wang, T. H.; Tomson, M. B.; Bedient, P. B. Environ. Toxicol. Chem. 1985, 4, 721-726.
78. Pignatello, J. J.; Martinson, M. M.; Steiert, J. G.; Carlson, R. E.; Crawford, R. L. Appl. Environ. Microbiol. 1983, 46, 1024-1031
79. Haller, H. D. J. Water Pollut. Contr. Fed. 1978, 50, 2771-2777.
80. Atlas, R. M. Microbiol. Rev. 1981, 45, 180-209.
81. Hogan, J. A.; Toffoli, G. R.; Miller, F. C.; Hunter, J. V.; Finstein, M. S. Proc. Int. Conf. Physiochem. Biol. Detoxific. of Haz. Wastes, 1988.
82. Delwiche, C. C. In Soil Biochemistry; McLaren, A. D.; Peterson, G. H., Eds.; Marcel Dekker: New York, NY, 1967; pp 173-193.
83. Clarke, P. H. In Evolution in the Microbial World; Cambridge University Press: London, England, 1974; pp 183-217.
84. Pitter, P. Water Res. 1976, 10, 231-235.
85. Boatman, R. J.; Cunningham, S. L.; Ziegler, D. A. Environ. Toxicol. Chem. 1986, 5, 233-243.
86. Niemi, G. J.; Veith, G. D.; Regal, R. R.; Vaishnav, D. D. Environ. Toxicol. Chem. 1987, 6, 515-527.
87. Boethling, R. S. Environ. Toxicol. Chem. 1986, 5, 797-806.
88. Howard, P. H.; Hueber, A. E.; Boethling, R. S. Environ. Toxicol. Chem. 1987, 6, 1-10.
89. Felsot, A. S.; Dzantor, E. K. This volume.

RECEIVED January 22, 1990

Appendix

Key Chemical Structures

This volume contains references to members of several classes of herbicides, insecticides, fungicides, and microbial inhibitors. The following appendix displays common names, chemical names, and chemical structures for most of the chemicals discussed in the preceding chapters.

For general information on the properties, uses, environmental fate, and microbial degradation of pesticides, the following references would be useful:

Pesticides

Beste, C. E., Ed. *Herbicide Handbook*; Weed Science Society of America: Champaign, IL, 5th edition, 1983.

Hartley, D.; Kidd, H., Eds. *The Agrochemical Handbook;* Royal Society of Chemistry: Nottingham, England, 1987.

Hassell, K. A. *The Chemistry of Pesticides: Their Metabolism, Mode of Action and Uses in Crop Protection;* Verlag–Chemie: Weinheim, Federal Republic of Germany, 1982.

Matsumura, F. *Toxicology of Insecticides;* Plenum Press: New York, 2nd edition, 1985.

Poplyk, J., Ed. *Farm Chemicals Handbook '89;* Meister Publishing Company: Willoughby, OH, 1989.

Pesticide Microbiology

Hill, I. R.; Wright, S. J. L., Eds. *Pesticide Microbiology;* Academic Press: London, 1978.

Lal, R., Ed. *Insecticide Microbiology*; Springer–Verlag: Berlin, 1984.

Microbial Ecology

Alexander, M. *Introduction to Soil Microbiology;* Wiley: New York, 1977.

Campbell, R. *Microbial Ecology;* Blackwell Scientific: Oxford, England, 1983.

Ellwood, D. C., Ed. *Contemporary Microbial Ecology;* Academic Press: London, 1980.

Environmental Chemistry

Haque, R.; Freed. V. H., Eds. *Environmental Dynamics of Pesticides;* Plenum Press: New York, 1975.

Sawhney, B. L.; Brown, K., Eds. *Reactions and Movement of Organic Chemicals in Soil;* SSSA Special Publication No. 22; Soil Science Society of America: Madison, WI, 1989.

Tinsley, I. J. *Chemical Concepts in Pollutant Behavior;* Wiley: New York, 1979.

PHENOXYALKANOIC HERBICIDES

2,4-D	2,4-dichlorophenoxyacetic acid 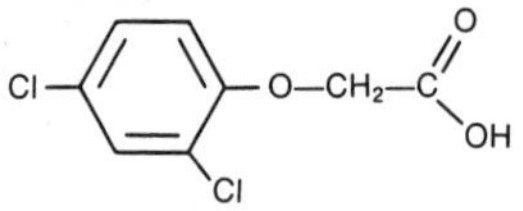
2,4,5-T	2,4,5-trichlorophenoxyacetic acid 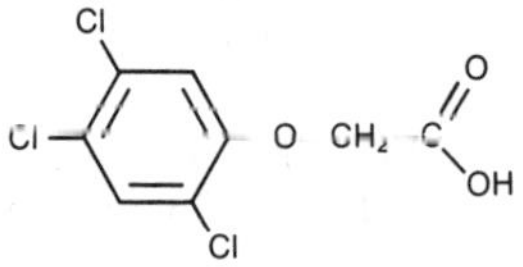
MCPA	4-chloro-2-methylphenoxyacetic acid 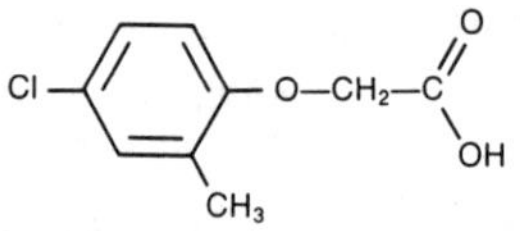
MCPB	4-(4-chloro-2-methylphenoxy) butanoic acid 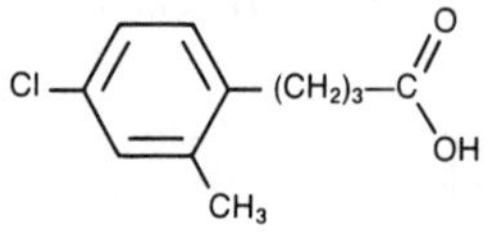

CARBAMOTHIOATE HERBICIDES

EPTC	*S*-ethyl dipropylcarbamothioate 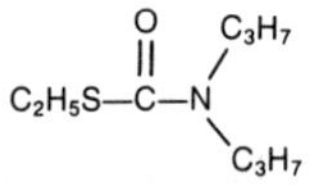
Butylate	*S*-ethyl bis(2-methylpropyl) carbamothioate 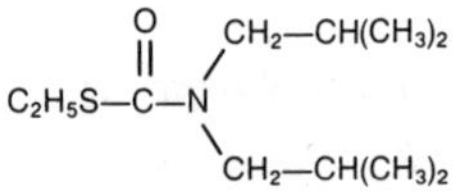
Vernolate	*S*-propyl dipropylcarbamothiate 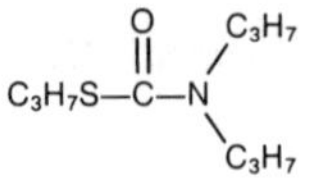

CARBAMOTHIOATE HERBICIDES (CONT.)

Pebulate	*S*-propyl butylethylcarbamothioate
Cycloate	*S*-ethyl *N*-ethylthiocyclohexanecarbamate
Molinate	*S*-ethyl hexahydro-1*H*-azepine-1-carbothioate

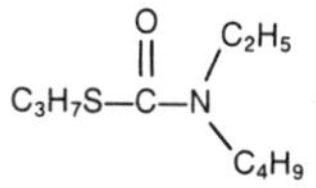

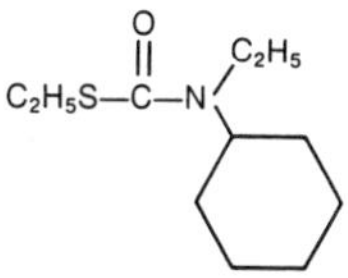

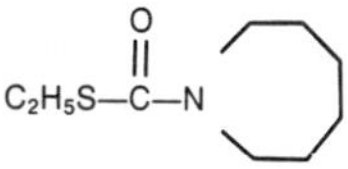

TRIAZINE HERBICIDES

Atrazine	2-chloro-4-(ethylamino)-6-(isopropylamino)-*s*-triazine
Cyanazine	2-[[4-chloro-6-(ethylamino)-*s*-triazin-2-yl]amino]-2-methylpropionitrile
Metribuzin	4-amino-6-*tert*-butyl-3-(methylthio)-*as*-triazin-5(4*H*)-one

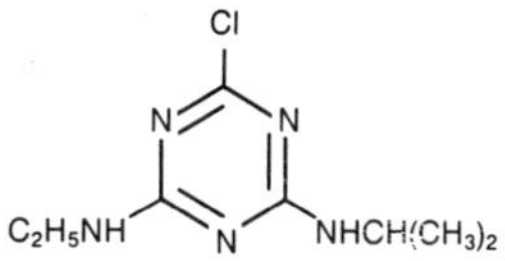

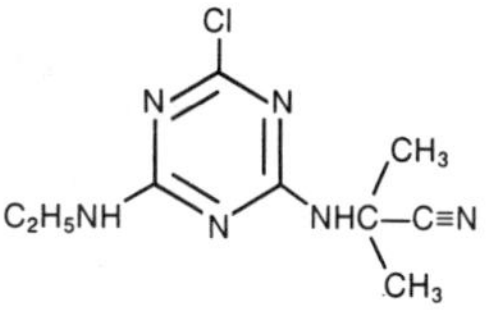

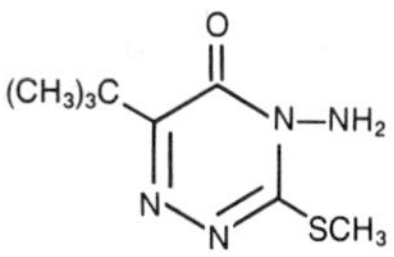

MISCELLANEOUS HERBICIDES

Chlorpropham	1-methylethyl(3-chlorophenyl) carbamate	$NH-C(=O)-OCH(CH_3)_2$; Cl
Diphenamid	*N,N*-dimethyl-2,2-diphenylacetamide	$CH-C(=O)-N(CH_3)_2$
Trifluralin	2,6-dinitro-*N,N*-dipropyl-4-(trifluoromethyl) benzamine	C_3H_7, N, C_3H_7; NO_2, NO_2; CF_3
Metolachlor	2-chloro-*N*-(2-ethyl-6-methylphenyl)-*N*-(2-methoxy-1-methylethyl)acetamide	CH_3; CH_3; $CH-CH_2-OCH_3$; N; $C(=O)-CH_2Cl$; C_2H_5
Alachlor	2-chloro-*N*-(2,6-diethylphenyl)-*N*-(methoxymethyl)acetamide	C_2H_5; CH_2OCH_3; N; $C(=O)-CH_2Cl$; C_2H_5

CARBAMATE INSECTICIDES

Carbaryl	1-naphthalenyl methylcarbamate	$O-C(=O)-NHCH_3$

CARBAMATE INSECTICIDES (CONT.)

Carbofuran	2,3-dihydro-2,2-dimethyl-7-benzofuranyl methylcarbamate	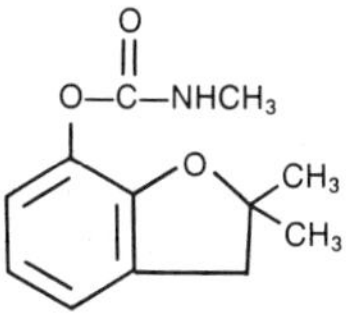
Bendiocarb	2,2-dimethyl-1,3-benzodioxol-4-yl methylcarbamate	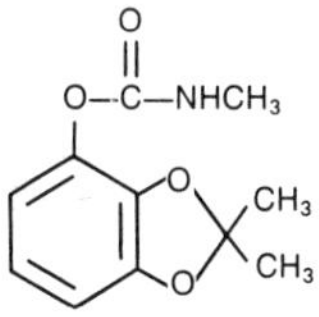
Trimethacarb	3,4,5-trimethylphenyl methylcarbamate & 2,3,5-trimethylphenyl methylcarbamate	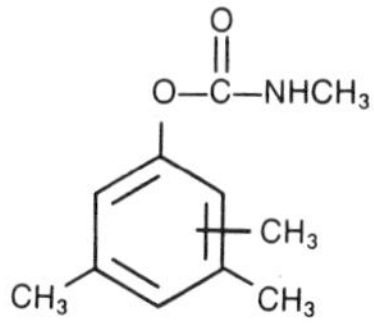
Cloethocarb	2-(2-chloro-1-methoxyethoxy) phenyl methylcarbamate	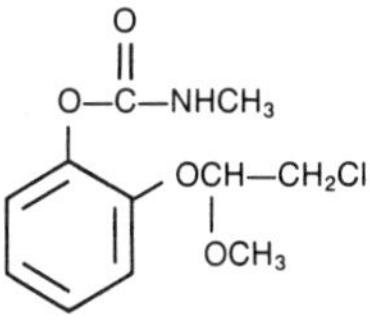
Aldicarb	2-methyl-2-(methylthio) propionaldehyde *O*-[(methylamino)carbonyl]oxime	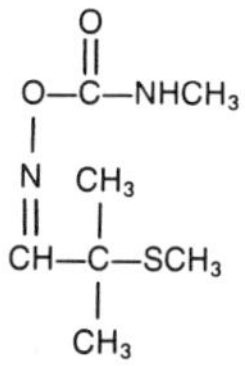

ORGANOPHOSPHORUS INSECTICIDES

Parathion	*O,O*-diethyl *O*-(4-nitrophenyl) phosphorothioate	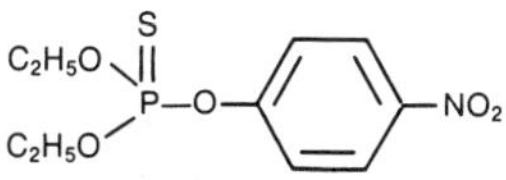

ORGANOPHOSPHORUS INSECTICIDES (CONT.)

Isofenphos	1-methylethyl 2-[[ethoxy[(1-methylethyl)amino] phosphinothioyl] oxy]benzoate 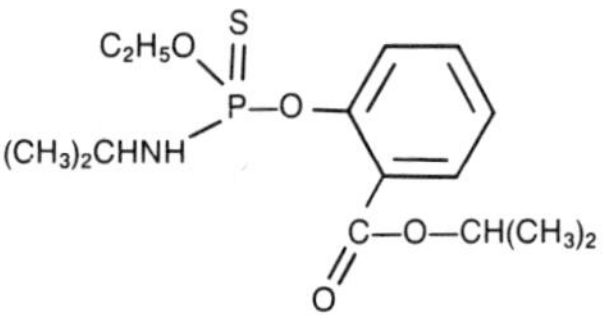
Fensulfothion	*O,O*-diethyl *O*-[4-(methylsulfinyl) phenyl]phosphorothioate 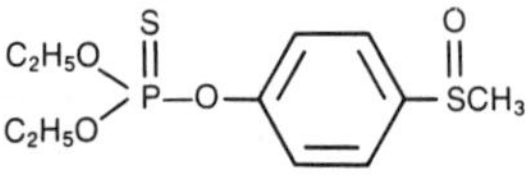
Chlorpyrifos	*O,O*-diethyl *O*-(3,5,6-trichloro-2-pyridyl)phosphorothioate 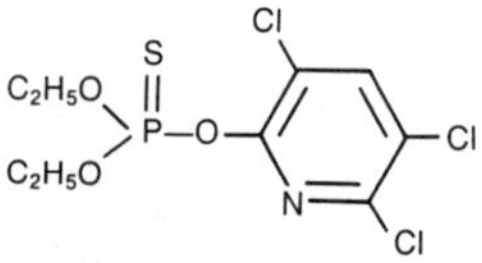
Diazinon	*O,O*-diethyl *O*-(6-methyl-2-(1-methylethyl) 4-pyrimidinyl phosphorothioate 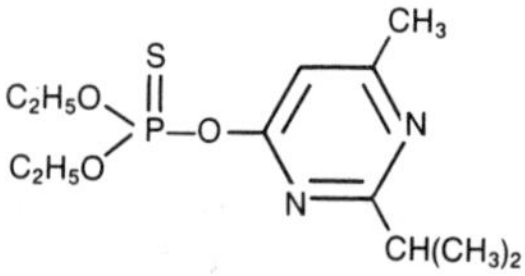
Dowco 429 (Butathiofos)	*O,O*-diethyl *O*-[2-(1,1-dimethylethyl)- 5-pyrimidinyl] phosphorothioate 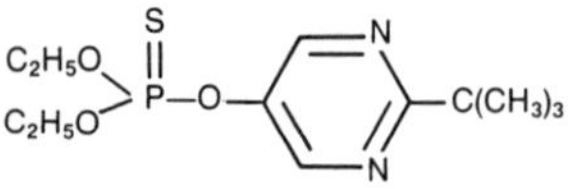
Fonofos	*O*-ethyl *S*-phenyl ethylphosphonodithioate 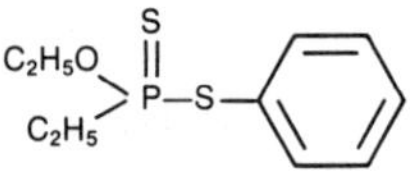
Phorate	*O,O*-diethyl *S*-[(ethylthio)methyl] phosphorodithioate 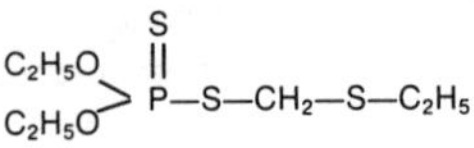

ORGANOPHOSPHORUS INSECTICIDES (CONT.)

Disulfoton — *O,O*-diethyl *S*-[2-(ethylthio)ethyl] phosphorodithioate

$(C_2H_5O)_2P(=S)-S-CH_2-CH_2-S-C_2H_5$

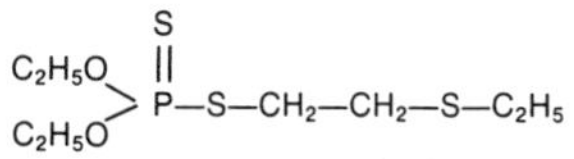

Terbufos — *S*-[[(1,1-dimethylethyl)thio]methyl] *O,O*-diethyl phosphorodithioate

$(C_2H_5O)_2P(=S)-S-CH_2-S-C(CH_3)_3$

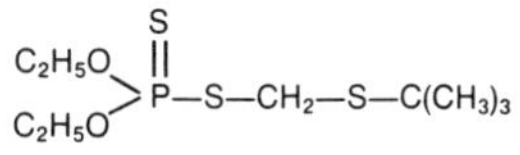

Ethoprop(hos) — *O*-ethyl *S,S*-dipropyl phosphorodithioate

$C_2H_5O-P(=O)(SC_3H_7)_2$

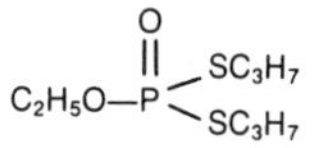

PYRETHROID INSECTICIDES

Tefluthrin — 2,3,5,6-tetrafluoro-4-methylphenyl *cis*-3-(z-2-chloro-3,3,3-trifluoro-prop-1-enyl)-2,2-dimethylcyclo-propane-carboxylate

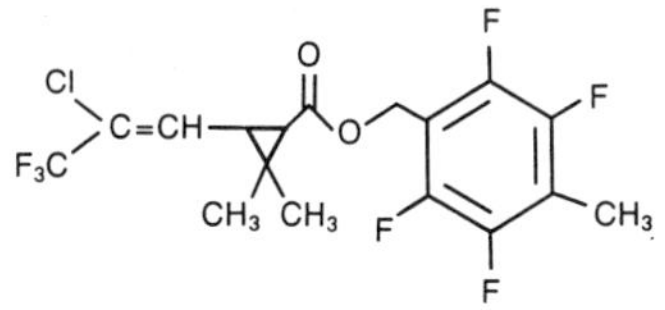

FUNGICIDES

Iprodione — 3-(3,5-dichlorophenyl)-*N*-(1-methylethyl)- 2,4-dioxo-1-imidazolidinecarboxamide

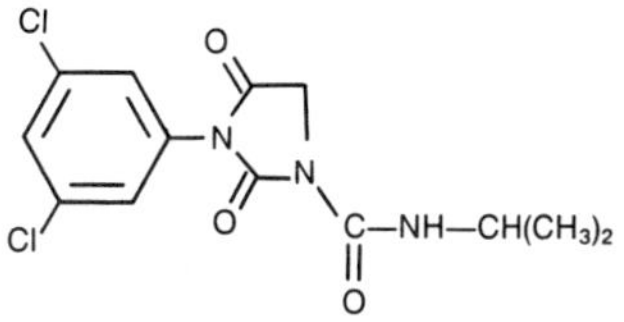

Vinclozolin — 3-(3,5-dichlorophenyl)-5-ethenyl-5-methyl-2,4-oxazolidinedione

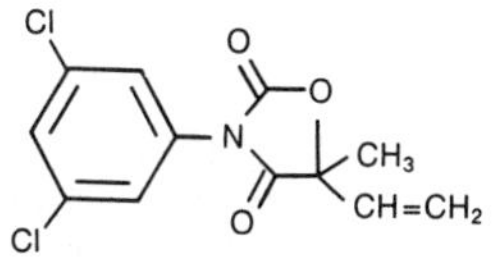

Myclozolin — 3-(3,5-dichlorophenyl)-5-methoxy-methyl-5-methyl-2,4-oxazolidine-dione

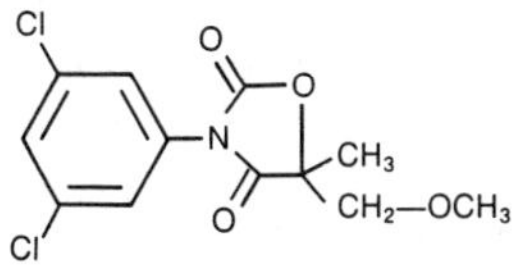

FUNGICIDES (CONT.)

Procymidone	3-(3,5-dichlorophenyl)-1,5-dimethyl-3-azabicyclo [3.1.0]hexane-2,4-dione
Carbendazim (MBC)	2-(methoxy carbamoylamino)-benzimidazole

MICROBIAL INHIBITORS

Dietholate (R33865)	*O,O*-diethyl *O*-phenyl phosphorothioate
R251005 (SC-0058)	*S*-ethyl-*N,N*-bis-(3-chloroallyl) carbamothioate
Chloramphenicol	[*R*-(*R**,*R**)]-2,2-dichloro-*N*-[2-hydroxy-1-(hydroxymethyl)-2-(4-nitrophenyl)ethyl] acetamide
Cycloheximide	[1*S*-[1α(*S**),3α,5β]]-4-[2-(3,5-dimethyl-2-oxocyclohexyl)-2-hydroxyethyl]-2,6-piperidinedione

Note: The microbial inhibitors listed above have been used in the field or laboratory to inhibit the enhanced biodegradation of certain pesticides and extend pesticide persistence. No attempt has been made to separate them based on their mode of action.

RECEIVED March 9, 1990

Author Index

Abou-Assaf, N., 98
Aharonson, N., 113
Ankumah, R. O., 98
Avidov, E., 113
Chapman, R. A., 82
Coats, Joel R., 68,128
Dick, W. A., 98
Drost, Dirk C., 222
Dzantor, E. K., 249
Felsot, A. S., 192,249
Harris, C. R., 82
Harvey, Robert G., 214
Hsu, Joanna K., 222
Karns, Jeffrey S., 141
Katan, J., 113
Keaster, Armon J., 240
Konopka, A. E., 153
Kremer, Robert J., 240
Lafond, Guy P., 14
Martin, Alex R., 23
McClung, G., 98
Moorman, Thomas B., 167
Racke, Kenneth D., 1,68,269
Reed, Joel P., 240
Rodebush, James E., 222
Roeth, Fred W., 23
Shea, Patrick J., 23
Skipper, Horace D., 37
Smith, Allan E., 14
Somasundaram, L., 128
Spain, J. C., 14
Tollefson, J. J., 192
Turco, R. F., 153
Walker, Allan, 53
Welch, Sarah J., 53
Wilson, Robert G., 23
Yarden, O., 113

Affiliation Index

Agriculture Canada, 14
Clemson University, 37
DowElanco, 1,68,269
ICI Americas, Inc., 222
Illinois Natural History Survey, 192,249
Institute of Horticultural Research, 53
Iowa State University, 68,128,192
Ohio State University, 98
Purdue University, 153
The Hebrew University of Jerusalem, 113
The Volcani Center, 113
U. S. Air Force, 181
U.S. Department of Agriculture, 98,141,167,240
University of Missouri, 240
University of Nebraska, 23
University of Wisconsin, 214

Subject Index

A

Abiotic mechanisms for degradation of herbicides in soil, discussion, 15
Abiotic transformation processes of pesticide dissipation in soil
  hydrolytic reactions, 5
  role in pesticide degradation, 4–5
Accelerated pesticide degradation rate, characterization, 273–274
Acridine orange direct counting, 156
Adaptation
  characterization, 181
  definition, 168
  mechanisms, 168–174
  microbial, *See* Microbial adaptation
  occurrence, 168

Adaptation in subsurface environments
effect of degrader growth, 173,175
effect of inherent biodegradability, 175
effect of nutrient limitations, 173,175
effect of soil sample depth on pesticide degradation, 173,174*f*
factors influencing degradation, 173,175
Adaptive enzyme theory, explanation of microbial involvement in enhanced pesticide degradation, 99
Advances in understanding of enhanced pesticide degradation, 273–277
Agrochemical retail facilities, accumulation of hazardous chemicals, 250
Alachlor
enhancing biodegradation for detoxification of waste in soil, 252–265
extraction and analysis, 257
structure and systematic name, 286
Alachlor-degrading microorganisms
bacteria, 260,261*t*
enrichment and isolation, 254,255*f*,256
fungi, 260–261
screening procedures, 256-257
Aldicarb, 287
4-Amino-6-*tert*-butyl-3-(methylthio)-*as*-triazin-5(4*H*)-one, *See* Metribuzin, 285
Antimicrobial agents, effect on enhanced pesticide degradation, 118*f*,119
Aquatic systems
chemicals that cause adaptation, 182*t*
microbial adaptation, 181–188
Aryl acylamidase, induction by pesticides, 168,170
Atrazine
effect on enhanced degradation, 33,34*t*
structure and systematic name, 285

B

Bacterial degraders, assessment of degradative capabilities, 120,121*f*,122*t*
Bacterial genetics
generalized structure of hypothetical catabolic transposon, 144,145*f*
mechanisms of gene exchange in bacteria, 144,145*f*,146
types of DNA found in bacteria cells, 143–144,145*f*
Bacterial growth, phases, 6–7
Bacteriophages, description, 144
Bendiocarb, 287
Bioaugmentation, cleanup of pesticide wastes, 251
Biodegradation
accelerated rate, 273–274
Biodegradation—*Continued*
cross-adaptation, 274
prediction of rate increases, 181–182
relation to pest damage and yield, 276–277
requirements, 168
Biomass, microbial, *See* Microbial biomass
Bioremediation
importance, 279
in situ remediation method, 176,177*f*
Biostimulation, cleanup of pesticide wastes, 251
Butathiofos, 288
Butylate
cross-adaptation, 44,45*f*,46*t*
degradation in soil, 222
effect of multiple applications, 41–42,43*f*
effect of repeated applications on weed control, 224
fluorescein diacetate assay, 244–245
performance in butylate-history soils, 39*t*,40
performance in vernolate-history soil, 40*t*
structure and systematic name, 284

C

Carbamate pesticides
control of rootworms, 69
cross enhancement factors, 93*t*,94
names, 286–287
structures, 286–287
Carbamothioate herbicides
advantages, 38
degradation
in soil, 222
mechanism in mammals, 99
mechanism is plants, 99
effect of cultural practices and chemicals on persistence in soil, 222–237
enhanced biodegradation in South Carolina soils, 37–50
enhanced biodegradation with repeated use, 225,226*t*
names and abbreviations, 284–285
persistence in soil, 222–223,226*t*
structures, 284–285
systems allowing continued use despite enhanced biodegradation, 214–221
Carbaryl, 286
Carbendazim
behavior in problem soils, 114–115
effect of fungicides and antimicrobial agents, 118*f*,119
effect of soil disinfestation on enhanced degradation, 116,117–118*f*,119
enhanced degradation in history and nonhistory soils, 114,115*f*
enhanced degradation rate, 123,124*t*

Carbendazim—*Continued*
pathways of degradation and enzymatic reactions associated with enhanced degradation, 123,125
structure and systematic name, 290
Carbofuran
cross enhancement factors, 93,94*t*
cumulative plots of mineralization in history and nonhistory soils, 72,73*f*
degradation
in companion soils, 72,74,75*t*,76
effect of hydrolysis product, 130,134*t*,137
effect of previous applications on disappearance rate, 82
enhanced degradation
by microorganisms, 82
cross-adaptations, 78
microbiology, 76–78
relation to pest control failure, 69–70
enhanced microbial activity to degrade carbofuran in soil
effect of moisture, 87,88*t*
effect of temperature, 87,88*t*
effect of treatment intensity, 87,88*t*
enhancement factors
effect of concentration, 87*t*
effect of soil moisture, 87*t*
effect of temperature, 86*t*
metabolite assay, 245*t*
microbial toxicity, 136*t*
mobility in soils, 137,138*t*
observed loss of efficacy, 9
proposed conversion pathway, 153,154*f*
response of microbial populations in soils enhanced for its degradation, 153–164
structure and systematic name, 287
typical carbofuran conversion, 153,154*f*
Carbofuran phenol, 130,137–138
Catabolism, description, 5–6
Chance mutation theory, explanation of microbial involvement in enhanced pesticide degradation, 99
Chloramphenicol, structure and systematic name, 290
Chlorinated hydrocarbon insecticides
control of corn pests, 68–69
evidence of rootworm control failures, 69
See also specific insecticides
2-Chloro-*N*-(2,6-diethylphenyl)-*N*-(methoxymethyl)acetamide, *See* Alachlor
2-Chloro-4-(ethylamino)-6-(isopropylamino)-*s*-triazine, *See* Atrazine
2-[[4-Chloro-6-(ethylamino)-*s*-triazin-2-yl]amino]-2-methylpropionitrile, *See* Cyanazine, 285
2-Chloro-*N*-(2-ethyl-6-methylphenyl)-*N*-(2-methoxy-1-methylethyl)acetamide, *See* Metolachlor
2-(2-Chloro-1-methoxyethoxy)phenyl methylcarbamate, *See* Cloethocarb
(4-Chloro-2-methylphenoxy)acetic acid, *See* MCPA
4-(4-Chloro-2-methylphenoxy)butanoic acid, *See* MCPB
4-(4-Chloro-2-methylphenoxy)butyric acid, *See* MCPB
2-(4-Chloro-2-methylphenoxy)propionic acid, *See* Mecoprop, 14–15
Chlorpropham, 286
Chlorpyrifos
degradation in companion soils, 72,74,75*t*,76
enhanced degradation
cross-adaptation, 78–79
effect of hydrolysis product on degradation, 134*t*
relation to mobility, 137–138
microbial toxicity, 136*t*
mobility in soils, 137,138*t*
structure and systematic name, 288
Cleanup technologies for pesticide wastes, 250–251
Cloethocarb
degradation in companion soils, 72,74,75*t*,76
structure and systematic name, 287
Colorimetric test, identification of rapid-degrading soils, 58,59*t*
Cometabolism
definition, 172
description, 6
occurrence, 172–173
Commercial perspective on managing enhanced biodegradation
impact of chemical extenders, 233,234*t*,235,236–237*t*
impact of climatic and edaphic conditions, 231,233
Conjugation, definition, 146
Corn pest management, 68
Corn rootworm, control, 3
Corn rootworm insecticides, pressure to use same product, 193
Cross-adaptation, definition, 44
Cross enhancement
among dicarboximide fungicides, 58–59, 60*f*,61*t*
definition, 93
discovery, 93
effect of treatment intensity, 94,95*t*
factors for carbamate pesticides, 93*t*,94
factors for organophosphorus insecticides and carbofuran, 93,94*t*
Cultural practices and chemicals affecting carbamothioate herbicide persistence in soil, 222–237

Cyanazine, structure and systematic name, 285
Cycloate, 285
Cycloheximide, 290

D

Dalapon, mechanism of bacterial degradation, 147
Death phase, description, 6
Degradation
  description, 4–5
  microbial adaptation, 6–8
Degradative cross-adaptations, influencing factors, 274
Diazinon
  enhanced degradation
    effect of hydrolysis product, 130,133*t*
    effect of mobility, 137–138
  microbial toxicity, 136*t*
  mobility in soils, 137,138*t*
  structure and systematic name, 288
Dicarboximide fungicides
  changes in effectiveness, 54
  failure to control white rot disease, 53–54
[*R*-(*R**,*R**)]-2,2-Dichloro-*N*-[2-hydroxy-1-(hydroxymethyl)-2-(4-nitrophenyl)-ethyl]acetamide, *See* Chloramphenicol
(2,4-Dichlorophenoxy)acetic acid (2,4-D)
  bacterial degradation
    mechanism, 147
    role of soil microorganisms, 15–17
  development, 14–15
  enhanced degradation
    effect of hydrolysis product, 130,131*f*,132*t*
    effect of mobility, 137–138
    effect of repeat treatments, 17–18,19*t*,20
  microbial toxicity, 136*t*
  mobility in soils, 137,138*t*
  structure, 284
4-(2,4-Dichlorophenoxy)butyric acid (2,4-DB), development, 14–15
2-(2,4-Dichlorophenoxy)propionic acid, *See* Dichlorprop, 14–15
3-(3,5-Dichlorophenyl)-1,5-dimethyl-3-aza-bicyclo[3.1.0]hexane-2,4-dione, *See* Procymidone
3-(3,5-Dichlorophenyl)-5-ethenyl-5-methyl-2,4-oxazolidinedione, *See* Vinclozolin
(*RS*)-3-(3,5-Dichlorophenyl)-5-ethenyl-5-methyl-2,4-oxazolidinedione, *See* Vinclozolin
3-(3,5-Dichlorophenyl)-*N*-isopropyl-2,4-dioxoimdazolidinecarboxamide, *See* Iprodione
3-(3,5-Dichlorophenyl)-5-(methoxymethyl)-5-methyl-1,3-oxazolidine-2,4-dione, *See* Myclozolin
3-(3,5-Dichlorophenyl)-5-(methoxymethyl)-5-methyl-2,4-oxazolidinedione, *See* Myclozolin
3-(3,5-Dichlorophenyl)-*N*-(1-methylethyl)-2,4-dioxo-1-imidazolidinecarboxamide, *See* Iprodione
Dichlorprop, development, 14–15
Dietholate
  effect on EPTC degradation in soil, 24,26,29*f*,33
  performance in butylate-history soils, 39*t*,40
  performance in vernolate-history soil, 40*t*
  structure and systematic name, 290
O,O-Diethyl *O*-[2-(1,1-dimethylethyl)-5-pyrimidinyl] phosphorothioate, *See* Butathiofos, 288
O,O-Diethyl *S*-[2-(ethylthio)ethyl] phosphorodithioate, *See* Disulfoton, 289
O,O-Diethyl *S*-[(ethylthio)methyl] phosphorodithioate, *See* Phorate
O,O-Diethyl *O*-[(6-methyl-2-(1-methylethyl)-4-pyrimidinyl] phosphorothioate, *See* Diazinon
O,O-Diethyl *O*-[4-(methylsulfinyl)phenyl] phosphorothioate, *See* Fensulfothion, 288
O,O-Diethyl *O*-(4-nitrophenyl) phosphorothioate, *See* Parathion
O,O-Diethyl *O*-phenyl phosphorothioate, *See* Dietholate
O,O-Diethyl *O*-(3,4,6-trichloro-2-pyridyl) phosphorothioate, *See* Chlorpyrifos
2,3-Dihydro-2,2-dimethyl-7-benzofuranyl methylcarbamate, *See* Carbofuran
2,2-Dimethyl-1,3-benzodioxol-4-yl methylcarbamate, *See* Bendiocarb, 287
N,N-Dimethyl-2,2-diphenylacetamide, *See* Diphenamid
S-[[(1,1-Dimethylethyl)thio]methyl] *O,O*-diethyl phosphorodithioate, *See* Terbufos
{1*S*-[1α(*S*),3α,5β]}-4-[2-(3,5-Dimethyl-2-oxocyclohexyl)-2-hydroxyethyl]-2,6-piperidinedione, *See* Cycloheximide, 290
2,5-Dinitro-*N,N*-dipropyl-4-(trifluoromethyl)benzamine, *See* Trifluralin, 286
Diphenamid
  behavior in problem soils, 114–115
  enhanced degradation
    by soil microorganisms, 37
    effect of fungicides and antimicrobial agents, 118*f*,119
    effect of soil disinfestation, 116,118*f*,119
    in history and nonhistory soils, 114,115*f*
    degradation rate, 123,124*t*
  pathways of degradation and enzymatic reactions associated with enhanced degradation, 123,125,126*f*

Diphenamid—*Continued*
structure and systematic name, 286
Disulfoton, 289
DNA, types found in bacterial cells, 143–144,145*f*
DNA–DNA hybridization technique, use in study of enhanced pesticide degradation, 149,150*f*,151

E

Enhanced biodegradation
apparent elimination, 225,227–228,230*t*
definition, 44
effect of cultural practices
herbicide rotation, 227–229,230*t*
herbicide tank mixtures, 229,231,232*t*
weed development, 227,228,230*t*
influencing factors, 114,128
management, 231,233–237
Enhanced biodegradation of carbamothioate herbicides in Nebraska soils
application(s) necessary to induce enhanced degradation, 30,31*f*
confirmation of enhanced degradation, 25–26,27–28*f*
controlled studies, 30,31–32*f*,33
coping with enhanced degradation, 33,34–35*t*
cross-adaptation, 26,28*f*
degradation of EPTC in soils, 25,26,27*f*,28*f*
effect of atrazine, 33,34*t*
effect of carbamothioate extender on efficacy, 26,29*f*
effect of dietholate, 33
effect of herbicide rotation, 34,35*t*
field loss of EPTC in treated and untreated soils, 26,27*f*
sulfoxide-induced enhanced degradation, 30,32*f*
Enhanced biodegradation of carbamothioate herbicides in South Carolina soils
CO_2 evolution, 44,45*f*
cross-adaptation, 40,41,44–48
degradation studies, 44–48
evidence, 39
factors
herbicide rotation, 42,44*t*
multiple applications, 41,42*t*,43*f*
previous soil exposure, 47,49*t*
soil moisture, 46
storage time, 46–47,48*f*
field research, 39–44
role of plasmids, 49
Enhanced biodegradation of dicarboximide fungicides in soil
color test for identification of rapid-degrading soils, 58,59*t*
Enhanced biodegradation of dicarboximide fungicides in soil—*Continued*
ease of induction and spread, 61–62,63*f*,64
effect of microbial inhibitors, 64,65*f*
evidence in commercial fields, 57*t*,58
field performance and preliminary observations with iprodione, 54*t*,55
possibilities of cross enhancement, 58–59,60*f*,61*t*
preliminary studies of microbial relationships, 64,66
Enhanced biodegradation of EPTC in soil
discovery, 98–99
effect of antibiotics, 103*t*
effect of multiple applications, 102*t*,103,104*f*
metabolic studies
in soil, 108–109
procedure, 101–102
preparation of microbial isolate, 101
results, 102–103*t*,104*f*,105*t*
use of isolate JE1, 103–112
microbial involvement, 99
procedures for study, 100–101
rhodanese activity of soils, 103,105*t*
studies of EPTC metabolism by JE1
autoradiogram of metabolites, 105,107*f*
fractionation of ^{14}C label during metabolism, 103,105,106*f*
GC of metabolites, 105,107*f*,108
proposed metabolic pathways, 109,110*f*,111
Enhanced biodegradation of insecticides in Midwestern corn soils
cross-adaptations, 78–79
degradation in companion soils, 72,74,75*t*,76
evidence, 69–70
experimental approach for study, 70
experimental definition, 68
microbiology, 76–78
rapid insecticide degradation assay, 70,71*t*,72,73*f*
Enhanced biodegradation of insecticides in soil
discovery, 83
enhanced microbial activity to degrade insecticides
effect of moisture, 87,88*t*
effect of soil treatment, 90,91*t*,92,93
effect of temperature, 87,88*t*
effect of treatment intensity, 87,88*t*
field-generated activity, 84,85*t*,86
laboratory-generated activity, 83*t*,84
longevity, 89*t*
enhancement factors
effect of insecticide concentration, 87*t*
effect of soil moisture, 87*t*
effect of temperature, 86*t*

Enhanced biodegradation of insecticides in soil—*Continued*
interaction between commercial formulations and enhanced soil microbial populations, 89,90*t*
methods of coping with phenomenon, 192–211
Enhanced biodegradation of *p*-nitrophenol
biodegradation in river water and sediment, 185,186*f*
changes in microbial mass during adaptation, 187,188*f*
duration of adaptation in river samples, 185,186*f*,187
effect of previous treatments, 187,188*f*
typical degradation curve, 184*f*,185
Enhanced biodegradation of pesticides
cause, 113
concern in agricultural community, 240
definition, 98
description, 9
discovery, 53,98
effect on pesticide design, 278
enzymatic reactions, 122
factors
ability of pesticides to induce genes coding for degradative enzymes, 168
fungicides and antimicrobial agents, 118*f*,119
microbial metabolism of pesticides, 129
microbial toxicity, 136*t*
pesticide availability, 129
presence of pesticide or hydrolysis product in soils, 136–137,138*t*,139
pesticide toxicity, 129
soil disinfestation, 116,117–118*f*,119
impact, 113, 269–273
induction mechanisms, 114,116
methods of coping, 192–211
microbial context, 277
microbial ecology of degraders, 279
microbial studies, 119–120,121*f*,,122*t*
molecular genetics, 142–151
pathways, 122
persistence, 122–123
potential of metabolites to condition soils, 129–135
predictive tools, 278
regional difference of incidence, 240
relationship of persistence and control, 278
relationship to pest damage and yield, 276–277
research challenges and opportunities, 278–279
response of microbial populations to carbofuran, 153–164
role of soil microorganisms, 8–9,141
spectrometric methods for prediction and study, 241–246
use in bioremediation, 279
Enhanced biodegradation of pesticides—*Continued*
see also specific compounds or compound classes
Enhanced biodegradation of phenoxyalkanoic acid herbicides, 17–18,19*t*,20
Enhanced soil microbial activity
effect of chemical biocide treatment of soil, 92*t*,93
effect of insecticide form on effectiveness, 89,90*t*
effect of thermal treatment of soil, 90,91*t*,92
field generation, 84,85*t*,86
laboratory generation, 83*t*,84
Enhancing biodegradation for detoxification of herbicide waste in soil
effect of concentration and formulation, 253–254,258,259*t*,260
effect of nutrient amendments and microbial inoculum, 257,261–265
effect of soil dilution, 253,258*t*
experimental materials and methods, 253–257
soil dilution experimental procedure, 253
Enzymatic assays
detection of pesticides, 242
detection of urease activity in problem soils, 242,243*t*
Enzyme activities in enhanced soil, general assay, 243,244*t*
EPTC
applications, 3,23
controlled degradation, 30,31–32*f*,33
control of annual grasses in crops, 214
cross-adaptation, 26,28*f*
degradation in soil, 222
effect of carbamothioate extender on efficacy, 26,29*f*
effect of corn stover on performance, 24
effect of cultural practices and chemicals on persistence, 222–237
effect of multiple applications on efficacy, 41,42*t*,43
effect of repeated applications on weed control, 223
effect of rotation on efficacy, 42,44*t*
enhanced degradation
confirmation, 25–26,27–28*f*
methods of coping, 33,34–35*t*
discovery, 98–99,214
failure of herbicidal activity, 223
fluorescein diacetate assay, 244–245
prediction of persistence studies, 222
shattercane control, 23–24,25*t*
structure and systematic name, 284
use of extenders to restore activity, 223
EPTC plus dietholate mixture
commercialization, 214

EPTC plus dietholate mixture—*Continued*
enhanced biodegradation, 214–215
Ethoprop
cumulative plots of mineralization in history and nonhistory soils, 72,73*f*
effect of use history on mineralization, 71*t*,72
structure and systematic name, 289
S-Ethyl *N,N*-bis(3-chloroallyl)-carbamothioate (SC–0058)
abbreviation, 290
effect on EPTC degradation, 33
structure, 290
S-Ethyl bis(2-methylpropyl)carbamothioate, *See* Butylate
S-Ethyl dipropylcarbamothioate, *See* EPTC
O-Ethyl *S,S*-dipropyl phosphorodithioate, *See* Ethoprop
S-Ethyl *N*-ethylthiocyclohexanecarbamate, *See* Cycloate, 285
S-Ethyl hexahydro-1*H*-azepine-1-carbothioate, *See* Molinate, 285
O-Ethyl *S*-phenyl ethylphosphonodithioate, *See* Fonofos
Exponential phase, description, 6
Extender
definition, 24
description, 214

F

Fensulfothion, 288
Field-generated enhanced soil microbial activity
enhancement factors for treated soil, 84,85*t*,86
microplot development, 84
Field plot technique, 224
Fluorescein diacetate assay, determination of EPTC and butylate, 244–245
Fluorescence spectrophotometry, use in biochemical analyses, 241
Fonofos
cross-adaptations for enhanced degradation, 78–79
cumulative plots of mineralization in history and nonhistory soils, 72,73*f*
degradation in companion soils, 72,74,75*t*,76
effect of use history on mineralization, 71*t*,72
effect on EPTC degradation, 33
microbiology of enhanced degradation, 76–78
structure and systematic name, 288
Fungal degraders, assessment of degradative capabilities, 120
Fungicides
applications, 3
effect on enhanced pesticide degradation, 118*f*,119
examples, 3
names, 289–290
structures, 289–290

G

GC, procedure, 225
Gene exchange, mechanisms in bacteria, 144,145*f*,146
Gene transfer and expression, adaptation mechanism, 168,170
Groundwater, contamination with pesticide, 167
Groundwater aquifers, importance, 167

H

Hazardous chemicals, accumulation in agrochemical retail facilities, 250
Herbicide(s)
application time, 3
names, 284–286
phenoxyalkanoic acid, *see* Phenoxyalkanoic acid herbicides
preemergence examples, 3
structures, 284–286
Herbicide rotation, effect on enhanced degradation, 34,35*t*
Herbicide tank mixtures, elimination or inhibition of enhanced biodegradation, 229,231,232*t*
Herbicide waste, enhancing biodegradation for detoxification in soil, 249–265
Hydrolytic reactions, description, 5

I

Immunoassays
definition, 242
detection of pesticides, 242
Impact of enhanced pesticide biodegradation
advances in understanding enhanced pesticide degradation, 273–277
effect on pesticide manufacturers, 270–272
effect on pesticide users, 270
effect on research and extension community, 272–273
quantification, 269–270
Insecticide(s)
enhanced biodegradation in Midwestern corn soils, 68–79
importance of persistence and bioavailability, 193

Insecticide(s)—*Continued*
names, 286–289
structures, 286–289
Insecticide degradation assay, rapid, *See* Rapid insecticide degradation assay
In situ remediation method, use in cleaning up contaminated soil, 176,177*f*
Iprodione
control of white rot disease, 3,54
cross enhancement, 59,61*t*
degradation factors
microbial inhibitors, 64,65*t*
multiple applications, 55,56*f*
pretreatment concentration, 62,63*f*,64
soil pH,55,57,60*f*
enhanced degradation
ease of induction and spread, 61–62,63*f*,,64
evidence in commercial fields, 57*t*,58
preliminary observations, 55
preliminary studies of microbial relationships, 64,66
field performance, 54*t*,55
metabolite assay, 246
structure and systematic name, 289
transfer of degrading ability, 62,63*f*
IR spectrophotometry, use in identification of organic compounds, 241
Isofenphos
cumulative plots of mineralization in history and nonhistory soils, 72,73*f*
degradation in companion soils, 72,74,75*t*,76
enhanced degradation
effect of hydrolysis products, 135*t*
effect of mobility, 137–138
effect of use history on mineralization, 71*t*,72
microbiology, 76–78
metabolite assay, 245,246*t*
microbial toxicity, 136*t*
mobility in soils, 137,138*t*
structure and systematic name, 288

L

Laboratory-generated enhanced soil microbial activity
design of laboratory experiment, 83
enhancement factors for treated soils, 83,84*f*
Lag phase
cause, 16
description, 6
role in microbiological soil degradation, 16

M

Managing enhanced biodegradation, commercial perspective, *See* Commercial perspective on managing enhanced biodegradation
MCPA
development, 14–15
degradation
effect of repeat treatments, 17–18,19*t*,20
importance of soil microorganisms, 15–17
structure and systematic name, 284
MCPB
development, 14–15
structure and systematic name, 284
Mechanisms of adaptation
dynamics of degrader populations, 170–174
gene transfer and expression, 168,170
Mechanisms of pesticide dissipation in soil
abiotic transformation processes, 4–5
microbiological transformation processes, 5–6
transport processes, 4
Mecoprop, 14–15
Metabolite assay
carbofuran, 245*t*
iprodione, 246
isofenphos, 245,246*t*
vinclozolin, 246
Methods for coping with enhanced biodegradation of soil insecticides
analytical methods, 200
chemical rotation procedure
field evaluation, 196–197,198*t*,199
laboratory evaluation, 196,197*t*
chemical rotation studies
field testing, 203–207
laboratory testing, 203,205*f*
controlled-release formulations
field testing, 199,201*t*,,211
laboratory testing, 199,205*f*,207–210
cultivation–time studies, 200,201*t*
cultivation–time treatment procedure, 194–195
experimental design, 194–200
long-term management strategies, 193
operational strategies, 193–194
studies of adult suppression to prevent larval damage, 195–196,200,201–202*t*,203
technological strategies, 193–194
2-(Methoxycarbamoylamino)benzimidazole, *See* Carbendazim
1-Methylethyl (3-chlorophenyl)carbamate, *See* Chlorpropham, 286
1-Methylethyl 2-[[ethoxy[(1-methylethyl)-amino]phosphinothioyl]oxy]benzoate, *See* Isofenphos

2-Methyl-2-(methylthio)propionaldehyde
O-[(methylamino)carbonyl]oxime,
See Aldicarb, 287
Metolachlor
enhancing biodegradation for
detoxification of waste in soil, 252–265
structure and systematic name, 286
Metribuzin, 285
Microbial adaptation
evidence, 168
for pesticide degradation, 6–8
mechanisms, 168–174
patterns of pesticide biodegradation, 168,169*f*
to pesticides in subsurface environments,
173,174*f*,175
use in bioremediation as treatment method,
175–176,177*f*
Microbial adaptation in aquatic ecosystems
chemicals that cause adaptation, 182*t*
enhanced degradation of *p*-nitrophenol,
185–188
patterns, 182–183,184*f*,185
Microbial biomass
definition, 160
effect of carbofuran on size,
160,161*f*,162
Microbial context of enhanced pesticide
biodegradation
microevolutionary nature of
adaptation, 277
pervasive nature, 277
Microbial inhibitors
names and abbreviations, 290
structures, 290
Microbial isolate JE1, metabolic studies,
103,105,106*f*,107
Microbial metabolism of pesticides
catabolism, 5–6
cometabolism, 6
Microbial or enzymatic inhibitors, use
to restore herbicidal activity
in problem soils, 38
Microbial population(s)
biomass estimations, 160,161*f*,162–163,164*f*
ecological implications of growth pattern, 6
plate count media preparation, 155–156
response to carbofuran in soils enhanced
for its degradation
attempts to understand impact, 153,155*t*
changes in bacteria numbers in soil, 156*t*
[^{14}C]CO_2 evolution, 156,157–158*f*
percentage distribution of ^{14}C
radioactivity, 156,159*f*,160
Microbial substrate adaptation
key factors, 7–8
microrevolutionary nature, 7
Microbiological mechanisms for degradation
of herbicides in soil
effect of concentration, 16–17
Microbiological mechanisms for degradation
of herbicides in soil—*Continued*
effect of repeat treatments, 17
importance of soil microorganisms, 15–16
influencing factors, 17
lag phase, 16
Microbiological transformation processes,
microbiological degradation, 5–6
Microbiology of enhanced insecticide
degradation, discussion, 76
Mobility of pesticides and hydrolysis
products in soils
determination
by R_f values, 137,138*t*
by soil thin-layer chromatography, 137
importance for enhanced degradation,
136–138
influence of soil characteristics, 139
Molecular genetics of pesticide degradation
by soil bacteria
adaptation mechanism of microbial
communities to yield pesticide-
degrading populations, 142
bacterial genetics, 143–146
future of research, 148–149,150*f*,151
role of extrachromosomal elements, 142–143
Molinate, 285
Monoclonal antibodies, use in study of
enhanced pesticide degradation, 149
Monod equation, prediction of microbial
population growth, 170–171
Movement of dissolved or particulate-sorbed
pesticide in water, description, 4
Myclozolin
cross enhancement, 59,60*f*,61*t*
field performance, 54
structure and systematic name, 289

N

1-Naphthalenyl methylcarbamate, *See*
Carbaryl, 286
p-Nitrophenol, enhanced degradation, 184–188
Nonproblem soils, definition, 240

O

Operational strategies for coping with
enhanced biodegradation
description, 193
examples, 193–194
Organic compounds, factors controlling fate
and transport in aquatic systems, 181
Organic matter of soil
fractions, 2
turnover times, 2

Organophosphorus insecticides
control of rootworms, 69
cross enhancement factors, 93,94t
names, 287–289
structures, 287–289

P

Parathion
enhanced degradation
effect of hydrolysis product, 130,133t
effect of mobility, 137–138
mechanism of bacterial degradation, 148
microbial toxicity, 136t
mobility in soils, 137,138t
structure and systematic name, 287
Patterns of adaptation
factors influencing length of lag phase, 183,184f,185
phases, 182–183
Pebulate, 285
Pest damage, relationship to enhanced pesticide degradation, 276–277
Pesticide(s)
active ingredients, 2
adverse health effects, 249–250
degradation in companion soils, 72,74,75t,76
diversity, 113
importance, 2
mobility in soils, 136–137,138t,139
requirement, 3
role in soil microbial ecosystem, 1
world market, 2
Pesticide biodegradation, enhanced, *See* Enhanced biodegradation of pesticides
Pesticide contamination of groundwater, routes, 167–168
Pesticide-degrading biomass
incorporation of ^{14}C, 162
methods of estimation, 162–163
size as indicated by fumigation, 163,164f
Pesticide-degrading microbial populations
characterization, 275–276
detection, 274–275
isolation, 275
Pesticide-degrading plasmids, use in cleanup technologies, 252
Pesticide dissipation in soil, mechanisms, 3–6
Pesticide Impact Assessment Program, sponsors of degradation research, 272
Pesticide incorporation efficiency, effect on pesticide persistence, 227
Pesticide manufacturers, impact of enhanced pesticide biodegradation, 270–272
Pesticide persistence, effect of cultural practices and chemicals, 222–237
Pesticide users, impact of enhanced pesticide biodegradation, 270
Pesticide wastes, cleanup technologies, 250–251
Phenoxyalkanoic acid herbicides
abiotic mechanisms for soil degradation, 15
bacterial metabolism, 147
development, 14–15
effects of long-term field applications on rate of microbial degradation, 15–20
enhanced degradation, 17–18
microbiological mechanisms for soil degradation, 15–17
names and abbreviations, 284
structures, 284
Phorate
cumulative plots of mineralization in history and nonhistory soils, 72,73f
effect of use history on mineralization, 71t,72
structure and systematic name, 288
Plasmid(s), role in pesticide degradation by soil bacteria, 146–148
Plasmid-mediated transfer of pesticide-degrading genes, role in adaptation mechanism, 170
Population growth
factors influencing growth–degradation relationships, 172–173
limited growth from pesticide degradation, 172t
measurement, 171–172
prediction using Monod equation, 170–171
role in adaptation mechanism, 170–171
theoretical amounts of carbon source necessary to support active microbial biomass, 171,174f
Problem soils, definition, 240
Procymidone
cross enhancement, 59,60f,61t
field performance, 54
structure and systematic name, 290
S-Propyl butylethylcarbamothioate, *See* Pebulate, 285
S-Propyl dipropylcarbamothiate, *See* Vernolate
Pyrethroid insecticides
names, 289
structures, 289

R

Rapid-degrading soils, identification using colorimetric test, 58,59t
Rapid insecticide degradation assay
cumulative plots of insecticide mineralization in history and nonhistory soils, 70

Rapid insecticide degradation assay—*Continued*
effect of use history on cumulative mineralization in soils, 71*t*,72
procedure, 71
soil collection, 70
Research and extension community, impact of enhanced pesticide biodegradation, 272–273
Rootworm control, 68–69
Rootworm control in pesticide-history soils, 243*t*
Rotation of herbicides, elimination or inhibition of enhanced biodegradation, 228–229,230*t*

S

Shattercane
control with EPTC, 23–24,25*t*
possible development of resistance to EPTC, 24
Silvex, development, 14–15
Soil
composition, 2
definition, 2
description, 2
ecosystem, 2
importance, 2
Soil actinomycete population, function, 7
Soil bacteria
initiators of degradation, 2
growth rates, 6–7
population function, 7
Soil characteristics, effect on pesticide mobility, 139
Soil disinfestation, effect on enhanced pesticide degradation, 116,117–118*f*,119
Soil ecosystem
composition, 2
energetic input, 7
Soil extraction, procedure, 225
Soil fungi
initiators of degradation, 2
population function, 7
Soil insect pest control, example, 3
Soil microorganisms
populations, 7
role in globally significant nutrient cycles, 2
role in pesticide degradation, 8–9
Soil moisture, effect on efficacy of pesticides, 46
Soil pH, effect on enhanced degradation of dicarboximide fungicides, 55,57,60*f*
Solarization, effect on enhanced microbial activity, 91*t*,92
Spectrophotometric methodologies for predicting and studying enhanced degradation
advantages, 240
applications, 242
description of spectrometer, 241
determination of enzyme activities in enhanced soil, 243,244*t*
determination of urease activity in problem soils, 242,243*t*
fluorescein diacetate assay, 244–245
metabolite assays
carbofuran, 245*t*
iprodione and vinclozolin, 246
isofenphos, 245,246*t*
types of methods, 241
Stationary phase, description, 6
Substrate addition method, definition of biomass response, 162
Systems allowing continued use of pesticides despite enhanced biodegradation
biodegradation of EPTC, dietholate, and SC–0058, 219,220*t*
effectiveness of crop rotations, 220–221
effectiveness of dietholate vs. SC–0058, 218*t*,219
effectiveness of herbicide rotation, 220–221
field studies
methodology, 215
results for location not previously treated with EPTC, 215,216*t*
results for location previously treated with EPTC, 216*t*
results for location previously treated with EPTC plus dietholate, 217*t*
6-year herbicide rotation study, 217,218*t*
laboratory studies, 219–220*t*

T

Technological strategies for coping with enhanced biodegradation
description, 193
examples, 194
Tefluthrin, 289
Terbufos
cumulative plots of mineralization in history and nonhistory soils, 72,73*f*
effect of use history on mineralization, 71*t*,72
structure and systematic name, 289
2,3,4,6-Tetrafluoro-4-methylphenyl *cis*-3-(*Z*-2-chloro-3,3,3-trifluoroprop-1-enyl)-2,2-dimethylcylcopropane-carboxylate, *See* Tefluthrin, 289

Tetramethylthiuram disulfide, effect on enhanced pesticide degradation, 118f,119
Thermal treatment of soil, effect on enhanced microbial activity, 91t
Transduction, definition, 144,146
Transport processes of pesticide dissipation in soil
movement of dissolved or particulate-sorbed pesticides in water, 4
volatilization, 4
Transposons
description, 144,145f
role in pesticide degradation by soil bacteria, 146–148
Triazine herbicides
names, 285
structures, 285
(2,4,5-Trichlorophenoxy)acetic acid (2,4,5-T)
development, 14–15
enhanced degradation
effect of hydrolysis products, 130,132t
effect of mobility, 137–138
effect of repeat treatments, 17–18,19t,20
importance of soil microorganisms, 15–17
mechanism of bacterial degradation, 147
microbial toxicity, 136t
mobility in soils, 137,138t
structure, 284
2-(2,4,5-Trichlorophenoxy)propionic acid, *See* Silvex
Trichloropyridinol, effect on enhanced microbial activity, 92t,93
Trifluralin, 286
Trimethacarb, 287
2,3,5-Trimethylphenyl methylcarbamate, *See* Trimethacarb, 287
3,4,5-Trimethylphenyl methylcarbamate, *See* Trimethacarb, 287
Triphenyltin acetate, effect on enhanced pesticide degradation, 118f,119

U

Urease activity in problem soils
effect of pesticide amendment, 242,243t
general assay, 242,243t
UV–visible spectrophotometry, description, 241

V

Vernolate
cross-adaptation, 44,46t
effect of multiple applications on efficacy, 40t
structure and systematic name, 284
Vinclozolin
control of white rot disease, 54
cross enhancement, 59,61t
degradation factors
microbial inhibitors, 64,65t
multiple applications, 55,56f
soil pH, 55,57,60f
enhanced degradation
evidence in commercial fields, 57t,58
preliminary studies of microbial relationships, 64,66
field performance, 54
metabolite assay, 246
structure and systematic name, 289
Volatilization, description, 4

W

Weeds, effect on pesticide persistence, 227-228,230t

Production: BethAnn Pratt-Dewey
Indexing: Deborah H. Steiner
Acquisition: Cheryl Shanks

Elements typeset by Hot Type Ltd., Washington, DC
Printed and bound by Maple Press, York, PA

Paper meets minimum requirements of American National Standard for Information Sciences—Permanence of Paper for Printed Library Materials, ANSI Z39.48–1984 ♾